Analog Electronics Applications

FUNDAMENTALS OF DESIGN AND ANALYSIS

Analog Electronics
Applications

FUNDAMENTALS OF DESIGN AND ANALYSIS

Hernando Lautaro Fernandez-Canque
Glasglow Caledonian University, Scotland, United Kingdom

CRC Press
Taylor & Francis Group
Boca Raton London New York

CRC Press is an imprint of the
Taylor & Francis Group, an **informa** business

CRC Press
Taylor & Francis Group
6000 Broken Sound Parkway NW, Suite 300
Boca Raton, FL 33487-2742

First issued in paperback 2019

© 2017 by Taylor & Francis Group, LLC
CRC Press is an imprint of Taylor & Francis Group, an Informa business

No claim to original U.S. Government works

ISBN-13: 978-1-4987-1495-2 (hbk)
ISBN-13: 978-0-367-87249-6 (pbk)

Visit the Taylor & Francis Web site at
http://www.taylorandfrancis.com

and the CRC Press Web site at
http://www.crcpress.com

To my wife Vicky

and our children Lalo, Bruni, Lia, and Mandela

Contents

Preface...xix
Acknowledgments...xxi
Author xxiii

Chapter 1   Analog Electronics Applications and Design 1

   1.1   Introduction to Analog Electronics 1
   1.2   Analog Signals.. 1
   1.3   Analog Systems 2
   1.4   Application and Design of Analog Systems............................... 3
         1.4.1   Customer Requirements 3
         1.4.2   Top-Level Specifications 4
         1.4.3   System Design Approach 4
                 1.4.3.1   Top-Level Design.. 5
                 1.4.3.2   Detailed Design 5
         1.4.4   Technology Choice.. 5
                 1.4.4.1   System Testing.. 6
                 1.4.4.2   Social and Environmental Implications 6
                 1.4.4.3   Documentation.. 6
         1.4.5   Distortion and Noise... 6
                 1.4.5.1   Noise 7
                 1.4.5.2   Distortion 7
         1.4.6   Electronic Design Aids 8
   1.5   Key Points.. 9

Chapter 2   Electric Circuits.. 11

   2.1   Introduction 11
   2.2   Units... 11
         2.2.1   Unit of Charge 12
         2.2.2   Unit of Force.. 12
         2.2.3   Unit of Energy 12
         2.2.4   Unit of Power... 13
         2.2.5   Unit of Electric Voltage... 13
         2.2.6   Unit of Resistance and Conductance........................... 13
   2.3   Concept of Electric Charge and Current 15
   2.4   Movement of Electrons and Electric Current in a Circuit....... 16
         2.4.1   Circuit.. 16
         2.4.2   Electromotive Force 17
         2.4.3   Source... 17
         2.4.4   Load.. 17

2.5   Passive Components: Resistance, Inductance, and
        Capacitance 17
        2.5.1   Resistance 17
                2.5.1.1   Resistors Connected in Series and
                          Parallel 19
                2.5.1.2   Resistors Connected in Series 19
                2.5.1.3   Resistor Connected in Parallel 19
                2.5.1.4   Special Case 20
        2.5.2   Capacitance 21
                2.5.2.1   Capacitors in Parallel 24
                2.5.2.2   Capacitors in Series 24
        2.5.3   Inductors 26
                2.5.3.1   Inductors in Series 27
                2.5.3.2   Inductors in Parallel 28
                2.5.3.3   Energy Storage W in an Inductor 29
        2.5.4   Application: Inductive Proximity Sensors 29
2.6   Active Components of a Circuit: Sources 29
        2.6.1   Ideal Voltage Source 30
        2.6.2   Practical Voltage Source 30
        2.6.3   Voltage Sources Connected in Series 31
        2.6.4   Voltage Sources Connected in Parallel 31
                2.6.4.1   Ideal Current Source 31
                2.6.4.2   Practical Current Source 31
2.7   Electric Circuits/Networks 32
        2.7.1   Selection of Components 34
2.8   Key Points 34

Chapter 3   Circuit Analysis 35

3.1   Concept of Steady State and Transient Solutions 35
3.2   DC Circuits 36
        3.2.1   Kirchhoff's Current Law Applied to Electric
                Circuits: KCL 36
        3.2.2   Kirchhoff's Voltage Law Applied to Electric
                Circuits: KVL 41
                3.2.2.1   The Voltage and Current Divider 58
3.3   AC Circuits 60
        3.3.1   Origin of Phasor Domain 62
        3.3.2   Application of Kirchhoff's Law to ac Circuits 64
                3.3.2.1   Impedance Z 64
                3.3.2.2   Impedance of an Inductor 65
                3.3.2.3   Impedance of a Capacitance 66
                3.3.2.4   Impedance of a Resistance 67
                3.3.2.5   Reactance X 68
                3.3.2.6   Polar–Cartesian Forms 69

3.3.2.7 Cartesian to Polar Form 69
3.3.2.8 Polar to Cartesian 69
3.3.2.9 Phasor Diagrams 70
3.4 Key Points 76

Chapter 4 Diodes 77

4.1 Introduction 77
4.2 Semiconductor Material 77
4.2.1 Conductivity and Energy Bands in Semiconductors 77
4.2.2 Doping 79
4.3 p–n Junction 80
4.4 Diode Current–Voltage Characteristics I–V 81
4.4.1 Forward Bias 81
4.4.2 Reverse Bias 82
4.5 Different Types of Diodes 83
4.5.1 Semiconductor Diodes 83
4.5.2 Zener Diodes 84
4.5.3 Avalanche Diodes 84
4.5.4 Light-Emitting Diodes 84
4.5.5 Tunnel Diodes 84
4.5.6 Gunn Diodes 84
4.5.7 Peltier Diodes 84
4.5.8 Photodiodes 85
4.5.9 Solar Cell 85
4.5.10 Schottky Diodes 85
4.6 Diode Applications 85
4.6.1 Rectification 85
4.6.2 Half-Wave Rectifiers 86
4.6.3 Full-Wave Rectifiers 87
4.6.4 Single-Phase Bridge Rectifier Circuit 87
4.6.5 Diode as Voltage Limiter 92
4.6.6 Voltage Doubler 93
4.7 Testing Diodes 94
4.8 Key points 94
Reference 94

Chapter 5 Bipolar Junction Transistor 95

5.1 Introduction 95
5.2 Bipolar Junction Transistor 95
5.3 BJT Characteristics 98
5.3.1 Transistor Configurations 98
5.3.2 Input Characteristics 98
5.3.3 Output Characteristics 98

5.3.4   Data for a Typical NPN Transistor........................... 100
5.3.5   Rating and Selection of Operating Point................. 101
5.4   Gain Parameters of BJT: Relationship of α and β
        Parameters 102
5.4.1   Common Base Connection..................................... 102
5.4.2   Common Emitter Configuration............................. 103
5.5   Testing Transistors... 106
5.6   Efficient BJT as Amplification Device.................... 106
5.6.1   Emitter Injection Efficiency η 108
5.6.2   Base Transport Factor χ 109
5.6.3   Punch-Through.. 111
5.7   Key Points... 111
Reference.. 111

Chapter 6   Field Effect Transistors 113

6.1   Introduction 113
6.2   Fabrication of FET 113
6.3   Different Types of FET 114
6.3.1   Insulated-Gate FETs... 114
6.3.2   Junction Gate FETs 115
6.3.3   FET Circuit Symbols.. 115
6.4   MOS FET 115
6.4.1   The MOS IGFET... 115
6.4.2   MOS Operation 116
6.4.3   Accumulation 117
6.4.4   Depletion 118
6.4.5   Strong Inversion 119
6.4.6   Threshold Voltage, V_T................................... 120
6.5   JFET Operation 120
6.6   Static Characteristics of FET 123
6.6.1   Input Characteristics 123
6.6.2   Output Characteristics... 123
6.6.3   Transfer Characteristics... 123
6.7   Current–Voltage Characteristics............................... 123
6.7.1   Saturation 124
6.8   Key Points... 126

Chapter 7   Bipolar Junction Transistor Biasing 129

7.1   Introduction 129
7.2   Load Line 129
7.2.1   Cut Off.. 132
7.2.2   Saturation 132
7.3   Biasing a BJT.. 133
7.3.1   Fixed Bias... 133
7.3.2   Auto Bias 137

	7.3.3	Collector-Feedback Bias	141
	7.3.4	Two Sources	143
7.4	Key Points		144

Chapter 8 Modeling Transistors 145

8.1	Introduction		145
8.2	Hybrid h Parameters		145
	8.2.1	h Parameters Common Emitter Configuration	146
8.3	Admittance Y Parameters		148
8.4	General Three-Parameter Model		149
8.5	T-Equivalent Two-Parameter Model		149
	8.5.1	AC-Simple T-Parameters Transistor Model	149
	8.5.2	Emitter Resistance, r_e (Small Signal)	150
8.6	Mutual Conductance Model		151
8.7	Key Points		152

Chapter 9 Small-Signal Analysis of an Amplifier under Different Models 153

9.1	Introduction		153
9.2	Analysis of Transistor Amplifiers Using h Parameters		153
	9.2.1	Current Gain, G_i	154
	9.2.2	Input Impedance, Z_i	155
	9.2.3	Voltage Gain, G_v	156
	9.2.4	Output Impedance, Z_o	156
	9.2.5	Power Gain, G_p	157
9.3	Small-Signal Practical CE Amplifier under h-Parameter Model		158
	9.3.1	Fixed Bias Amplifier	158
	9.3.2	Auto Bias Amplifier	159
		9.3.2.1 Voltage Gain	161
		9.3.2.2 Current Gain	161
		9.3.2.3 Power Gain	162
9.4	Analysis of Transistor Amplifiers Using Y Parameters		165
	9.4.1	Voltage Gain, G_v	166
	9.4.2	Current Gain, G_i	166
	9.4.3	Power Gain, G_p	166
9.5	Analysis of Transistor Amplifiers Using the General Three-Parameter Model		168
	9.5.1	Voltage Gain, G_v	168
	9.5.2	Current Gain, G_i	169
	9.5.3	Power Gain, G_p	170
9.6	Analysis of Transistor Amplifiers Using the T-Model Two-Parameters		172
	9.6.1	Voltage Gain, G_v	172
		9.6.1.1 CE Amplifier Using T Parameters	172
	9.6.2	Current Gain, G_i	173

        9.6.3   Input Impedance, Z_i 173
        9.6.4   Output Impedance, Z_o 174
                9.6.4.1   Effect of a Load Resistance 175
                9.6.4.2   Effect of the Source Resistance 175
        9.6.5   Power Gain, G_p 177
    9.7   Key Points 178

Chapter 10 Amplifiers Frequency Response 179

    10.1   Introduction 179
    10.2   Half-Power Gain, Concept of 3 dB 179
    10.3   CE Amplifier at Low Frequency 182
        10.3.1   Voltage Gain at Low Frequency 182
        10.3.2   Low-Frequency Cutoff 183
        10.3.3   Phase Change at Low Frequency 184
                10.3.3.1   Effect of the Other Capacitance on the
                          Low-Frequency Response 184
    10.4   CE Amplifier at High Frequency 185
        10.4.1   Voltage Gain at High Frequency 185
        10.4.2   High-Frequency Cutoff 187
        10.4.3   Phase Change at High Frequency 187
    10.5   Total Frequency Response 187
    10.6   Key Points 189

Chapter 11 Common Collector Amplifier/Emitter Follower 191

    11.1   Voltage Gain, Gv_{CC} 191
    11.2   Current Gain, Gi_{CC} 194
    11.3   Input Impedance, Zi_{CC} 194
    11.4   Power Gain, Gp_{CC} 195
    11.5   Key Points 196

Chapter 12 Common Base Amplifier 197

    12.1   Common Base Amplifiers under h Parameters 197
    12.2   Voltage Gain, Gv_{CB} 198
    12.3   Current Gain, Gi_{CB} 199
    12.4   Power Gain, Gp_{CB} 199
    12.5   Key Points 200

Chapter 13 Common Emitter Amplifier in Cascade 201

    13.1   Introduction 201
    13.2   Overall Gain of Amplifiers in Cascade 201
        13.2.1   Voltage Gain of the Last Stage n 202
        13.2.2   Voltage Gain for the (n − 1) Stages 202
        13.2.3   "Voltage" Gain of the Source 203

13.3   Frequency Response 205
     13.3.1   Low-Frequency Cutoff 205
         13.3.1.1   Frequency Cutoff of Last Stage 205
         13.3.1.2   Frequency Cutoff for 1 to (n − 1) Stages 206
         13.3.1.3   Frequency Cutoff Input Circuit
                  First Stage 206
     13.3.2   High-Frequency Cutoff 208
         13.3.2.1   Frequency Cutoff of Last Stage 208
         13.3.2.2   High-Frequency Cutoff for 1 to (n − 1)
                  Stages 209
         13.3.2.3   Frequency Cutoff Input Circuit
                  First Stage 209
13.4   Key Points..212

Chapter 14   Field Effect Transistor Biasing......................................213
14.1   Introduction 213
14.2   MOSFET Biasing 213
     14.2.1   Depletion MOSFET 213
     14.2.2   Enhancement MOSFET 214
         14.2.2.1   Drain-Feedback Bias 216
14.3   JFET Biasing 217
     14.3.1   Self-Bias 217
     14.3.2   Voltage Divider Bias....................................219
14.4   Key Points..220

Chapter 15   Field Effect Transistor as Amplifiers 223
15.1   Introduction 223
15.2   Common Source Amplifier 223
     15.2.1   AC Voltage Gain at Medium Frequencies...............223
     15.2.2   AC Voltage Gain at High Frequencies 227
     15.2.3   Input Impedance at High Frequency...............228
15.3   Common Drain (Source Follower) 229
     15.3.1   AC Voltage Gain at Medium Frequency 229
     15.3.2   AC Voltage Gain at High Frequency......................232
     15.3.3   Input Impedance at High Frequency......................233
     15.3.4   Output Impedance at High Frequency 234
15.4   Key Points..235

Chapter 16   Transfer Function and Bode Diagrams 237
16.1   Introduction 237
16.2   Transfer Functions...237
     16.2.1   Examples of Transfer Functions Passive
             Components...237

16.2.2 First-Order Low-Pass Transfer Function...................238
16.2.3 First-Order High-Pass Transfer Function................239
16.3 Bode Plots...243
16.3.1 Magnitude Bode Plot...243
16.3.2 Transfer Function = TF = K (Constant)....................244
16.3.3 Transfer Function = TF = j2πf...............................245
16.3.4 Transfer Function = TF = $(1 + jf/f_p)$ 245
16.4 Idealized Bode Plots...245
16.4.1 Magnitude Asymptotic..245
16.4.2 Bode Plot Asymptotic Phase 246
16.5 Construction of Bode Plots..247
16.6 Bode Plot Example 247
16.7 Key Points...250

Chapter 17 Feedback in Amplifiers 251

17.1 Introduction 251
17.2 Negative Feedback..251
17.2.1 Series Voltage NFB..253
17.2.1.1 Series Voltage NFB Voltage Gain 254
17.2.1.2 Series Voltage NFB Effects on Input
           Impedance 254
17.2.1.3 Series Voltage NFB Effects on Output
           Impedance 255
17.2.1.4 Series Voltage NFB Effects on
           Frequency Response 255
17.2.1.5 Series Voltage NFB Effects on Internal
           Distortion 257
17.2.1.6 Series Voltage NFB Current Gain 257
17.2.2 Series Current NFB..258
17.2.2.1 Series Current NFB Voltage Gain 259
17.2.2.2 Series Current NFB Current Gain 260
17.2.2.3 Series Current NFB Effects on Input
           Impedance 260
17.2.2.4 Series Current NFB Effects on Output
           Impedance 260
17.2.3 Parallel Voltage NFB 260
17.2.3.1 Parallel Voltage NFB Voltage Gain..........261
17.2.3.2 Parallel Voltage NFB Current Gain 261
17.2.3.3 Parallel Voltage NFB Effects on Input
           Impedance 262
17.2.3.4 Parallel Voltage NFB Effects on
           Output Impedance 262
17.2.4 Parallel Current NFB 263
17.2.4.1 Parallel Current NFB Current Gain 263

17.2.4.2 Parallel Current NFB Effects on Input
Impedance 264
17.2.4.3 Parallel Current NFB Effects on
Output Impedance 264
17.2.4.4 Parallel Current NFB Voltage Gain 265
17.3 Key Points 269

Chapter 18 Differential Amplifiers 271

18.1 Introduction 271
18.2 Single Input Voltage 271
18.3 Differential-Mode Voltage Gain 274
18.4 Common Mode Voltage Gain: Effect on Noise 274
18.5 Common Mode Rejection Ratio 276
18.6 Key Points 278

Chapter 19 Operational Amplifiers 281

19.1 Introduction 281
19.2 Op-Amps Characteristics 283
19.2.1 Offset Null Circuit 284
19.2.2 Compensation Circuit 285
19.2.3 Slew Rate 285
19.3 Op-Amp Gain 285
19.4 Inverting Amplifier 285
19.4.1 Inverting Amplifier Voltage Gain 285
19.4.2 Virtual Earth 287
19.4.3 Input Impedance 289
19.4.4 Output Impedance 290
19.4.5 Bias Equalization 290
19.5 Noninverting Amplifier 292
19.5.1 Noninverting Amplifier Voltage Gain 292
19.5.2 Input Impedance 295
19.5.3 Output Impedance 295
19.5.4 Voltage Follower 296
19.6 The Differential Amplifier 296
19.6.1 Differential Amplifier Voltage Gain 297
19.6.1.1 Op-Amp as Comparator 299
19.6.1.2 The Summing Amplifier 300
19.7 Op-Amp Frequency Response 301
19.8 Key Points 303

Chapter 20 Filters 305

20.1 Introduction 305
20.2 Low-Pass Filter Responses 308

20.2.1   Single First-Order Low-Pass Filter 308
20.2.2   Second-Order Low-Pass Filters 310
20.2.3   Sallen and Key Low-Pass Filter 310
20.3   High-Pass Filter Response.......................................311
20.3.1   First-Order High-Pass Filter.................................311
20.3.2   Second-Order Filters 312
20.3.3   Sallen and Key High-Pass Filter 313
20.4   Band-Pass Filter Response 314
20.5   Band-Stop Response..314
20.6   Fourth-Order Response 315
20.7   Filter Response Characteristics 316
20.8   Filter Design using Standard Tables...............................316
20.8.1   Bessel..318
20.8.1.1   First Second-Order Filter.........................318
20.8.1.2   Second Second-Order Filter 318
20.8.2   Butterworth 319
20.8.3   Chebyshev 320
20.8.3.1   First Second-Order Filter.........................320
20.8.3.2   Second Second-Order Filter 321
20.9   Key Points..322
References 322

Chapter 21   Applications of Analog Electronics 323

21.1   Introduction 323
21.2   Simulation in Circuit Applications...............................323
21.3   Selection of Components and Circuit Elements in an
          Application 324
21.3.1   Transistors 325
21.3.2   Resistors 325
21.3.3   Capacitors..326
21.3.4   Inductors...326
21.3.5   Nominal Preferred Values......................................326
21.4   Building and Realization of a Circuit.............................327
21.5   Testing–Troubleshooting 329
21.5.1   Testing 329
21.5.2   Troubleshooting...329
21.6   Analog Electronic Applications Examples...........................330
21.6.1   DC Power Supply 330
21.6.1.1   Description......................................330
21.6.1.2   The Transformer Stage 330
21.6.1.3   Rectifier Stage..................................331
21.6.1.4   Filter Smoothing Stage 331
21.6.1.5   Regulator Stage..................................332
21.6.2   Audio Amplifier 334

           21.6.2.1 Description...334

           21.6.2.2 Design of a Class A Amplifier.................334

      21.6.3 Sallen and Key Second-Order Butterworth
Low-Pass Filter...341

           21.6.3.1 Description...341

      21.6.4 Automatic Switch on of Lamp in the Dark
Using a BJT...342

           21.6.4.1 Description...342

      21.6.5 Automatic Switch on of a Lamp in the Dark
Using Op-Amp 343

           21.6.5.1 Description...343

      21.6.6 Automatic Switch-On of Lamp in the Presence
of Light Using a BJT..344

           21.6.6.1 Description...344

      21.6.7 Automatic Switch-On of Lamp in the Presence
of Light Using an Op-Amp 345

           21.6.7.1 Description...345

      21.6.8 Humidity Detector...346

           21.6.8.1 Description...346

      21.6.9 Delay Switch..347

           21.6.9.1 Description...347

      21.6.10 Smoke Detector 348

           21.6.10.1 Description...348

      21.6.11 Multivibrators and the 555 Timer............................348

           21.6.11.1 Astable Multivibrator...............................348

      21.6.12 Humidity Detector Using 555 Timer........................351

           21.6.12.1 Description...351

      21.6.13 Two-Tone Musical Instrument Using 555 Timers 352

           21.6.13.1 Description...352

      21.6.14 Automatic Switch-On of Lamp in the Presence
of Light Using a 555 Timer 352

           21.6.14.1 Description...352

  21.7 Key Points..354

  References 354

Chapter 22 Future Trend of Analog Electronics..355

  22.1 Introduction 355

  22.2 Reconfigurable Analog Circuitry 356

  22.3 Analog Devices at High Power..358

  22.4 Future Advances in Applications of Analog Electronics 359

      22.4.1 Nanotechnology...359

      22.4.2 New Analog Building Block Working Voltage
Under 1 V 359

      22.4.3 Analog Electronics Sees a Revival in the Music
Industry 359

          22.4.4   New Type of Analog Computing 360
          22.4.5   Biotechnology..360
    22.5 Key points..360
    References 361

Chapter 23 Computer-Aided Simulation of Practical Assignment....................365

    23.1 Introduction 365
    23.2 First Introduction to Using Pspice...................................366
          23.2.1   Circuit Creation/Schematics....................................367
          23.2.2   To Connect the Components 367
          23.2.3   Names to Elements...368
          23.2.4   Running a Simulation...369
          23.2.5   DC Analysis 370
    23.3 Practical Assignments Using Pspice 370
          23.3.1   Assignment 1. DC Networks: Bias Point Analysis... 371
                   23.3.1.1   Experiment 23.1..371
                   23.3.1.2   Experiment 23.2..372
                   23.3.1.3   Experiment 23.3..372
          23.3.2   Assignment 2. AC Networks: AC Sweep 374
                   23.3.2.1   Experiment 23.4..374
                   23.3.2.2   Experiment 23.5..374
                   23.3.2.3   Experiment 23.6..376
          23.3.3   Assignment 3. BJT Operating Point, Q, Stability 377
                   23.3.3.1   Experiment 23.7..378
          23.3.4   Assignment 4. BJT Amplifier Analysis....................380
                   23.3.4.1   Experiment 23.8..381
                   23.3.4.2   Experiment 23.9..382
          23.3.5   Assignment 5. FET Amplifier Analysis and
                   Differential Amplifier 386
                   23.3.5.1   Experiment 23.10......................................386
                   23.3.5.2   Experiment 23.11 387
          23.3.6   Assignment 6. Active Filter–Power Amplifier 391
                   23.3.6.1   Experiment 23.12......................................391
                   23.3.6.2   Experiment 23.13......................................392
                   23.3.6.3   Experiment 23.14 393
                   23.3.6.4   Experiment 23.15......................................396
    23.4 Key Points..396
    References 396

Index...397

Preface

I have organized the material presented in this book in a way that can be readily used and effortlessly understood. The material presented is based on topics taught over a number of years. As a result, it has been tried and tested.

The book has a hierarchical and mixed method of presenting the learning material combining top-down and bottom-up approaches.

The book starts by examining an electronic system as a top-down approach to provide opportunity for the learner to acquire an understanding of what is globally required from electronics systems. Students may not be asked to achieve complete design projects in their early years of study, but it is worthwhile to encourage an understanding of the nature of global tasks required in analyzing a complete electronic system. This book includes in Chapter 1, a top-down approach to look at some methodologies used for electronic system design indicating the usefulness of embracing a methodical, rational method.

The book then switches to a bottom-up approach to include detailed circuit analysis leading students to gain a good understanding of how circuits operate. This will help in comprehending the material presented and will also contribute to a more enjoyable learning experience.

Some students may be familiar with the material provided in the early parts of the book. However, many students will benefit from going through these topics more formally, and the book provides notes and guidance to ensure that students are familiar with the fundamentals of analog electronics material before progressing.

The book is written to cover material for the first 2 years of an electrical and electronic engineering university degree. The book is also beneficial for all the years of a technician grade study and it can be used by students of science related degrees at any level to acquire acknowledge in analog electronics. Some courses may, if appropriate, ignore the material on semiconductor physics and concentrate on the external characteristics of semiconductor devices. Students of electronics or electrical engineering will need to study the physics-related material but students of other disciplines may find it more appropriate to omit these sections.

The book goes from a detailed explanation of fundamental to advanced concepts required to perform and design some applications of an analog electronic system.

Once the idea of amplification is clearly explained, the learning moves to electronic amplifiers of various characteristics. The book presents various models for transistors used in analog electronics to investigate the characteristics of electronic amplifiers. This permits a detailed analysis of voltage, current, and power gains to be determined as well as loading effects, input, and output impedance to be investigated. The concepts of frequency response and bandwidth are then discussed in more detail, and material is included to analyze differential amplifiers leading to the concept and realization of an operational amplifier.

Op-amps are important to all students and they will all benefit from lectures and practical work in this area. The book includes an in-depth analysis of negative feedback and deduces the effects of negative feedback on voltage and current gains, input and output impedances, and frequency response. The material in each chapter is presented with an emphasis on design and possible practical applications. Chapter 21 includes various common basic applications of analog electronics systems. Where students have a particular specialist interest in a specific topic, he or she may choose to perform the applications and laboratory experiments presented in Chapters 21 and 23, thus making the practical work more directly relevant to their course of study.

Once the general principles of feedback have been established, the book explains the derivation of the transfer function of a generalized feedback system. It presents a comprehensive treatment of different types of negative feedback. This is then used to design and analyze a range of electronic circuits that make use of negative feedback. Chapters include examples tailored to illustrate and consolidate the learning material presented in a chapter.

This book is beneficial to electrical and electronics engineering students in year 1 and 2 of their course. For advanced students, the book offers a good source of reference and revision on how detailed deductions of the different expressions used in electronic circuits are deduced. The book is also useful for all years of a degree in engineering where electronics is not the major subject.

The manner in which the material is presented prepares students to grasp the fundamentals of analog electronics and to use the knowledge acquired to perform basic design work as well as realize some basic applications.

A large section of the book is dedicated to details on realization and implementation of practical applications of an analog electronic system. It also includes many practical assignments to be performed using computer-aided packages tailored to the material covered in the book.

The book includes over 300 figures to illustrate the topics covered, examples, and applications. Over 100 specific tailored examples are included enhancing the material covered in this book. Calculations on examples are accurate to two decimal places. Each chapter includes a key point of the material covered.

In general, the book covers the fundamentals of analog electronics with an enhanced projection for design, analysis, and practical applications.

Pspice is registered trademark of Microsim Corporation.

Hernando L. Fernandez-Canque
Glasgow

Acknowledgments

I thank my wife, Vicky Grandon, for her support and assistance with the preparation of this book. Special thanks to my students who, through the years, have allowed me to guide them and observe how they come up with both queries and unexpected wrong results that needed to be corrected. These have helped me produce teaching material that avoids misconceptions in analysis of electronic circuits.

I received great support from the technician team at Glasgow Caledonian University Department of Engineering in setting up laboratory experiments maintaining software packages used in this book. Thanks to Tony Floyd for his help in the photographic material included. Thanks to my colleagues at Glasgow Caledonian University School of Engineering and Computing for their encouragement, discussion, and suggestions. Thanks to Sorin Hintea and colleagues of Cluj-Napoca University for joint work and publications in advanced analog electronic systems.

Finally, I acknowledge with gratitude the unconditional support and patience of my family and friends.

Author

Hernando L. Fernandez-Canque, PhD, MEng, BSc(Hons), Ing, MIEE, MIET, ILTM, Ceng, FHEA, was born in Arica, Chile. He earned his first degree at the University of Chile, Faculty of Engineering, Science and Mathematics, where he also taught electronics. In 1975, he moved to the United Kingdom where he earned a BEng in electrical and electronics engineering at the University of Glasgow, an MEng in solid state electronics at UMIST Manchester, and a PhD from Sheffield University with a thesis on capacitance and conductance studies on sputtered amorphous hydrogenated silicon Schottky barrier diodes. As part of his industrial experience, he has worked in the Sheffield Information Technology Centre, National Semiconductor Ltd., as director of the Intelligent Technologies Research Centre ITRC at Glasgow Caledonian University, and in various European projects and consultancies, including work with the Scottish Parasitic Diagnostic Laboratory, CERN, Moscow Academy of Science, Oviedo University, and Cluj-Napoca University where he is visiting professor. He has lectured at the University of Chile, the University of Sheffield in the United Kingdom, the University of Oviedo in Spain, the University of Cluj-Napoca in Romania, and as senior lecturer taught analog electronics for over 28 years at Glasgow Caledonian University in Scotland. He has published over 100 research publications in international journals, and at conferences, and book chapters on image processing, intelligent systems, analog electronic design, amorphous silicon and education in engineering. He is a fellow of the Higher Education Academic UK, a chartered engineer with the Engineering Council UK, and a member of the Institute of Engineering and Technology.

1 Analog Electronics Applications and Design

1.1 INTRODUCTION TO ANALOG ELECTRONICS

Humans experience sound, vision, smell, hearing, or any external physical stimuli in form. In general, the natural physical world is analog. A physical stimulus can be transformed into an electronic signal using a transducer; and outputs of transducers are, in general, analog electronic signals. In this sense, humans cannot totally eliminate the manipulation of analog signals. With the advancement in semiconductor technology it has become possible to create millions of efficient electronic components of small physical size that utilize very little energy to operate as integrated circuits. The availability of integrated circuits with huge numbers of components provide the potential for creating robust reliable electronic applications that can embed ways of testing the circuit for functionality. This in turn can make the manufacture of applications easier.

Transistors can be made to operate in two states, saturation and cut off, to provide output voltages at two defined levels (zero and one). Subsequently, digital electronics was born and the manipulation of signals was made easier at an ultrafast rate and the number of signals to manipulate increased exponentially. Machines can communicate with each other in digital form but when human interference is required analog systems continue to be necessary.

Analog electronics has been replaced by digital electronics in many fields, media and communication in particular, but with the advancement of digital electronics it is necessary to match these advances with developments in analog electronics. Trends of pulses in digital form at a very high speed resemble an analog signal. To obtain a detailed analysis of circuitry manipulating these ultrafast signals, analog electronic theory is needed. Analog electronics is still a vibrant part of electronics to accompany and complement digital electronics and electrical power.

1.2 ANALOG SIGNALS

An electronic analog signal provides the variation of a quantity such as voltage or current, with time, in a continuous fashion. In an analog signal waveform there is no discontinuity of values and there is no abrupt or large variation from one value to the next. A digital waveform differs from an analog waveform in that the digital waveform has only well-defined steps values kept at different periods of time. The transition between the two steps is assumed to be instantaneous. Figure 1.1 shows a sample of an analog signal and a binary digital signal.

In the analog waveform represented in Figure 1.1a the analog voltage value at times t_1, t_2, t_3, and t_4 are different values of voltage. At any particular time, values

1

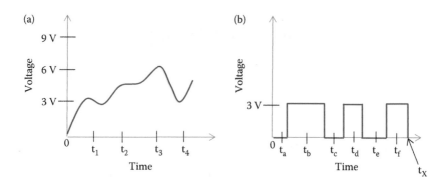

FIGURE 1.1   Example of (a) analog and (b) digital signals.

of voltages in the proximity are close in value to the value at this particular time and show a continuous change. In the digital waveform represented in Figure 1.1b, there are only two values of voltages: 0 or 3 V. Therefore, at different times the voltages can be only 0 or 3 V. The voltages at times t_a, t_c, and t_e are 0 V and the voltage at times t_b, t_d, and t_f are 3 V. The value at t_X shows a discontinuity because in theory it can be 0 or 3 V and it can change instantaneously from 0 to 3 V. Then an *analog signal* is any continuous signal for which the time-varying characteristic of the signal (usually voltage, current, or power) is represented as a function of time or some other time-varying quantity.

Electronic systems that can manipulate analog signals are termed analog systems.

1.3   ANALOG SYSTEMS

An electronic system is an assembly of electronic devices and components connected in order to respond to a defined input or inputs so as to produce the desired output or outputs. Electronic systems have characteristics that accept the input or inputs and manipulate them to create the desired output signal or signals. A good electronic system is one that has been designed to solve a particular problem in the most appropriate and efficient way.

In this chapter, we look at various approaches to system design and note the benefits of adopting a methodical, rational approach. We also consider the various stages of system design, briefly discuss the choice of the technology to be used, and consider various automated design tools.

Many systems take as their input some physical quantities and produce as their output a variation in these physical quantities. Figure 1.2 shows a block diagram of such system.

In an electronic system, we normally sense the physical input quantity using some form of sensor, transducer, or actuator that generates an electrical signal correlated to the physical input. In the same way, in a system the output quantity is usually uninhibited by a sensor, transducer, actuator, or display that is controlled by the electronic output from the electronic system. Considering the sensors, transducers, and actuators, the inputs and outputs will be related to the external physical quantities

FIGURE 1.2 Electronics system as part of a complete system.

concerned. By allowing for the sensors, transducers, and actuators to be external to the electronic system, the input and output of an electronic system will be electronic signals.

In many cases, electronic systems are used mainly because they deliver an economical solution, in some cases electronic systems provide the only solution. The processing and operations required by an electronic system will vary with the nature of the input signals and output signals and the required global function. The processing and operations required by an electronic system can include: amplification, addition, subtraction, integration, differentiation, filtering, counting, timing, signal generation, etc.

A designer of an electronic system has to be able to analyze a circuit so as to predict its electronic behavior. To design and construct the circuits included in an electronic system, the designer should also be able to select suitable components in terms of value, tolerance, voltage, current, power rating, and cost.

In order to create or implement an analog electronic application it is necessary to be able to analyze circuits used. Design and implementation of an application of analog electronics goes hand in hand. In both cases, knowledge of the fundamentals of electronics is required. Knowing the fundamentals of electronics will allow a designer or practitioner to predict the behavior of a circuit or modify an existing circuit as well as create and improve an electronic system.

A design methodology will be a concept that students can become familiar with as they learn about existing systems and how they work. Complex systems can be made easier to understand if broken down. A range of steps is necessary to successfully conclude the realization of an electronic design system.

1.4 APPLICATION AND DESIGN OF ANALOG SYSTEMS

An electronic system can be represented as a closed structure for which all the inputs and outputs are known. In practice, we select and choose to enclose a component, or group of components, that are of specific interest to us.

1.4.1 CUSTOMER REQUIREMENTS

An important part of the process of designing an application is determining precisely the requirements of the user or customer. The final user of the electronic system will, in many cases, articulate their requirements in nontechnical or imprecise terms. It is imperative at this stage to clarify exactly what the system will do and at what range of variables. The designer and customer must agree on what is required from the system. This agreement will create a contract wherein the designer will design an electronic system to satisfy this contract. The designer will convert these requirements into the technical system *design specifications*.

1.4.2 Top-Level Specifications

A top-level specification considers the system as a global entity. A top-level specification looks at characteristics that are features of the entire system, rather than of individual components. These properties are often complex in nature, and may relate to several diverse aspects of the system. The top-level specifications determine what the system is to do without taking into consideration how to do it. This task will produce a global arrangement for the system. The top-level specifications must define precisely what the system should do in response to all possible inputs, and must also establish any restrictions on the design of the electronic system. The top-level specifications usually include the production of a system block diagram, identifying all the inputs and outputs of the system and input and outputs of each block.

1.4.3 System Design Approach

In a systematic approach, a complex system is simplified by dividing it into a number of smaller building blocks. These building blocks are then themselves subdivided, the process can be repeated until the various parts have devolved into building blocks that are sufficiently simple to be easily understood. Partitioning is a process of dividing a complex system into a series of modules or subsystems to aid its design and implementation.

In a system consisting of a series of subsystems, the output of each subsystem represents the input of the next. This process is illustrated in Figure 1.3. The modules within the system can then be defined in terms of the characteristics of the inputs and outputs that they represent.

Depending on the system two approaches can be used. Either you start with the global system at the top level and go down in levels until the design is completed (*top-down*) or you start at the component bottom level and build up until the system is completed (*bottom-up*) (Figure 1.4).

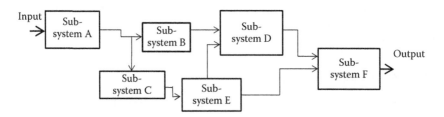

FIGURE 1.3 Partitioning of a complex system into subsystems.

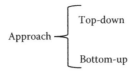

FIGURE 1.4   Two approaches to design an electronic system.

The top-down approach leads to a study of the way in which complex systems may be represented diagrammatically, and how complex arrangements may be divided into a set of less complex modules.

1.4.3.1   Top-Level Design

The top-level design determines a way of implementing the top-level specification. It involves the proposal of one or more solutions to the problem, then assesses and decides on one solution to be implemented. It involves identifying the techniques to be used to produce function described in the specifications. This does not involve detailed design of circuits or generation of programs, but basically determining their overall process of operation. It also requires a decision on whether the proposed solution includes system partitioning. It includes a decision on how the system will be implemented, whether the system will use discrete components or integrated circuit, the device technology to be used, etc. Any decision required for a global implementation is part of this process.

1.4.3.2   Detailed Design

As the function and principles of operation of the system or subsystems are known from the top-level design, the detailed circuitry or software can be designed. The detailed design will produce the final circuits at the level of components and devices and their connections. The designer must consider all aspects affecting the system such as speed of operation, power dissipation, time taken to complete the design and realization, and testability. The designer must be aware of all the significant factors affecting the performance of the design and needs to have a good knowledge of the variation of parameters of fabricated devices.

1.4.4   Technology Choice

Determining the right technology and methodology of implementation for an electronic system can make the difference whether a project succeeds or fails. After the specification of the design has been completed it is possible to decide on the technology to be used according to the nature of the project. There are a variety of choices for implementing an electronic system, such as analog system, digital, discrete components, standard integrated circuit, programmable logic array, microprocessor, software, hardware, bipolar technology, field effect transistor (FET) technology, and others.

If the technology used is based on integrated circuits, then all tests are done in software and the system has to behave perfectly within specification. It must be working according to specification before it is committed to fabrication. If the technology choice is to use discrete components a prototype is required in most cases. Tests will then be done in hardware before the system is committed to production. In the case of using discrete components, the selection of components is important. One important difference to consider among components is the variation in physical size of components and the specifications for a given component type. Electrical components vary in size depending on their power rating rather than value, and low-power devices are small. Manufacturers produce only certain values of passive components. The values of passive components available adhere to a defined sequence known as nominal preferred values (npv).

In many cases, the component's values obtained from a particular design calculation will not coincide with an npv component. In this case, the designer could use a higher tolerance component in order to obtain the required value, or use a variable component, but using a variable component, the production cost will be increased. The nearest npv of component could be used, but then it is essential to make sure that the design specifications are satisfied. Worst-case analysis should be performed for a designed circuit to ensure that design specifications are fulfilled for all values of every component, especially if the circuit is to be manufactured in a large number of units. The worst-case analysis should consider the tolerance of components used.

1.4.4.1 System Testing

After the system is constructed a detailed and comprehensive testing of the whole electronic system under various conditions is required to make sure the system performed as expected.

1.4.4.2 Social and Environmental Implications

A designer has to consider the impact of the system on the environment and any social implications. There are various regulations in existence that must be adhered to. Some are provided by professional bodies, geographical groups of nations, or by certain restrictions within the area where the electronic system will be used.

1.4.4.3 Documentation

A design is never conclusively completed. It can go to production to satisfy specification, but in the future it may be intended to improve it. New components may be found with better properties, or it may need adaptation to a new application; or a new upgraded version may be required. Any modification to the electronic system or new way of testing the device will require the original information accumulated when it was first designed. A design is not complete without detailed documentation of the design process, problem, and results. Documentation records of a design job are always required.

1.4.5 Distortion and Noise

There are many electromagnetic signals accessible in the space surroundings and electronic systems. These waves are emitted from other electronic systems, general communication systems, natural electromagnetic phenomenon, and switches from other electromechanic systems. An electronic system in many cases contains

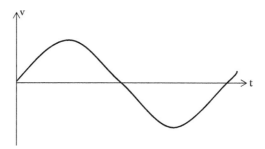

FIGURE 1.5   Pure sinusoidal original signal.

an amplifier and a small noise signal can be amplified affecting the functioning of the electronic system that is being implemented. Internal signals are also generated by components or other modules used in the system to be implemented at certain current and voltages level or different frequencies. All these signals are undesired noise to the system to be implemented.

The noise particularly affects an analog electronic system as noise can increase or reduce a value of a voltage and current used at a particular point of the system that can change the required response of the system.

In electronic systems, in general, and analog electronic systems, in particular, the *distortion* and *noise* can limit the performance of an electronic system. The designer should design a system in such a way that these effects are eliminated or minimized.

Below there are some familiar examples of common noise and distortion found in analog electronic systems. We can illustrate noise and distortion assuming a pure sinusoidal signal as shown in Figure 1.5.

1.4.5.1   Noise

Noise can occur due to fluctuation of the signal as a consequence of variation within the system or/and external effects of the environment. Figure 1.6 shows the sinusoidal signal of Figure 1.5 with noise added.

1.4.5.2   Distortion

Figure 1.7 shows examples of sinusoidal waveform including distortion due to clipping, crossover, and harmonic distortion.

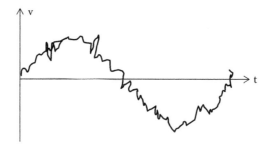

FIGURE 1.6   Sinusoidal signal including noise.

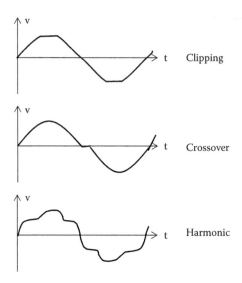

FIGURE 1.7 Example of the sinusoidal waveform including clipping, crossover, and harmonic distortion.

Clipping is the elimination of part of the signal that can happen due to improper biasing. Crossover distortion is the problem of wrong synchronization of more than one part of the signal. Harmonic distortion occurs when signals of different frequencies are added or subtracted from the original signal creating a response that contains frequencies that are not required. One of the major tasks of the electronic designer is to eliminate or reduce the magnitude of noise and distortion.

1.4.6 ELECTRONIC DESIGN AIDS

There are a number of electronic design aid tools that you can use to simplify the system design process. It is possible to use tools to

- Produce circuit and block diagrams
- Perform circuit simulation
- Perform placement of circuit board layout
- Automatically place components and route connections
- Perform testing
- Predict behavior

It is important to note that the use of the electronic design aid tools *do not remove the requirement to understand the details of circuit design* or the operation of electronic components. A computer-aided design software to simulate circuit can simplify the task of circuit analysis, but it is always necessary to have a detailed understanding of the basic principles that rule currents and voltages in circuits. *The circuit designer must always know what to expect from the simulator,* and should always be able to predict the signal response. The simulator is then used to investigate the operation in more details for different component values or different variation of the input. A circuit simulator can investigate a worst-case analysis where all the components are varied according to their tolerances.

The electronic design aid tools complement the designer skills; they do not replace them.

1.5   KEY POINTS

- An analog signal is a continuous signal for which the time-varying feature of the signal is represented as a continuous function of time.
- Systems may be described in terms of the nature of their inputs, their outputs, and the relationship between them.
- Systems sense external physical quantities using *sensors or transducers* and affect external quantities using *actuators or transducers.*
- Within a system we often use electrical signals to represent physical quantities. The nature of these signals varies with the application.
- Complex systems are often represented by block diagrams, which often partition systems into smaller, more manageable modules.
- Complex systems are more easily designed or understood through the use of a systematic, top-down approach.
- Major components of the design process include the following: the customer requirements; top-level specification; top-level design; detailed design; module testing; and system testing.
- Top-down design is normally followed by bottom-up testing.
- At an early stage, it is necessary to decide on the method of implementation to be adopted.
- Technological decisions include a choice between analog and digital techniques, between integrated and discrete approaches, and between various device families.
- A large number of automated tools are available to simplify the design process.
- The electronic design aid tools complement the designer skills; they do not replace them.

2 Electric Circuits

2.1 INTRODUCTION

This chapter introduces concepts to help students progress through the early parts of circuit analysis and prepare them for the more detailed analysis that follow. For some students this would be a beneficial revision of material seen before. This chapter defines the basic characteristics and components of electrical circuits. It is expected that on completion of the chapter students will understand the concept of units, current, voltage, power, sources, and electrical loads. They will also be able to describe the characteristics of circuit components and the differences between ideal and practical voltage and current sources. They will be able to understand the relation between voltage and current in a resistance provided by Ohm's law and its application to resistive circuits. The characteristics and relationship between voltage and current in active and passive components will also be studied.

Resistors, inductors, and capacitors are the most widely used electronic components, an understanding of their operation and characteristic is necessary for everyone studying analogue electronics.

2.2 UNITS

There are some basic units used in analogue electronics. It is important to use them correctly in order to avoid having to recalculate a design. Until the last century different systems of units were used by different institutions, laboratories, or countries. This created confusion for many students and researchers as different results were published for the same experiments. A conference in units was organized to unify criterion on units. The conference concluded with an International System of Units or SI system. The SI system was formally introduced in 1960 and has been accepted by many countries and institutions as their system of measurements. Throughout this book we will use the SI system of units.

The SI system selects six physical quantities as the basis of their units. All other units are derived from these basic units. The six basic physical quantities selected are indicated in Table 2.1.

You will encounter a much larger number of derived units. All derived units are related to the six basic units and can be expressed in terms of the basic quantities: mass, length, time, electric current, absolute temperature, and luminous intensity. For example, a unit of area can be expressed as the square of length, that is, square meters. Other common examples are shown below.

Volume	V	m^3
Velocity	v	$m/s = ms^{-1}$
Acceleration	a	m/s^{-2}

TABLE 2.1
Six Physical Quantities Selected in SI System of Units

Quantity	Names	Symbol
Mass	Kilogram	kg
Length	Meter	M
Time	Second	S
Electric current	Ampere	A
Absolute temperature	Kelvin	K
Luminous intensity	Candela	Cd

In electrical and electronic engineering there are some common units used in different types of physical quantities, devices, components, and circuits. Here are short definitions of some of the units to be used in this book.

2.2.1 UNIT OF CHARGE

The unit of charge is the *coulomb* (C). The coulomb is defined as the quantity of electricity that flows in a circuit when a current of 1 ampere (A) is maintained for 1 second (s). One Coulomb is equal to 6.24×10^{18} electrons. The electric charge (Q) can then be expressed in terms of time and current as

$$Q = I \cdot t \, [C] \tag{2.1}$$

where I is the current in amperes, t is the time in seconds, and Q is the charge in coulomb.

2.2.2 UNIT OF FORCE

The unit of force is the *newton* (N). A newton is defined as the force that, when applied to a mass of 1 kilogram (kg), gives it an acceleration of 1 m/s². The force (F) can then be expressed in terms of mass and acceleration as

$$F = m \cdot a \, [N] \tag{2.2}$$

where m is the mass in kilograms, a is the acceleration in meter per second squared, and F is the force in newton.

2.2.3 UNIT OF ENERGY

The unit of work or energy is the *joule* (J). The joule is defined as energy transferred when a force of 1 N is applied through a distance of 1 m in the direction of the force. The energy (E) can then be expressed in terms of force and length as

$$E = F \cdot l \, [J] \tag{2.3}$$

where F is the force in newton, l is the distance in meters moved by the body in the direction of the force, and E is the energy in joules.

2.2.4 UNIT OF POWER

The unit of power is the *watt* (W). Power is defined as the rate of transferring energy. The power (P) can then be expressed in terms of energy and time as

$$P = \frac{E}{t} [W] \tag{2.4}$$

where E is the energy transferred in joules, t is the time in seconds, and P is the power in watts.

Note that using the definition of power above, the energy (E) can also be expressed in terms of power and time as

$$E = P \cdot t [J] \tag{2.5}$$

This provides energy expressed in watts per second, when dealing with large amounts of energy consumption, the unit used is the kilowatt hour (kW h) where

$$1\,kW\,h = 1000\,W\,h = 1000 \times 3600\,W\,s \quad \text{or} \quad J = 3{,}600{,}000\,J$$

2.2.5 UNIT OF ELECTRIC VOLTAGE

The unit of electric voltage is the *volt* (V). One volt is defined as the difference in potential between two points in a circuit when carrying a current of 1 A dissipates 1 W. The voltage (V) can be expressed in terms of power and current as

$$V = \frac{P}{I} = Volts = \frac{Watt}{Amperes} = \frac{Joule/seconds}{Amperes} = \frac{Joule}{Amperes \cdot seconds} = \frac{Joule}{Coulombs} \tag{2.6}$$

where P is the power in watts, I is the current in amperes, and V is the voltage in volts.

2.2.6 UNIT OF RESISTANCE AND CONDUCTANCE

The unit of electric resistance is the *ohm* (Ω). One ohm is defined as the resistance between two points in a circuit as 1 V applied at the two points provides a current of 1 A. The resistance (R) can then be expressed in terms of voltage and current as

$$R = \frac{V}{I} [\Omega] \tag{2.7}$$

where V is the potential difference across the two points containing the resistance in volts, I is the current flowing between the two points in amperes, and R is the resistance in ohms.

The reciprocal of resistance is called *conductance* and is measured in siemens (S). The conductance (G) can then be expressed in terms of the resistance as

$$G = \frac{1}{R}[S] \tag{2.8}$$

where R is the resistance in ohms and G is the conductance in siemens.

The electric power from a direct current (dc) can also be expressed in terms of voltage and currents as

$$P = VI = I^2R = \frac{V^2}{R}[W] \tag{2.9}$$

EXAMPLE 2.1

An electric heater has a resistance of 28 Ω. If this heater is connected to a dc power supply of 98 V, calculate the current that flows through this heater and its electric power dissipation.

SOLUTION

$$\text{Current} = I = \frac{V}{R} = \frac{98}{28} = 3.5\,A$$

$$\text{Power dissipation} = P = VI = 98 \times 3.5 = 343\,W$$

EXAMPLE 2.2

An electric iron has a dc current of 7 A flowing to its resistance of 25 Ω. Calculate the potential difference across its resistance and the power dissipated by its resistance.

SOLUTION

The potential difference, V_i, across its resistance is

$$\text{Potential difference} = V_i = IR = 7 \times 25 = 175\,V$$

The power, P_i, dissipated by its resistance is

$$\text{Power dissipation} = P_i = VI = 175 \times 7 = 1.285k\,[W]$$

or

$$P_i = \frac{V_i^2}{R} = \frac{175^2}{25} = 1.285k\,[W]$$

or

$$P_i = I^2R = 7^2 \times 25 = 1.285 \text{k[W]}$$

EXAMPLE 2.3

Calculate the voltage supply and the power dissipated by an electric heater of 10 Ω resistance when a current of 5 A flows through it. If the heater is connected to the voltage supply for 10 h determine the energy used and cost assuming that the value of a unit of electricity is \$0.5.

SOLUTION

The voltage of the power supply is

$$\text{Voltage of supply} = V = IR = 10 \times 5 = 50\,\text{V}$$

$$\text{Power dissipation} = P = VI = 50 \times 5 = 250\,\text{W}$$

or

$$\text{Power dissipation} = P = I^2R = 5^2 \times 10 = 250 \text{[W]} = 0.25 \text{k[W]}$$

Unit of electricity = 1 kW h
Then the cost of energy used per hour is

$$\text{Cost of energy in 1h} = \text{Power used} \times \text{time} \times \text{cost of electricity unit}$$

$$= 0.25\,\text{k} \times 1 \times \$0.5 = \$0.125/\text{kW h}$$

Total cost of the electricity for the use of this heater for 10 h is equal to

$$\text{Total cost} = \text{No. of hours} \times \text{Cost of unit of electricity} = 10 \times 0.125 = \$1.25$$

2.3 CONCEPT OF ELECTRIC CHARGE AND CURRENT

In our universe all matter consists of atoms. These atoms contain particles classified according to their charge as electrons, protons, and neutrons (Figure 2.1).
A short definition in terms of type of charge for these particles is

An *electron* is an elementary particle charged with a small and constant quantity of negative electricity.
A *proton* is an elementary particle charged with a small and constant quantity of positive electricity.
A *neutron* is uncharged.

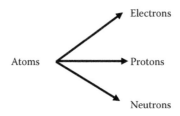

FIGURE 2.1   Classification of particles in atoms according to their charge.

If an atom has excess electrons, it is said to be negatively charged. If an atom has excess protons, it is said to be positively charged. A charged atom is called an ion. A body containing a number of ionised atoms is also said to be electrically charged.

2.4   MOVEMENT OF ELECTRONS AND ELECTRIC CURRENT IN A CIRCUIT

Any movement of charges creates an electric current. Conventionally, current is defined in terms of movement of electrons. Electrons vibrate within the atoms, but their movement is in different directions that can cancel each other's movement, adding up to a total movement of all electrons equal to zero. As a consequence, in a material with no external energy applied, there is no current circulating through it. If an ammeter is connected to measure their total current it will indicate 0 A. But if an external energy is applied to a material making electrons move in a concerted motion in one direction then a current is produced. Electrons have a certain potential energy moving freely from one energy level to another and this movement, when undertaken in a concerted manner, is called an electric current flow.

For convenience the point of high potential is termed the positive and the point of low potential is termed the negative, hence conventionally a current is said to flow from positive to negative. Because electrons are assumed to have negative charge, conventional current flow is in the opposite direction to that of an electron flow. This can create some confusion in the analysis of circuits. Later in the book we will develop a way of avoiding any confusion in determining value or direction of currents in active or passive components.

In any practical application, it is required that the current flow continuously for as long as it is needed. In order to obtain a current in a circuit it is necessary that *two conditions must be fulfilled*:

1. There must be an entire circuit around which the electrons may move.
2. There must be a driving stimulus to cause the continuous flow.

These two conditions will lead to the concept of *circuit* and *electromotive force (emf)*.

2.4.1   CIRCUIT

The path required to allow electrons to move to fulfill condition 1. This path will generate what is known as the electric circuit.

2.4.2 ELECTROMOTIVE FORCE

The *driven* stimulus required to fulfill condition 2 is known as the *electromotive force (emf)*. Each time a charge passes through the source energy provided by the *emf* the continuous current flow is maintained.

2.4.3 SOURCE

An element that provides electrical energy supplied to a circuit is termed a *source*. The *emf* in a circuit is provided by source energy such as a battery or a generator and is measured in volts. A change in electric potential between two points in an electric circuit is called a potential difference.

2.4.4 LOAD

When a current is established in a circuit, some elements in this circuit will absorb or convert the electrical energy supplied by the source. An element that absorbs and/or converts the electrical energy supplied by the source is termed a *load*.

We can classify electrical elements in a circuit according to whether they provide energy to the circuit as active components or whether they absorb or convert energy to a circuit as passive components.

2.5 PASSIVE COMPONENTS: RESISTANCE, INDUCTANCE, AND CAPACITANCE

Passive components in an electric circuit are components that do not add energy to the circuit. They usually dissipate or transform energy. There are different types of passive components according to their relationship between their voltages and currents.

The potential difference (V) across the terminals of circuit component is proportional to the current (I) flowing between them.

$$V \text{ proportional to } I$$

This proportionality varies for different components. We can identify the type of passive component according to its relationship between the voltage and the current. Some of the most common types of components found in an electric circuit are resistors, capacitors, and inductors. In this section we can look at some of the characteristics of these three types of passive components.

2.5.1 RESISTANCE

The *resistance* is an electrical characteristic that describes how a component opposes the circulation of current. If the ratio of voltage to current provides a constant value then the passive component is known as a resistance or resistor. In a resistance, the relationship between voltage and current is defined in *Ohm's law*. This states that the flow of electric current through a resistance is proportional to the applied voltage. Ohm's law can be expressed as

$$V = R \cdot I \quad \text{or} \quad I = \frac{V}{R} \qquad (2.10)$$

Let us illustrate Ohm's law with an example.

EXAMPLE 2.4

Let us draw a graph of current versus voltage when a variable dc voltage source is connected to a resistance of 5 Ω. Assume that the source voltage can be changed between 0 and 8 V in increments of 2 V.

SOLUTION

A sketch of the final circuit diagram is shown in Figure 2.2.
For each voltage, we take I = V/R to calculate the current in the circuit
Let us work it out for a value of 2 V on the battery

$$I = V/R = 2/5 = 0.4\,A$$

Similarly, by changing the values of the source to create voltages of 4, 6, 8, and 10 V the current can be calculated as tabulated in Table 2.2.
A graphical diagram of the voltage against current is shown in Figure 2.3.
This graph shows the relationship between voltage and current in a resistor for this example.

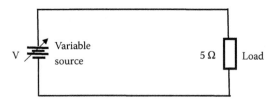

FIGURE 2.2   Circuit diagram of voltage source and a resistor to illustrate Ohm's law.

TABLE 2.2
Voltages and Currents for Ohm's Law for Example 2.4

Voltage (V)	Current (A)
0	0
2	0.4
4	0.8
6	1.2
8	1.6

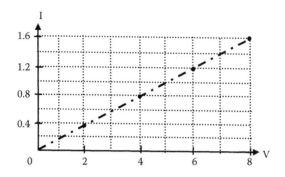

FIGURE 2.3 Diagram of I–V Ohm's law example.

2.5.1.1 Resistors Connected in Series and Parallel

There are many ways of connecting resistors. Some connections of resistors are widely used and a name is given to them as they repeat themselves in different systems. You can find connections such as Star, Delta, Pi, Tee, and others in many systems. Let us look at two common way of connecting resistors: series and parallel.

2.5.1.2 Resistors Connected in Series

If the *current* passing through resistors is the *same* then these resistors are connected in series. Figure 2.4 illustrates an example of four resistors in series.

It is possible to reduce all resistors connected in series as one equivalent resistor R_T. Figure 2.5 shows the equivalent resistor R_T of n resistors connected in series.

For n resistors connected in series this equivalent resistance R_T is given by

$$R_T = R_1 + R_2 + R_3 + \cdots + R_n \tag{2.11}$$

2.5.1.3 Resistor Connected in Parallel

If the *voltage* across connected resistors is the *same* then these resistors are connected in parallel. Figure 2.6 shows an example of five resistors connected in parallel.

It is possible to reduce all resistors connected in parallel as one equivalent resistor R_T. Figure 2.7 shows n resistor in parallel and its equivalent resistor R_T.

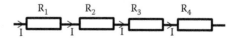

FIGURE 2.4 Resistors in series.

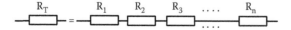

FIGURE 2.5 Equivalent resistor for n resistors connected in series.

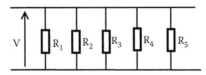

FIGURE 2.6   Resistor in parallel.

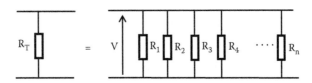

FIGURE 2.7   Equivalent resistor for n resistors connected in parallel.

For n resistors connected in parallel the equivalent resistance R_T is given by

$$\frac{1}{R_T} = \frac{1}{R_1} + \frac{1}{R_2} + \frac{1}{R_3} + \cdots + \frac{1}{R_n} \qquad (2.12)$$

2.5.1.4   Special Case

For the special case of *two resistors connected in parallel* as indicated in Figure 2.8 an easy reduction can be found

$$\frac{1}{R_T} = \frac{1}{R_1} + \frac{1}{R_2} = \frac{R_2 + R_1}{R_1 \cdot R_2}$$

and

$$R_T = \frac{R_1 \cdot R_2}{R_1 + R_2} \qquad (2.13)$$

Two resistances in parallel can be reduced to one resistor equivalent with value equal to the product divided by the sum values of the two resistors. This can be used for more than two resistors in parallel by taking two resistors at a time.

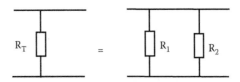

FIGURE 2.8   Equivalent resistor for two resistors in parallel.

EXAMPLE 2.5

Three resistors $R_1 = 100\ \Omega$, $R_2 = 150\ \Omega$, and $R_3 = 300\ \Omega$ are connected in parallel. Resistor R_3 has a current I_3 of 30 mA. Calculate the value of the supply voltage, V_s, across the resistances and the supply current I_s.

SOLUTION

The potential difference across R_3 is

$$V_{R3} = R_3 \times I_3 = 300 \times 30 \times 10^{-3} = 9\,V$$

As the resistors are connected in parallel, they will be connected to 9 V and $V_s = 9$ V.

The three resistors are in parallel and can be reduced into one, R_T, as

$$\frac{1}{R_T} = \frac{1}{R_1} + \frac{1}{R_2} + \frac{1}{R_3} = \frac{1}{100} + \frac{1}{150} + \frac{1}{300} = \frac{6+4+2}{600} = \frac{12}{600}$$

and $R_T = 50\ \Omega$.

Then the supply current, I_s, is

$$I_S = \frac{V_S}{R_T} = \frac{9}{50} = 180\,mA$$

2.5.2 CAPACITANCE

A capacitor is a component that can store charges. The effect of storing charges in a capacitor makes the relationship between voltage and current more complicated and creates an electric field. A capacitor consists of two conducting surfaces separated by an insulating layer called a dielectric. Figure 2.9 shows different types of capacitor fabrication.

Capacitors are better understood if we consider a simple capacitor consisting of two parallel plate capacitor as the one shown in Figure 2.10.

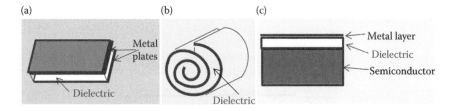

FIGURE 2.9 Capacitor fabrication. Consisting in two conductors separated by an insulator: (a) Parallel plates capacitor. (b) Tubular capacitor made from metal foil and flexible dielectric material. (c) Integrated circuit capacitor.

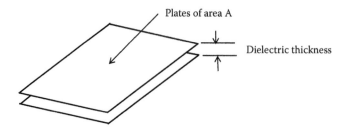

FIGURE 2.10   Parallel plate capacitor.

Electrons flowing around a circuit can produce a positive charge on one side of the capacitor and an equal negative charge on the other. This results in an electric field between the two plates (Figure 2.11).

The charge stored in a capacitor is directly proportional to the voltage across it. The constant of proportionality is the capacitance C. Then, we can write the relationship among the capacitance C, the charge Q, and the voltage V as

$$C = \frac{Q}{V} \qquad (2.14)$$

The physical dimensions of a capacitor effects the amount of charge to be stored, then the physical dimension of a capacitor affects the capacitance of this component. The capacitance of a parallel plate capacitor is proportional to the surface area of the plates and inversely proportional to their separation. The constant of proportionality is the permittivity of the dielectric (ε). We can write an equation for the capacitor in terms of its physical properties as

$$C = \frac{\varepsilon A}{d} \qquad (2.15)$$

If the charge is measured in coulombs and the voltage in volts, the capacitance is given in farads.

The permittivity ε is a quantity that indicates the electrical characteristic of a material. The permittivity is normally expressed as the product of the absolute value ε_0 (8.85×10^{-12} F/m) and the relative permittivity ε_r as

$$\varepsilon = \varepsilon_0 \cdot \varepsilon_r \qquad (2.16)$$

Positive charge Q

Plates of area A

Dielectric thickness = d

Voltage V

Negative charge −Q

Electric field

FIGURE 2.11   Electric field in parallel plates capacitor.

and the capacitance can be expressed as

$$C = \frac{\varepsilon_0 \varepsilon_r A}{d} \qquad (2.17)$$

The charge on the plates of the capacitor produces an electric field E, which is proportional to the voltage applied to it and inversely proportional to the distance of separation between the plates d. Then,

$$E = \frac{V}{d} \qquad (2.18)$$

The stored charge in a capacitor produces a force between positive and negative charges and is described in terms of the *electric flux* linking them, which is measured in coulombs.

Flux density D given by the flux per unit area can be expressed as

$$D = \frac{Q}{A} \qquad (2.19)$$

where Q is the charge in the capacitor and A is the area of the capacitor plates.

The relationship between voltage (v_c) and current (i_c) in a capacitor is very different compared to a resistor. The current–voltage relationship can be expressed in terms of its voltage or current.

The voltage on a capacitor, v_c, is given by

$$v_c = \frac{1}{C} \int i_c \, dt \qquad (2.20)$$

where i_c is the current through the capacitor integrated as a function of time and C is the capacitance.

The current into a capacitor i_c is given by

$$i_c = C \frac{dv_c}{dt} \qquad (2.21)$$

where dv_c/dt is the first derivative of the capacitor's voltage with time.

Combining Equations 2.14, 2.18, and 2.19 for a capacitor, it can be deduced that the permittivity of a dielectric within a capacitor is equal to the ratio of the electric flux density to the electric field strength.

As

$$\varepsilon = \frac{D}{E} \qquad (2.22)$$

When capacitors are used with sinusoidal signals the current leads the voltage by 90° as we will discuss in Chapter 3.

The energy (W) stored within a charged capacitor can be deduced in terms of their voltage or charge as

$$W = \frac{1}{2}CV^2 = \frac{1}{2}\frac{Q^2}{C} \qquad (2.23)$$

2.5.2.1 Capacitors in Parallel

Consider a voltage V applied across two capacitors connected in parallel as indicated in Figure 2.12.

The charge on each capacitor is $Q_1 = VC_1$ and $Q_2 = VC_2$ from Equation 2.14. If the two capacitors are replaced by an equivalent capacitor C_E, then the total charge Q of the single capacitor is equal to the sum of the charges of the two capacitors in parallel, Q_1 and Q_2.

Then the total charge Q is equal to $Q_1 + Q_2$. Using Equation 2.14 the charges can be written in terms of the capacitance and voltage as

$$Q = Q_1 + Q_2 = VC_E = VC_1 + VC_2 = V(C_1 + C_2)$$

Then, $C_E = C_1 + C_2$.

Expanding for n capacitors in parallel. The equivalent capacitance C_T of n capacitors connected in parallel is equal to the sum of all n capacitors. Then,

$$C_T = C_1 + C_2 + C_3 + \cdots + C_n \qquad (2.24)$$

2.5.2.2 Capacitors in Series

Consider a voltage V applied across two capacitors connected in series as indicated in Figure 2.13.

The capacitors are in series the current is the same, then charge passing through the capacitors is the same. The charge Q in each capacitor C_1, C_2, and an equivalent capacitor C_E is the same. The sum of the voltages across C_1 and C_2 add to the total voltage V applied across the two capacitors. Then, $V = V_1 + V_2$ or is written in terms of charge and capacitance using Equation 2.14:

$$V = V_1 + V_2 = \frac{Q}{C_E} = \frac{Q}{C_1} + \frac{Q}{C_2} = Q\left(\frac{1}{C_1} + \frac{1}{C_2}\right)$$

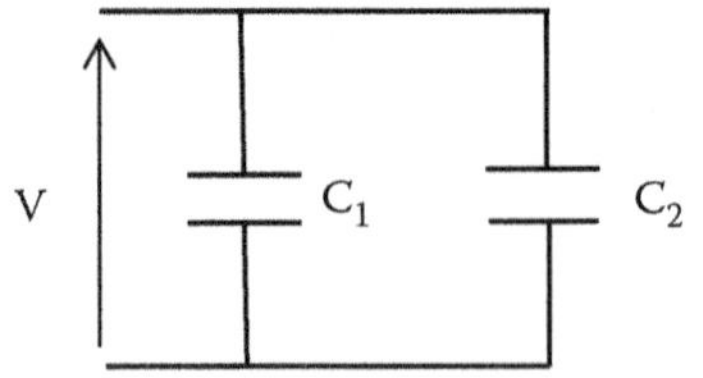

FIGURE 2.12   Two capacitors connected in parallel.

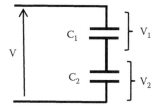

FIGURE 2.13   Two capacitors connected in series.

and

$$\frac{1}{C_E} = \frac{1}{C_1} + \frac{1}{C_2}$$

In general, the reciprocal of the equivalent capacitor C_T of n capacitors connected in series is equal to the sum of the n reciprocal capacitors.

$$\frac{1}{C_T} = \frac{1}{C_1} + \frac{1}{C_2} + \frac{1}{C_3} + \cdots + \frac{1}{C_n} \tag{2.25}$$

EXAMPLE 2.6

Calculate the equivalent capacitance of three 9 μF capacitors connected in series and then in parallel.

SOLUTION

Three 9 μF capacitors in series can be reduced to one equivalent capacitance C_{ST} as

$$\frac{1}{C_{ST}} = \frac{1}{9\,\mu F} + \frac{1}{9\,\mu F} + \frac{1}{9\,\mu F} = \frac{3}{9\,\mu F}$$

Then, $C_{ST} = 3$ μF.

And the three 9 μF capacitors connected in parallel can be reduced to one equivalent capacitance C_{PT} as

$$C_{PT} = 9\,\mu F + 9\,\mu F + 9\,\mu F = 27\,\mu F$$

EXAMPLE 2.7

Three capacitors $C_1 = 20$ μF, $C_2 = 30$ μF, and $C_3 = 50$ μF are connected in parallel to a dc voltage source $V_s = 110$ V. Calculate the total equivalent capacitance of the three capacitors in parallel C_T, the total charge Q_T, and the charge on each capacitor.

Solution

The total equivalent capacitance of the three capacitors connected in series is

$$C_T = C_1 + C_2 + C_3 = 20 \ \mu F + 30 \ \mu F + 50 \ \mu F = 100 \ \mu F$$

The total charge Q_T is

$$Q_T = C_T V_s = 100 \cdot 10^{-6} \times 110 = 11 \ mF$$

The charge on the 20 μF capacitor is $Q_1 = C_1 V_s = 20 \times 10^{-6} \times 110 = 2.2 \ mC$.
The charge on the 30 μF capacitor is $Q_2 = C_2 V_s = 30 \times 10^{-6} \times 110 = 3.3 \ mC$.
The charge on the 50 μF capacitor is $Q_3 = C_3 V_s = 50 \times 10^{-6} \times 110 = 5.5 \ mC$.

also $Q_T = Q_1 + Q_2 + Q_3 = 2.2 \ mC + 3.3 \ mC + 5.5 \ mC = 11 \ mC$ as before.

2.5.3 INDUCTORS

An inductor is a coil of conducting material. In the simple form, an inductor can be illustrated as in Figure 2.14.

A changing current in the conducting wire produces a changing magnetic field around the coil. This changing magnetic field induces an emf in the conductor within the field. Then when a current changes in a coil it induces an emf in the coil. This is known as self-inductance or commonly known as inductance. The voltage of the emf in the coil depends on the rate of change of the current with time. Then the voltage induced is proportional to the variation of current with time di/dt. The proportionally factor is the inductance L and is measured in Henry (H). We can write an expression for the voltage as

$$v_L = L \frac{di_L}{dt} \tag{2.26}$$

where the term di_L/dt indicates the first derivative of the inductor current with respect to time.

The inductance of a coil depends on its dimensions and the material around which it is formed.

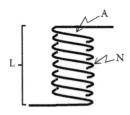

FIGURE 2.14   An inductor with air-filled coil.

The inductance of an air-filled coil can be expressed as

$$L = \frac{\mu_o A N^2}{l} \qquad (2.27)$$

where N is the number of turns in the coil, l is the length of the coil, A is the cross-sectional area of the coil, and μ_o is the permeability of the free space (air in this case).

The permeability μ is a measure of the ease with which magnetic flux can be established in a material. The permeability of free space μ_o is $4\pi \times 10^{-7}$ H/m. The ratio of the permeability of any substance to that of the free space is called the relative permeability, μ_r, a dimensionless number. The permeability can be expressed in terms of the relative permeability and the permeability of free space as

$$\mu = \mu_o \mu_r \qquad (2.28)$$

The relative permeability for most materials is nearly one, but in ferromagnetic materials such iron, nickel, cobalt, and their alloys μ_r is very high.

If a ferromagnetic material is used in the coil, instead of air, the inductance is greatly increased as the permeability is high. Figure 2.15 illustrates an inductor with ferromagnetic material filled.

The inductance of a ferromagnetic material-filled coil can be expressed in terms of its permeability as

$$L = \frac{\mu_o \mu_r A N^2}{l} \qquad (2.29)$$

Inductors and capacitors are elements with capacity to store energy. Inductors store energy within a magnetic field.

2.5.3.1 Inductors in Series

Inductors can be connected in series when the current passing through them is the same. Inductors that are not linked magnetically can be connected in series as indicated in Figure 2.16. The equivalent inductance, L_T, of n inductances connected is series is the sum of all inductances, that is, inductors can be combined the same way as for resistors.

$$L_T = L_1 + L_2 + L_3 + \cdots + L_n \qquad (2.30)$$

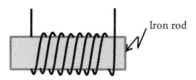

FIGURE 2.15   A coil wound on an iron rod.

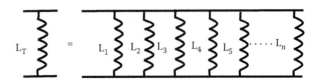

FIGURE 2.16 Inductors in series.

FIGURE 2.17 Inductors in parallel.

2.5.3.2 Inductors in Parallel

Inductors can be connected in parallel when the voltage across them is the same.

Inductors that are not linked magnetically can be connected in parallel as indicated in Figure 2.17.

The reciprocal equivalent inductance, L_T, of n inductances connected in parallel is the sum of all n reciprocal inductances in parallel.

$$\frac{1}{L_T} = \frac{1}{L_1} + \frac{1}{L_2} + \frac{1}{L_3} + \cdots + \frac{1}{L_n} \tag{2.31}$$

EXAMPLE 2.8

Calculate the equivalent inductance of three 12 mH inductors connected in series and then in parallel.

SOLUTION

Three 12 mH inductors connected in series can be reduced to one equivalent inductance L_{ST} as

$$L_{ST} = 12\,mH + 12\,mH + 12\,mH = 36\,mH$$

Then $L_{ST} = 36$ mH

And the three 12 mH inductors connected in parallel can be reduced to one equivalent inductance L_{PT} as

$$\frac{1}{L_{PT}} = \frac{1}{12\,mH} + \frac{1}{12\,mH} + \frac{1}{12\,mH} = \frac{3}{12\,mH}$$

Then $L_{PT} = 4$ mH

EXAMPLE 2.9

Calculate the inductance, L_1 that connected in parallel with $L_2 = 30$ mH provide a total equivalent inductance L_T of 12 mH.

SOLUTION

The total inductor L_T is equivalent to the parallel of L_1 and L_2 as

$$\frac{1}{L_T} = \frac{1}{L_1} + \frac{1}{L_2} = \frac{1}{12\,mH} = \frac{1}{L_1} + \frac{1}{30\,mH}$$

Then,

$$\frac{1}{L_1} = \frac{1}{12\,mH} - \frac{1}{30\,mH} = \frac{30\,mH - 12\,mH}{360(mH)^2} = \frac{18\,mH}{360(mH)^2}$$

and

$$L_1 = \frac{360\,mH}{18} = 20\,mH$$

2.5.3.3   Energy Storage W in an Inductor

The energy added to the magnetic field in increment of time dt, is the product of the voltage, the current and the increment time. Then the energy added ΔW is

$$\Delta W = vidt \tag{2.32}$$

Replacing the voltage as function of inductance, we have

$$\Delta W = L\frac{di}{dt}idt = Lidi \tag{2.33}$$

The current is increased from zero to I, we can calculate to total energy W as

$$W = \int_0^I idt = \frac{1}{2}LI^2 \tag{2.34}$$

2.5.4   APPLICATION: INDUCTIVE PROXIMITY SENSORS

Consider a fixed coil wrapped around a ferromagnetic material as indicated in Figure 2.18. If a ferromagnetic plate moves close to the fixed coil it will change the inductance of the fixed coil. The changes in inductance can be sensed indicating proximity between the ferromagnetic plate and the fixed coil.

2.6   ACTIVE COMPONENTS OF A CIRCUIT: SOURCES

For the electrons to move they need external energy to be added to them. This can come from different types of energy, such as solar cells, chemical reactions,

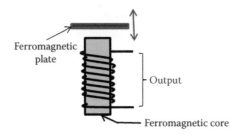

FIGURE 2.18 Proximity sensor application.

electro-mechanical conversion, etc. In circuit analysis this energy is concentrated and it is represented by an electrical source. There are voltage and current sources.

2.6.1 IDEAL VOLTAGE SOURCE

An ideal voltage source is a source that does not have any losses. In practice, it is possible to have sources with small losses close to ideal sources. An ideal source does not exist in practice, but the concept of ideal sources is a very useful concept in the analysis of electric circuits.

In an ideal voltage source the voltage level is maintained at all levels of current, this make the voltage independent of current drawn from it. Figure 2.19 shows the output characteristics of an ideal voltage source where the terminal voltage (V) is independent of the value of the current (I).

2.6.2 PRACTICAL VOLTAGE SOURCE

Practical voltage sources have some losses. A practical source can be represented by adding external passive components to an ideal source to represents losses in the source. The practical source is represented by an equivalent circuit consisting of an ideal source in series with a resistance. This resistance added to the ideal source is known as the internal resistance. Figure 2.20 shows a practical source and its output characteristics.

A practical voltage source has a terminal voltage, V, which will depend on the value of the load current I. As the current drawn from the source increases the losses increase and the output voltage level is reduced.

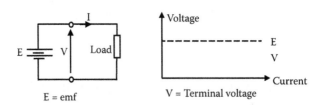

FIGURE 2.19 Ideal dc source and its output characteristics.

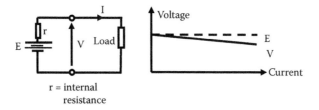

FIGURE 2.20   Practical dc source and its output characteristics.

2.6.3   VOLTAGE SOURCES CONNECTED IN SERIES

For voltage sources connected in series, the total emf is equal to the sum of the source's emfs (electromotive forces) and the total internal resistance is equal to the sum of the source's internal resistances.

Note that polarity must be considered when adding sources in series.

2.6.4   VOLTAGE SOURCES CONNECTED IN PARALLEL

For sources connected in parallel, assuming each source has the same emf and internal resistance, the total emf is equal to emf of one source and the total internal resistance of n cells is equal to internal resistance of one cell/n.

In practice, sources of different voltage are not connected in parallel as this may produce a large current circulating through the sources that may damage them.

2.6.4.1   Ideal Current Source

As in the case of ideal voltage source, the concept of ideal current source is a useful concept in the analysis of electric circuits. An ideal current source has constant current independent of voltage across the terminals.

Figure 2.21 shows an ideal dc current source and its output characteristics.

2.6.4.2   Practical Current Source

As in the case of a voltage source, a practical current source can be represented by adding external passive components to represent losses. In the case of a practical current source, an internal resistor is added in parallel to an ideal current source.

A practical current source has a terminal current, which depends on of the value of the load voltage V. Figure 2.22 shows a practical current source and its output characteristics.

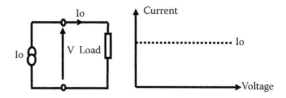

FIGURE 2.21   Ideal dc current source and its output characteristics.

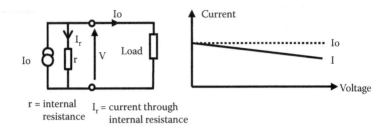

r = internal   I$_r$ = current through
   resistance   internal resistance

FIGURE 2.22   Practical current source and its output characteristics.

2.7   ELECTRIC CIRCUITS/NETWORKS

Now if we combine passive components and active components then we create an *electric circuit*. A circuit is a combination of active and passive elements joined at terminal points providing at least one closed path through which charge can flow.

Let us look at a very simple circuit or network consisting in only two elements: a passive element and an active element connected as shown in Figure 2.23.

The active component, in this case the voltage source, is known as the emf. The emf represents the driving influence that causes a current to flow. The energy transferred due to the passage of unit charge between two points in a circuit is known as the potential difference (pd).

The *emf* is always active in that it tends to produce an electric current in a circuit. A pd may be either passive or active. A network or a circuit is said to be passive if it contains no source of *emf*. A network or circuit is said to be active if it contains one or more of *emf.*

Now let us see what happens in this simple circuit shown in Figure 2.23, with the relationship between the polarity of voltages and the direction current in the passive and active component. As it was defined, the current is produced by movement of electrons and then the electron was defined as a negative charge; there is a difference in the current direction and polarities in passive and active components. This sometimes creates confusion in the analysis of circuits.

Let us create the circuit shown in Figure 2.23 from the beginning. First, we start with the emf source alone with the given polarity as shown in Figure 2.23. As this source is not connected, the current through it is zero. If we connect the source to a load as indicated in Figure 2.23 a current will circulate in the source from the negative terminal to the positive terminal, that is, from bottom to top of the page.

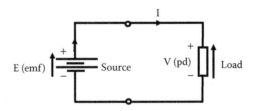

FIGURE 2.23   Two elements electric circuit.

The source is connected now to the passive component with the positive terminal to top of the component and negative terminal to the bottom of the passive component imposing the polarity in the passive component indicated in Figure 2.23. Now there is only one current in this circuit as there is only one path for the electrons to circulate. This current is imposed by the source and circulates clockwise. If we compare the polarity of the source and the direction of the current we see that in the source the current is circulating from negative polarity to positive polarity. In the passive component this not the case, the direction of the current is circulating from positive polarity to the negative polarity. Then, there is difference between circulation of current and polarity of voltages in active components compared to passive components. This sometimes creates confusion in solving a circuit but there are general methods of solving circuits that avoid this difficulty.

EXAMPLE 2.10

For the circuit shown in Figure 2.24 you are required to indicate, with arrows, the *direction of all currents*. Also indicate with arrows the *direction of all voltages*, the arrow for voltages go from the negative terminal to the positive terminal as indicated in Figure 2.24.

In this case, the active component imposes the direction of the current through R_1 in the direction from the right to the left of this page. Then, this current splits into two currents through R_2 and R_3 as indicated in the solution shown in Figure 2.25. Once the directions of the currents are determined, the polarities can be deduced. Figure 2.25 shows all voltages and currents.

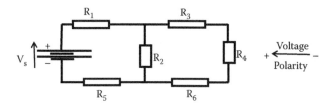

FIGURE 2.24 Electric circuit exercise in the direction of current and voltages for Example 2.10.

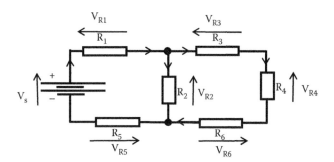

FIGURE 2.25 Solution of circuit indicated in Figure 2.24 indicating all voltages and current for Example 2.10.

In some cases it is difficult to determine the direction of the current, especially if there is more than one source. In general, if the direction of the current is not known to solve a circuit, we can assume a direction of the current in a component and stick to that assumption. Once the circuit is solved, if the current or voltage to be calculated comes out with a negative value, it indicates that the current circulates in the opposite direction to the assumed one. In the case of a voltage, a negative voltage indicates the polarities assumed are reverse.

2.7.1   SELECTION OF COMPONENTS

Electrical components can have a large variation in physical size for different components and also for a given component type. In Chapter 21, information on how to select electronic components is included.

2.8   KEY POINTS

- An electric current is a flow of charge.
- A voltage source produces an emf that can cause a current to flow.
- Ideal sources have zero internal resistance. Practical voltage sources are represented by an equivalent circuit including their internal resistance.
- Ohm's law indicates that the current in a resistor is directly proportional to the voltage across it.
- The equivalent resistance of n resistors connected in series is given by the sum of all n resistances.
- The reciprocal of an equivalent resistance of n resistors in parallel is equal to the sum of the reciprocals of the n resistances.
- A capacitor consists of two plates separated by a dielectric.
- The equivalent capacitance of n capacitors in parallel is equal to the sum of n individual capacitances.
- The reciprocal of an equivalent capacitance of n capacitors in series is equal to the sum of the reciprocals of the n individual capacitances.
- The energy stored in a capacitor is $1/2(CV^2)$ or $1/2(Q^2/C)$.
- Inductors store energy within a magnetic field.
- A wire carrying a current creates a magnetic field.
- Inductors can be made by coiling a wire; a ferromagnetic material will increase the inductance.
- The energy stored in an inductor is equal to $1/2(LI^2)$.

3 Circuit Analysis

3.1 CONCEPT OF STEADY STATE AND TRANSIENT SOLUTIONS

When an external energy is applied to a circuit, through either the connection of a voltage or current source, some components will not respond instantaneously to this stimulus. In particular, capacitors and inductors can absorb or dissipate energy in a finite time. This means that the total response of circuits consists of two parts: the transient response and the permanent or steady-state response. The total response then will be the sum of the transient response plus the steady-state response.

The transient response is a response that takes place in a very short period of time. It is the time for the circuit to settle before it reaches a permanent state. This transient response is a response to changes in the circuit; changes in a circuit can be due to the closing or opening switches or any sudden variation in a circuit. The transient response also depends on the state of a component previous to its connection to a circuit. For example, if a capacitor is connected to a circuit, and this capacitor has been precharged, the response will be different than if the connected capacitor was discharged. In circuit analysis this is known as the *initial condition* of a component. Usually it is an initial voltage for a capacitor or an initial current for an inductor. The transient response only remains for a very short period of time usually in the order of milliseconds and is the reaction of a circuit to instantaneous changes. The charging or discharging of a capacitor, and the energizing or de-energizing of an inductor will take time to settle into a permanent response.

The *steady-state response*, also known as permanent response, is the response of a circuit after the time engaged by the transient response has been completed. In this chapter, we will concentrate on the steady-state response of a circuit.

When an ac signal is applied to a circuit, voltages and currents change as the time varies. A solution of an ac circuit is time-dependent. If the signal is a periodic signal, it is possible to use a method based on the variation of the amplitude and phase of the response to make it artificially independent of the time variation. In this chapter we will discuss this method by introducing the concept of phasors and working in the frequency domain instead of the time domain.

Once a circuit is created then it is necessary to calculate or measure the different voltages and currents in the different components or part of the circuit to check its behavior. With the information of voltages and currents the electric power can be calculated. There are two important rules that can be used to *solve a circuit*, that is, calculation of voltages and currents. These regulations are known as *Kirchhoff's laws* in circuit analysis. The application of Kirchhoff's laws will solve a circuit no matter how many elements are in it or how complex it is.

One of the aims of this chapter is to understand the basic concept of Kirchhoff's laws associated with potential differences (pds) in closed circuits and electric currents in a circuit junction and their applications to solve dc and ac circuits.

3.2 DC CIRCUITS

A dc circuit is a circuit that consists of dc sources and passive components. The relationship between voltage and current in a particular component is given by the expression relating their voltage and current as indicated in Chapter 2. The relationship between a voltage and a current in a resistance is given by Ohm's law as indicated in Chapter 2. Under dc state a capacitor will be charged during its transient response and under steady response it will remain charged. Then, in a dc network the capacitor can be considered as an open circuit for the steady-state dc analysis. In the case of an inductor under transient response, the inductor will store a current. Under the steady-state regime this current will remain, then in a dc network the inductor can be considered as a closed circuit for the steady-state dc analysis.

If we want to calculate voltages and current in circuits, then we have to make use of Kirchhoff's laws. The relationship among voltages and currents in a circuit (not just a single component) is given by Kirchhoff's laws.

3.2.1 Kirchhoff's Current Law Applied to Electric Circuits: KCL

A junction also known in circuit analysis as *node* is a point in a circuit where three or more circuit elements are joined. The Kirchhoff's current law (KCL) applies to currents in a junction. KCL is a simple concept and it is based on the fact that in nodes the total number of electrons that remain in the node is zero, so electrons going into a node are just passing through and they come out of the node in the same number.

KCL can be defined as

The sum of the currents entering a node is equal to the sum of the currents leaving the same node.

It is also expressed assigning value to direction of current as

The algebraic sum of all the currents meeting at a common node is zero.

But in this case, it is assumed that the currents toward a junction are considered positive and those away from the same junction negative or vice versa.

The first definition may have certain advantages as this makes it easier to apply Kirchhoff's voltage law (KVL) in a similar way as KCL. Therefore, for the application of KCL the key words are *currents* and *node*.

Sum of currents entering a node = Sum of currents leaving the same node

Some common source of errors in applying this law is to use currents from different nodes, or not considering all currents in a node, or not identifying a node properly. The law applies to *all currents in a particular node*.

EXAMPLE 3.1

We can illustrate the application of KCL with an example. Let us identify nodes and apply KCL to the circuit shown in Figure 3.1.

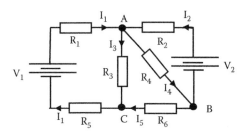

FIGURE 3.1 Circuit diagram for Example 3.1. Application of KCL.

KCL can be applied to any node. In this circuit we have three nodes. Nodes: A, B, and C

Node A

Currents entering node A: I_1 and I_2
Currents leaving node A: I_3 and I_4
Sum of currents entering node A $= I_1 + I_2$
Sum of currents leaving node A $= I_3 + I_4$

Therefore, by KCL

$$I_1 + I_2 = I_3 + I_4$$

(The sum of the currents entering a junction [*node*] is equal to the sum of the currents leaving the junction.)
Rearranging this equation we have

$$I_1 + I_2 - I_3 - I_4 = 0$$

If the currents toward a junction are considered positive and those away from the same junction negative, then the algebraic sum of all the currents meeting at a common junction is zero. This is another way of stating KCL.
Similarly, using KCL for nodes B and node C we can obtain equations that relate the existing currents in those nodes.

Node B

$$I_4 = I_2 + I_5$$

Node C

$$I_3 + I_5 = I_1$$

EXERCISE 3.1

Identify all nodes existing in the network shown in Figure 3.2. Without looking at the solution spend a few minutes to determine the number of nodes.

Solution

The circuit of Exercise 3.1 contains five nodes. In circuit analysis it is assumed that the wire connecting elements have zero resistance; hence, two or more points in a wire from the electrical point of view are connected to the same point. The solution to this exercise is given in Figure 3.3.

EXAMPLE 3.2

For the circuit shown in Figure 3.4 calculate the currents I_1, I_2, and I_s using KCL.

SOLUTION

At node B. Currents going into node B: 10 A; currents going out of node B: I_1 and 1 A.

By KCL at node B

$$10\,A = I_1 + 1\,A$$

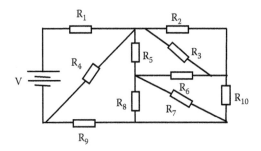

FIGURE 3.2   Circuit diagram for Exercise 3.1. Identification of nodes in order to apply KCL.

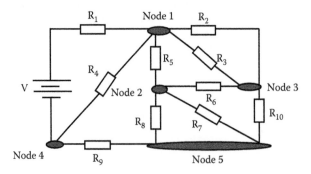

FIGURE 3.3   Circuit diagram showing all nodes as the solution to Exercise 3.1.

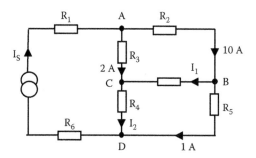

FIGURE 3.4   Circuit diagram for Example 3.2. Application of KCL.

Then,

$$I_1 = 10\,A - 1\,A = 9\,A$$

At node C. Currents going into node C: 2 A and I_1; currents going out of node C: I_2.
By KCL at node C

$$I_2 = I_1 + 2\,A = 9\,A + 2\,A = 11\,A$$

At node A. Currents going into node A: I_S; currents going out of node A: 2 and 10 A.
By KCL at node A

$$I_S = 2\,A + 10\,A = 12\,A$$

Also at node D. Currents going into node D: I_2 and 1 A; currents going out of node D: I_S.
By KCL at node D

$$I_S = I_2 + 1\,A = 11\,A + 1\,A = 12\,A$$

as indicated earlier.

EXAMPLE 3.3

For the circuit shown in Figure 3.5, calculate the currents I_1, I_2, I_3, I_4, and I_5 using KCL.

SOLUTION

By KCL at node A

$$30\,A = I_1 + 10\,A$$

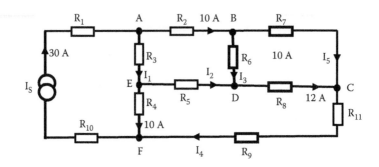

FIGURE 3.5   Circuit diagram for Example 3.3. Application of KCL.

Then

$$I_1 = 30\,A - 10\,A = 20\,A$$

By KCL at node E

$$I_1 = I_2 + 10\,A$$

Then

$$I_2 = I_1 - 10\,A = 20\,A - 10\,A = 10\,A$$

By KCL at node D

$$I_2 + I_3 = 12\,A$$

Then

$$I_3 = 12 - I_2 = 12 - 10 = 2\,A$$

By KCL at node F

$$30\,A = 10 + I_4$$

Then

$$I_4 = 20\,A$$

By KCL at node B

$$10\,A = I_3 + I_5$$

Then

$$I_5 = 10\,A - I_3 = 10\,A - 2\,A = 8\,A$$

3.2.2   Kirchhoff's Voltage Law Applied to Electric Circuits: KVL

A closed circuit, or a loop, is created by starting at a selected node and then tracing through a set of connected circuit elements in such a manner that we return to the original starting node without passing through any intermediate node more than once.

The KVL applies to voltages in a closed circuit. KVL is a simple concept to provide the relationship between voltages in a closed circuit taking into account their polarities.

KVL can be defined as

The sum of all the rises of potentials around any closed circuit equals the sum of the drops of potential in the same closed circuit.

KVL can also be expressed assigning polarities to the potential around a closed circuit. If the potential rises are considered positive and the potential drop as negative, then KVL can also be defined as:

The algebraic sum of the potential differences around a closed circuit is zero.

Therefore, for the application of KVL the key words are closed circuit and voltages.
KVL:

Sum of potential rises around a closed circuit = sum of potential drops in the same closed circuit.

KVL is a simple concept but in some cases the application of this law brings some small complications. In some circuits it is difficult to know the right polarity of a voltage in a component before the circuit is solved.

In Chapter 2, we indicated the difference with direction of voltages and currents in passive and active elements. In some cases, before we solve a circuit we do not know the direction of a current. This happens when there is more than one source in the circuit and each source will try to impose a direction of current in the circuit. In general, an approach to solve this problem is to assume direction of currents. Once the direction of currents is assumed, the direction of the voltages in passive components can be deduced. In these passive elements, the direction of the current is opposite to the direction of the arrow indicating the voltage. Let us look at one example to illustrate this.

EXAMPLE 3.4

In the circuit of Figure 3.6a we want to identify the direction arrows for voltages and currents.

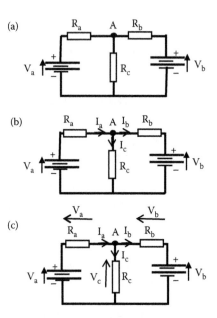

FIGURE 3.6 (a) Circuit diagram for Example 3.4. (b) Illustration of direction of currents in this circuit. (c) Polarities of voltages in the same circuit.

In this circuit we do not know the direction of the current passing through R_a. The two active components are competing, with source V_a trying to impose a current going from left to right (current in source V_a goes from negative to positive). On the other hand, source V_b tries to impose a current in the opposite direction. We can assume that the current in R_a (I_a) goes from left to right. Once this is assumed we can then deduce the other currents (I_b, I_c). Current passing through R_a reaches node A then splits into two currents as indicated in Figure 3.6b. Once the directions of the currents have been established, then we can deduce the direction of the arrows indicating the voltages (polarities). Figure 3.6c indicates voltages in passive components. The polarities of the active components are indicated by the positive and negative of their source terminals.

Now let us look at a way of applying KVL that makes it easy to decide the polarity issue. The law applies to a closed circuit. If we determine a direction of voltages in a closed loop then in this closed loop we have only two directions: clockwise or anticlockwise. By representing a voltage with an arrow with direction going from the negative terminal to the positive terminal, we can have all voltages in a closed circuit with their particular directions, as indicated in Figure 3.6c for Example 3.4. The application of this law is made easier because we can assume the "clockwise voltages" as potential rise then "anticlockwise" voltages must be potential drop or vice versa. For the application of this law it does not matter. If we determine the direction of the voltages then we can apply KVL in a similar manner as the application of the KCL. This is particularly beneficial in cases where we do not know the polarities of a voltage across a component. There is no confusion in relation to the sources as we know their polarities.

In summary, the proposed method is as follows: assuming direction of currents allows the deduction of the direction of voltages (remember in a passive component the current enters by the positive terminal); then the arrow indicating the voltage is in opposite direction to the current direction. The directions of the voltage due to sources are indicated by the polarities of the source. Example 3.5 illustrates the application of KVL in this way.

EXAMPLE 3.5

In the circuit shown in Figure 3.7 we have one closed circuit (loop) and five pds. The direction of the current is not known as it depends on the source that imposes the current. Source V_A will try to impose a clockwise current and source V_B will try to impose an anticlockwise current. Let us assume that the dominant source is V_A and that the current, I, circulates clockwise as indicated in Figure 3.7. Once this assumption is made we can deduce the direction of the pds in the other passive components. They are indicated as V_R, V_L, and V_C in Figure 3.7. The direction of the "voltage arrow" is in opposition to the direction of the current as they are passive components.

The pds can be potential drops or potential rises. Since we applied KVL in a closed circuit the direction of the arrows indicating the pds can be either clockwise or anticlockwise.

The directions of pds in Example 3.5 are shown in Figure 3.8.

We can determine the *potential drops* and the *potential rises* by inspecting the *direction of the arrow* of the pds. In this example we have the pd for V_A as having a clockwise direction, so we can take V_A as a potential rise.

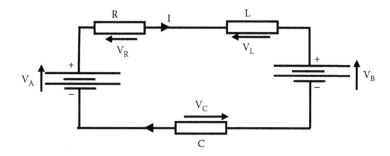

FIGURE 3.7   Circuit for KVL for Example 3.5.

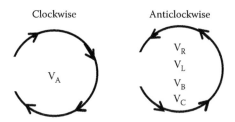

FIGURE 3.8   Clockwise and anticlockwise direction of voltages for Example 3.5.

Potential rise : V_A

The rest of the pds are anticlockwise, so we can take them as potential drops.

Potential drops : V_R, V_L, V_B, and V_C

Then, by applying KVL we have that

$$V_A = V_R + V_L + V_B + V_C$$

The sum of the rises of potentials around any closed circuit equals the sum of the drops of potential in that circuit. Or rearranging this equation we have

$$V_A - V_R - V_L - V_B - V_C = 0$$

The algebraic sum of the pds around a closed circuit is zero.

EXERCISE 3.2

Inspect the circuit shown in Figure 3.9 and determine the total number of closed circuits.

Without looking at the solution spend a few minutes to determine the number of closed circuits.

Solution

The solution to this exercise provides seven closed circuits; they can be expressed in term of nodes A, B, C, and D as

Closed circuit 1: ABCA
Closed circuit 2: CBDC

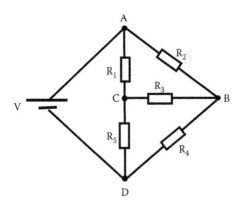

FIGURE 3.9   Circuit for Exercise 3.2. Determination of the number of closed circuits.

Closed circuit 3: ACDA
Closed circuit 4: ABDA
Closed circuit 5: ABDCA
Closed circuit 6: ACBDA
Closed circuit 7: ABCDA

EXAMPLE 3.6

For the circuit shown in Figure 3.10, calculate the value of voltages V_a, V_b, and V_c using KVL.

In circuit theory a node is formed when three or more branches meet. In this example the letters ABCDEGF and H indicate points in the circuit. Their only purpose is to identify the closed circuits.

Loop BCDEFB:

Clockwise: V_a, 10 V
Anticlockwise: 100 V
KVL: $V_a + 10 = 100$ therefore $V_a = 90$ V

Loop FEGHF:

Clockwise: 150 V
Anticlockwise: V_c, 20 V, V_a
KVL: $150 = V_c + 20 + V_a = V_c + 20 + 90$ therefore $V_c = 40$ V

Loop ABFHA:

Clockwise: 200 V
Anticlockwise: 150 V, 10 V, V_b
KVL: $200 = 150 V + 10 V + V_b$ therefore $V_b = 40$ V

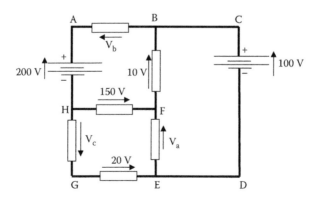

FIGURE 3.10   Circuit for KVL for Example 3.6.

EXAMPLE 3.7

For the circuit shown in Figure 3.11, calculate the value of voltages V_1, V_2, and V_3 using KVL.

In circuit theory a node is formed when three or more branches meet. In this example, the letters ABCDEGF and H indicate points in the circuit. Their only purpose is to identify the closed circuits.

Applying KVL to closed circuit BCHGEB, we can write

$$V_1 = 20\,V + 10\,V = 30\,V$$

Therefore, $V_1 = 30$ V

Applying KVL to closed circuit ABEDA, we can write

$$50 = V_2 + V_1 + 10$$

or

$$V_2 = 50 - V_1 - 10 = 50 - 30 - 10 = 10\,V$$

Therefore, $V_2 = 10$ V

Applying KVL to closed circuit DEGFD, we can write

$$8 + V_3 = V_2 + 10$$

or

$$V_3 = V_2 + 10 - 8 = 10 + 10 - 8 = 12\,V$$

Therefore, $V_3 = 12$ V

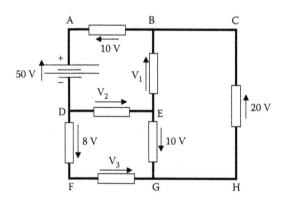

FIGURE 3.11   Circuit for KVL for Example 3.7.

EXAMPLE 3.8

In the circuit shown in Figure 3.12 the current passing through the 200 Ω resistor, I_1, is 20 mA. Calculate the current supply by the dc voltage source and the emf provided by the dc source.

SOLUTION

If $I_1 = 20$ mA then the voltage across the 200 Ω resistor, V_{200}, is

$$V_{200} = 200 \times I_1 = 200 \times 20 \times 10^{-3} = 4\,V$$

This is the same voltage across the combination of the three resistors 20, 100, and 40 Ω connected in series. Then,

$$I_L = \frac{4}{20 + 100 + 40} = 25 \times 10^{-3}\,A$$

Now by KCL

$$I_s = I_1 + I_L = 20 \times 10^{-3} + 25 \times 10^{-3} = 45 \times 10^{-3}\,A$$

and applying KVL in the loop formed by E_s and voltages across the 50, 200, and 90 Ω we can calculate E_s as

$$E_s = 50I_s + 200I_1 + 90I_s = 50 \times 25 \times 10^{-3} + 200 \times 20 \times 10^{-3} + 90 \times 25 \times 10^{-3} = 7.5\,V$$

EXAMPLE 3.9

Two pieces of semiconductor material having 20 and 30 Ω, respectively, are connected across a dc voltage source of emf = 12 V with an internal resistance of 3 Ω. Draw a circuit for this system and calculate the current drawn by the source, I_s, and the output terminal voltage of the source, V_s.

SOLUTION

An equivalent circuit of this system is drawn in Figure 3.13.

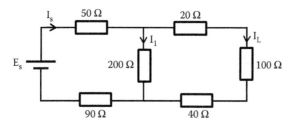

FIGURE 3.12 Circuit for Example 3.8.

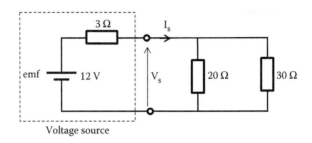

Voltage source

FIGURE 3.13   An equivalent circuit for system of Example 3.9.

Resistors 20 and 30 Ω can be reduced into one resistor as shown in Figure 3.14. The source current, I_s, can be calculated as

$$I_s = \frac{12}{3+12} = 0.8\,A$$

and applying KVL to circuit shown in Figure 3.14, we can write

$$12 = 3I_s + V_s$$

or

$$V_s = 12 - 3I_s = 12 - 3 \times 0.8 = 9.6\,V$$

EXAMPLE 3.10

A 15 Ω resistor is connected in parallel to a variable resistor, Rv. This parallel combination is connected in series with a piece of equipment, W, and the whole system connected to a dc voltage supply of 50 V dc. Draw an electric circuit for the system and determine the value of the variable resistor that makes the equipment W dissipate 60 W when a current of 3 A passes through W.

SOLUTION

Figure 3.15 shows a circuit for this example.

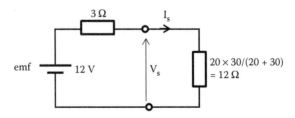

FIGURE 3.14   Simplified circuit of system from Example 3.9 with the 20 and 30 Ω resistors combined into one resistor.

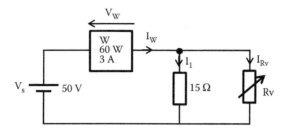

FIGURE 3.15   An equivalent circuit for system of Example 3.10.

The voltage across the equipment W at 3 A will be

$$V_W = \frac{P_W}{I_W} = \frac{60}{3} = 20\,V$$

Applying KVL to the loop that contains equipment W, we have

$$V_s = V_W + 15 \times I_1 = 50 = 20 + 15 \times I_1$$

or

$$I_1 = \frac{30}{15} = 2\,A$$

Applying KCL we can write

$$I_W = I_1 + I_{Rv} = 3 = 2 + I_{Rv}$$

or

$$I_{Rv} = 1\,A$$

The 15 Ω resistor and Rv resistor are in parallel. Hence, they have the same voltage across them:

$$V_{Rv} = 15 \times I_1 = 15 \times 2 = 30\,V$$

With the voltage across Rv and its current I_{rv} we can calculate the value of Rv as

$$Rv = \frac{V_{Rv}}{I_{Rv}} = \frac{30}{1} = 30\,\Omega$$

Application of KCL and KVL: Combining the application of KCL and KVL we can find the solution to any circuit no matter how complex or large it is. Let us look at the application of the two laws to solve a circuit. This can be illustrated in the following example.

EXAMPLE 3.11

Let us apply KCL and KVL to the circuit shown in Figure 3.16 to determine the currents I_1, I_2, and I_3.

Let us consider node A and apply KCL to this node. This will lead to the following equation:

$$I_1 + I_2 = I_3 \tag{3.1}$$

Now let us express the voltages in terms of their currents and resistances using Ohm's law. Select the *left-hand side (LHS) loop* alone as shown in Figure 3.17 and apply KVL to this loop.

Voltage across 2 Ω resistor is $2 \times I_1$ and voltage across 8 Ω resistor is $8 \times I_3$ by Ohm's law ($V = RI$).

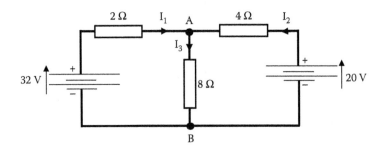

FIGURE 3.16 Circuit diagram for Example 3.11. Application of KVL and KCL.

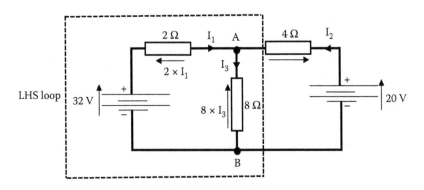

FIGURE 3.17 LHS loop circuit for Example 3.11.

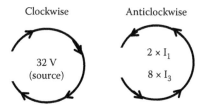

FIGURE 3.18 Clockwise and anticlockwise voltages for Example 3.11.

Identifying the voltage with clockwise or anticlockwise direction we have (Figure 3.18):

Clockwise voltages: 32 V (*source*)
Anticlockwise voltages: $2 \times I_1$ and $8 \times I_3$

Then, by KVL

$$32 = 2I_1 + 8I_3 \qquad (3.2)$$

Now let us consider the right-hand side (RHS) loop alone and apply KVL to this loop (Figure 3.19).
Direction of voltages (Figure 3.20):
By KVL

$$4I_2 + 8I_3 = 20 \qquad (3.3)$$

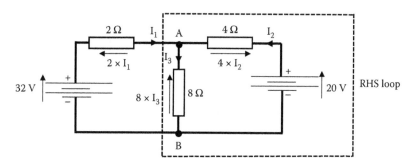

FIGURE 3.19 RHS loop circuit for Example 3.11.

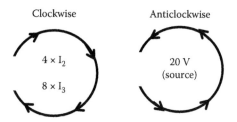

FIGURE 3.20 RHS loop clockwise and anticlockwise voltages for Example 3.11.

The system of Equations 3.1 through 3.3 has three unknowns: I_1, I_2, and I_3. By applying KCL and KVL we have obtained three *independent equations*: Equations 3.1 through 3.3. Therefore, the system has a solution. Solving this system of three simultaneous equations we can obtain the solution for I_1, I_2, and I_3.

The solutions are as follows: $I_1 = 4$ A, $I_2 = -1$ A, and $I_3 = 3$ A.

Note: To solve a system we need the same number of *independent equations* as the number of unknown equations.

As an exercise, you can solve the system of simultaneous equation: Equations 3.1 through 3.3 to obtain I_1, I_2, and I_3 to verify the solution given above.

"External" loop: Now let us consider the other closed circuit in Example 3.11, that is, the external loop (Figure 3.21).

Direction of voltages (Figure 3.22):

By applying KVL, we obtain

$$32 + 4I_2 = 20 + 2I_1 \qquad (3.4)$$

Now let us subtract Equation 3.3 from Equation 3.2

$$(\text{Equation } 3.2) \gg 32 = 2I_1 + 8I_3$$

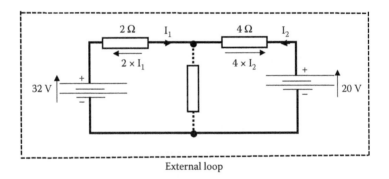

External loop

FIGURE 3.21   External loop for Example 3.1.

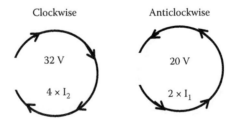

FIGURE 3.22   External loop. Clockwise and anticlockwise voltages.

$$(\text{Equation } 3.3) \gg 20 = 4I_2 + 8I_3$$

$$32 - 20 = 2I_1 - 4I_2$$

Equations 3.2 through 3.3 gives us

$$32 - 20 = 2I_1 - 4I_2$$

Rearranging this equation, we obtain

$$32 + 4I_2 = 20 + 2I_1$$

This is the same as Equation 3.4. Therefore, Equation 3.4 is not an independent equation!

Note: In order to obtain the solution of a system we need the same number *of independent equations* as the number of unknown equations.

The application of KCL and KVL will provide a large number of equations. The application of KCL will produce a number of equations equal to the number of nodes in a circuit. The application of KVL will produce a number of equations equal to the number of closed circuits in a network. It should be noted that not all these equations are independent. In the previous example we saw that some equations come from combining equations, hence not independent. To solve a circuit we need the same number of equations as the number of unknowns, but these equations have to be independent.

In circuit analysis there are various methods to acquire the required independent equations. The mesh method is one of them.

Mesh current method (General approach): In a network there can be many closed circuits. A mesh (or *window*) is a special type of closed circuit. It is a closed circuit that does not contain any other loops within it.

The mesh current method to solve a circuit is a general systematic approach that can be applied to any circuit. It is based is on the application of KVL in *meshes*. The mesh current method is based on KVL applied in a special way; it considers voltages in meshed as function of one current in the mesh. It is a systematic approach. Therefore, let us consider the mesh current method application in four steps.

The required steps to apply mesh current method are as follows:

   i. Determine the meshes "windows" in the network
  ii. Assign a mesh current (clockwise direction) in each mesh
 iii. Apply KVL in each mesh
 iv. Solve the system of equations

Let us look at an example to illustrate this method and follow the four steps to solve a circuit.

EXAMPLE 3.12

Solve the circuit shown in Figure 3.23 using the mesh current method.

> Step i: *Determine the meshes "windows" in the network.*
> This circuit has two "windows"—abda and bcdb
> (abcda is not a mesh, because the branch containing R_3 is part of other loop).
> Step ii: *Assign a mesh current (clockwise direction) in each mesh.*
> We can assign I_1 to loop abda and I_2 to loop bcdb as shown in Figure 3.24.
> Step iii: *Apply KVL in each mesh.*
> Apply *KVL* to loop abda and bcdb in term of mesh currents I_1 and I_2.
> > Voltage across R_1 is $R_1 \times I_1$
> > Voltage across R_2 is $R_2 \times I_2$
> > Voltage across R_3 is $R_3 \times (I_1 - I_2)$

Figure 3.25 indicates all voltage across the resistances in terms of currents passing through each resistor.
Applying KVL to mesh abda, we obtain

$$V_1 = R_1 I_1 + R_3 (I_1 - I_2) \tag{3.5}$$

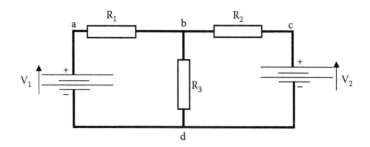

FIGURE 3.23   Circuit to be solved using the mesh current method in Example 3.12.

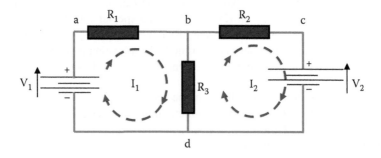

FIGURE 3.24   Circuit for mesh current method containing mesh currents I_1 and I_2.

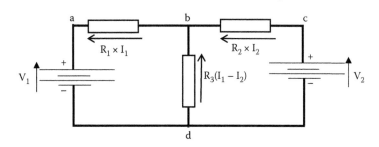

FIGURE 3.25 Meshes for Example 3.12 including the pds across the resistors.

Applying KVL to mesh bcdb, we obtain

$$R_3(I_1 - I_2) = R_2I_2 + V_2 \qquad (3.6)$$

Rearranging Equation 3.6, we can write the same equation as

$$-V_2 = R_3(I_2 - I_1) + R_2I_2$$

Note: Equations 3.5 and 3.6 can be written in a kind of mechanical way by leaving voltages due to sources on one side of the equation and on the other side of the equation the voltages due to passive components. Solving the brackets for Equations 3.5 and 3.6, we have

$$V_1 = (R_1 + R_3)I_1 - R_3I_2 \qquad (3.7)$$

and

$$-V_2 = -R_3I_1 + (R_2 + R_3)I_2 \qquad (3.8)$$

These equations can be written directly by examining Figure 3.24: In loop abda the arrow of I_1 and V_1 are in the same direction, then V_1 becomes positive on one side of Equation 3.7. In loop bcdb the arrow of I_2 and V_2 are in opposite directions, then V_2 becomes negative in Equation 3.8. On the other side of these equations, voltages corresponding to passive components can be written by adding resistance in the path of the current as positive voltage. When we encounter a resistance that involves two currents such as R_3 we add a negative voltage in terms of the other mesh current. For example, in loop abda (Equation 3.7) R_1 and R_3 are in the path of I_1, then we have a term $+(R_1 + R_3)I_1$, but R_3 is common to the other mesh in terms of I_2, then a term $-R_3I_2$ is added to this equation.

Step iv: *Solve the system of equations.*
    Solve the system of Equations 3.5 and 3.6 for I_1 and I_2.

The example above gives the value of current I_1 and I_2. This allows you to calculate the voltage across R_1 and R_2. To calculate the voltage across R_3 we need to know the current through R_3. The current through R_3 is the difference of I_1 and I_2

$$I_{R3} = (I_1 - I_2) \tag{3.9}$$

In general, the current through an element is the difference of the two and has the direction of the larger.

EXAMPLE 3.13

For the circuit shown in Figure 3.26 calculate the currents through the voltage sources using the mesh current method.

   Step i: Choose meshes—This circuit has two meshes: one formed by resistor of 1 Ω, 2 Ω, and the 4 V source; the other one is formed by resistor 2 Ω, 3 Ω, and the 5 V source.
   Step ii: Assign current I_1 in the mesh formed by resistor of 1 Ω, 2 Ω, and the 4 V source and I_2 in mesh formed by resistor 2 Ω, 3 Ω, and the 5 V source.
   Step iii: Applying KVL in each mesh, we have

$$4 = 1 \times I_1 + 2(I_1 - I_2) \tag{3.10}$$

$$-5 = 3 \times I_2 - 2(I_1 - I_2) \tag{3.11}$$

   Step iv: Solve the system of equations.

Solving the brackets and rearranging these equations, we have

$$4 = 3I_1 - 2I_2 \tag{3.12}$$

$$-5 = -2I_1 + 5I_2 \tag{3.13}$$

Multiplying Equation 3.12 by 2 and Equation 3.13 by 3, we have

$$8 = 6I_1 - 4I_2 \tag{3.14}$$

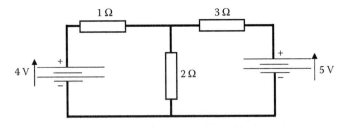

FIGURE 3.26   Circuit for Example 3.13. Mesh current method.

$$-15 = -6I_1 + 15I_2 \qquad (3.15)$$

Adding Equations 3.14 and 3.15 gives

$$-7 = 0 + 11I_2$$

and

$$I_2 = -\frac{7}{11}[A]$$

The negative sign means that the current has the opposite direction to the one assumed.

Now multiplying Equation 3.12 by 5 and Equation 3.13 by 2, we have

$$20 = 15I_1 - 10I_2 \qquad (3.16)$$

$$-10 = -4I_1 + 10I_2 \qquad (3.17)$$

Adding Equations 3.16 and 3.17 gives

$$10 = 11I_1$$

and

$$I_1 = \frac{10}{11}[A]$$

EXAMPLE 3.14

For the circuit shown in Figure 3.27, write a set of equations to solve I_1, I_2, and I_3 using the mesh current method.

SOLUTION

$$V_1 = R_1 I_1 + R_2(I_1 - I_3) + R_3(I_1 - I_2)$$

$$0 = R_4 I_2 + R_5(I_2 - I_3) + R_3(I_2 - I_1)$$

$$-V_2 = R_6 I_3 + R_2(I_3 - I_1) + R_5(I_3 - I_2)$$

or

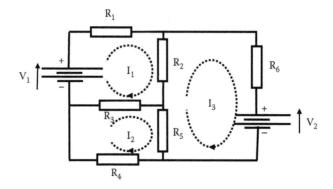

FIGURE 3.27   Circuit diagram for Example 3.14. Mesh current method.

$$V_1 = (R_1 + R_2 + R_3)I_1 - R_3I_2 - R_2I_3$$

$$0 = -R_3I_1 + (R_4 + R_5 + R_3)I_2 - R_5I_3$$

$$-V_2 = -R_3I_1 - R_5I_2 + (R_6 + R_5 + R_2)I_3$$

3.2.2.1   The Voltage and Current Divider

Combining resistor in series or in parallel, it is possible to split a voltage or current to obtain a different level of voltage or current; these circuits are known as voltage and current dividers.

3.2.2.1.1   Voltage Divider

The voltage divider also known as a potential divider consists of resistors in series connected to a voltage source. Figure 3.28 shows a potential divider of two voltages, V_1 and V_2.

The current through R_2 and R_1 is given by

$$I = \frac{V}{R_1 + R_2} \tag{3.18}$$

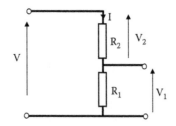

FIGURE 3.28   Circuit of a voltage or potential divider.

By Ohm's law, the voltage across R_1 is $V_1 = I \times R_1$, replacing current I from Equation 3.18 into the expression for V_1 gives

$$V_1 = I \times R_1 = \frac{VR_1}{R_1 + R_2}$$

Rearranging, we have

$$V_1 = \left(\frac{R_1}{R_1 + R_2} \right) V \qquad (3.19)$$

Equation 3.19 is known as the voltage divider rule, as the equation can be written directly by examining the circuit shown in Figure 3.28. The fraction term has as the denominator the sum of the two resistors $(R_1 + R_2)$ and the numerator is the resistor that provides the voltage V_1: (R_1). This fraction, $[R_1/(R_1 + R_2)]$, is multiplying the voltage V.

Similarly, the voltage across R_2 can be written as

$$V_2 = \left(\frac{R_2}{R_1 + R_2} \right) V \qquad (3.20)$$

3.2.2.1.2 *Current Divider*

The current divider consists of resistors connected in parallel with a current going into this parallel combination as shown in Figure 3.29.

The two resistors R_2 and R_1 can be combined into one resistor, R_T, as

$$R_T = \frac{R_1 R_2}{R_1 + R_2}$$

Then, by Ohm's law the voltage across R_T can be expressed as

$$V = R_T \times I = \frac{R_1 R_2}{R_1 + R_2} I$$

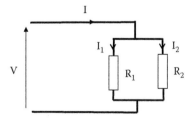

FIGURE 3.29 Circuit of a current divider.

By Ohm's law, the current through resistor R_1, I_1, can be obtained dividing the voltage V by resistance R_1 as

$$I_1 = \frac{V}{R_1} = \frac{R_1 R_2}{R_1 + R_2} \frac{I}{R_1}$$

Rearranging, we have

$$I_1 = \frac{R_2}{R_1 + R_2} I \qquad (3.21)$$

Equation 3.21 is known as the current divider rule. This equation can be written directly by examining the current divider circuit shown in Figure 3.23. In the fraction multiplying the current I, the denominator is the sum of the two resistors $(R_1 + R_2)$ and the numerator is the resistor parallel to R_1.

Similarly, the current through R_2 can be written as

$$I_2 = \left(\frac{R_1}{R_1 + R_2} \right) I \qquad (3.22)$$

3.3 AC CIRCUITS

Ac signal: As we saw in Chapter 2, the dc voltage or current in a source or circuit element keeps its value independent of time. There are electronic and electrical systems where the value of the voltages and current changes with time. These systems are therefore not dc systems. In this section, we look at some characteristics of one of these systems. It is possible to generate voltages and currents that change with time in a great variety of forms.

The example in Figure 3.30 shows signals that change with time but keep their polarities, that is, they do not alternate between negative and positive voltages or currents. These are usually called unidirectional signals.

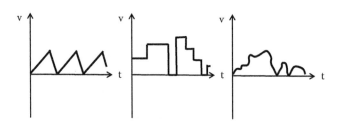

FIGURE 3.30   Example of unidirectional wave forms.

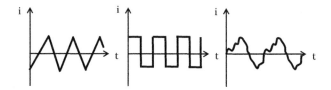

FIGURE 3.31   Example of alternating current waveforms.

There are signals that can alternate values between positive and negative voltages or currents. These are called alternating voltages or currents signals. In Figure 3.31, there are three examples of this type of alternating currents waveforms.

Of all the infinite number of signals that alternate between positive and negative currents, there is one that is commonly used in electrical and electronic systems. This is the sinusoidal shape voltage, known as the ac voltage or current. Figure 3.32 shows an ac signal voltage. This is an important form of alternating waveform used in analog electronics and a clear understanding of its nature and characteristics is essential. This section looks at the parameters that define such waveforms, and discusses the relationship between the various measures used to quantify them.

In this ac sinusoidal signal, E_p is the *peak or maximum voltage* and T is the time that the signal repeats itself known as the *period*.

The ac signal is the signal commonly generated in large quantities by electric plants. It has the advantage that it is relatively easy to generate. It is a periodic signal and also its shape can be represented by a mathematical expression in terms of a sine or a cosine function. Figure 3.33 shows the corresponding mathematical terms in the mathematical representation and the actual ac signal.

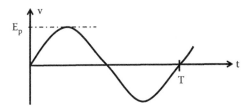

FIGURE 3.32   Sample of an ac waveform.

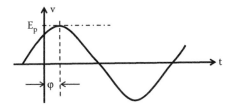

FIGURE 3.33   Representation of an ac signal as sinusoidal waveform.

We can write a mathematical expression for an ac signal as

$$v(t) = E_p \sin(\omega t + \varphi) \tag{3.23}$$

where E_p is the peak voltage, ω the angular frequency, and φ is the phase angle.

The frequency can be expressed in radian per second ω or in cycles per second Hz. The relationship between ω and f is given by

$$\omega = 2\pi f \tag{3.24}$$

It follows that the period T is given by

$$T = \frac{1}{f} = \frac{2\pi}{\omega} \tag{3.25}$$

3.3.1 ORIGIN OF PHASOR DOMAIN

We can now consider the behavior of circuits in response to sinusoidal voltages and currents. If an ac signal is applied to a circuit, voltages and currents values will be different at different times. In order to solve a circuit under ac condition we have to solve the circuit taking into account the different times. As the ac signal is a periodic signal that repeats itself and is regular in shape, theoreticians working on circuit analysis have developed an artificial way to solve ac circuit excluding the time dependence. This is achieved by working in the frequency domain instead of the time domain. It is necessary to transfer the circuit from the time domain to the frequency domain. This is achieved using an operant known as the *phasor*. The advantage of working with phasors is that circuits can be solved in the same way we solve dc circuits as time dependency is excluded.

In this section, we illustrate the characteristics and application of phasors to solve ac circuits.

Let us illustrate the origin of phasors with one example. Figure 3.34 shows a circuit containing a resistor, a capacitor, and an inductor connected to an ac voltage source.

Applying KVL we can write an equation for all voltages as a function of time (t)

$$v_s(t) = v_R(t) + v_L(t) + v_C(t) \tag{3.26}$$

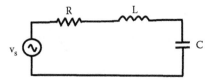

FIGURE 3.34   RLC ac circuit diagram to illustrate the origin of the phasor domain.

Assuming that the source has a voltage $v_s(t)$, an ac source equal to

$$v_s(t) = E_p \sin \omega t \tag{3.27}$$

Now as indicated in Chapter 2, the relation between voltages and currents in inductor and capacitor are

$$v_L(t) = L\frac{di}{dt} \tag{3.28}$$

$$v_c(t) = \frac{1}{C}\int i\,dt \tag{3.29}$$

Replacing the expression for voltages across the components of circuit shown in Figure 3.34 as a function of time in Equation 3.24, we can write

$$E_p \sin \omega t = Ri + L\frac{di}{dt} + \frac{1}{C}\int i\,dt \tag{3.30}$$

Solving this second-order differential equation we can obtain an expression for the current. The general solution of Equation 3.30 will have an expression that can be written as

$$i(t) = I_p \sin(\omega t + \varphi) \tag{3.31}$$

where I_p is the peak value of the current, ω the angular frequency, and φ is the phase of the current.

Now let us compare the expression for the current, $i(t)$, and the source voltage, $v_s(t)$, applied to the circuit. There are four main characteristics:

- Both signals are sinusoidal signals
- Both signals have the same frequency
- The signals have different amplitudes
- The signals have different phase φ

Thus when we want to solve an ac circuit, we already know that the current and voltage are sinusoidal and we also know that voltage and current have the same frequency. If the value of the voltage source is known, the problem to obtain the current and to solve an ac circuit is reduced to calculating the new amplitude of the current and the new phase of the current with respect to the voltage. These two quantities form *the phasor*.

By replacing voltages and currents by their phasors we can move from the time domain into the frequency domain and solve an ac circuit. Once the circuit is solved using phasors it can then move back to the time domain, if necessary.

Phasors are operand that behave as vectors, that is, they have a magnitude and an angle. Voltages and currents are not vectors, they are scalar quantities, but in the frequency domain, phasors give them the characteristics of vectors. Because phasors have magnitude and angle they can be represented mathematically by a complex number.

3.3.2 APPLICATION OF KIRCHHOFF'S LAW TO AC CIRCUITS

In order to solve an ac circuit it is easy to work in the frequency domain. In this section, we can see some examples of how to prepare the circuit to be analyzed in the frequency domain. By replacing the sources by their phasors and the passive components by their impedances, we can transfer a circuit from the time domain into the frequency domain.

For a voltage or a current, the phasor equivalent will contain only the amplitude and the phase angle. Table 3.1 shows some examples of how to transform an ac voltage source into a phasor.

3.3.2.1 Impedance Z

The impedance of a component or a circuit is defined as the ratio of its voltage to its current

$$Z = \frac{v}{i} \tag{3.32}$$

Since the impedance is the ratio of voltage to current, the unit to measure the impedance is ohms. In circuits containing only resistive elements the ratio of applied voltage to its current is given by its resistance. In ac circuits voltages and currents will be time-dependent and they will have amplitude and a phase angle. The impedance in general will have *amplitude* and an *angle* (phase). Impedances can be used in ac circuits in a similar way as resistances are used in dc circuits

Using the definition of impedance, we can calculate the impedances of inductors and capacitors. We can work in the time domain to obtain the value of amplitude and phase required to determine the impedance of capacitors and inductors.

TABLE 3.1
Example of Transferring Voltages from Time Domain to Phasor Domain

Time Domain (V)	Phasor
$v_s(t) = 10 \sin(100\,t + 30°)$	$V_s = 10\angle 30°$
$v_a(t) = 10 \sin(100\,t - 60°)$	$V_a = 10\angle -60°$
$v_b(t) = 10 \sin(200\,t + 30°)$	$V_b = 10\angle 30°$

Note: $\angle$ is used to represent the phasor angle.

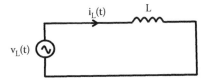

FIGURE 3.35   Ac circuit with a pure inductor.

In the circuit shown in Figure 3.35 a current $i_L(t)$ is applied to a pure inductor. Let us assume that a current is given by

$$i_L(t) = I_M \sin \omega t \tag{3.33}$$

This current passes through the inductor and we can deduce the voltage across the inductor.

We can calculate the impedance of this inductor by obtaining the voltage across the inductor as

$$v_L(t) = L\frac{di}{dt} = L\frac{d(I_M \sin \omega t)}{dt} = \omega L I_M \cos \omega t = \omega L I_M \sin(\omega t + 90) \tag{3.34}$$

Note that in an inductor the phase of the voltage is equal to 90°; this indicates that the voltage leads the current by 90° or that the current lags the voltage by 90°. Figure 3.36 shows the voltage and current in an inductor in the time domain.

3.3.2.2   Impedance of an Inductor

From the expression for the voltage across the inductor in the time domain indicated by Equation 3.34, we can write the phasor voltage for the inductor as

$$V_L = \omega L I_M \angle 90 \tag{3.35}$$

and from the expression for the current in the inductor in the time domain indicated by Equation 3.33, we can deduce the current phasor as

$$I_L = I_M \angle 0 \tag{3.36}$$

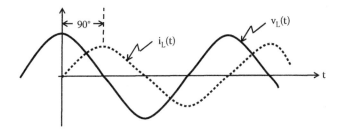

FIGURE 3.36   Voltage and current in a pure inductor in the time domain.

With the phasor voltage and phasor current we can calculate the impedance of this inductance as

$$Z_L = \frac{V_L}{I_L} = \frac{\omega L I_M \angle 90}{I_M \angle 0} = \omega L \angle 90 \qquad (3.37)$$

Impedances and phasors can be written as complex numbers in polar form as above or in rectangular form. The impedance of an inductor written in rectangular (or Cartesian) form is

$$Z_L = j\omega L [\Omega] \qquad (3.38)$$

where j indicates the imaginary part of a complex number (mathematicians use i to indicate imaginary parts of a number, but i can be confused as a current), then the impedance of an inductance is an imaginary number.

3.3.2.3   Impedance of a Capacitance

Similarly, we can deduce the impedance of a capacitor
    Let us assume a current

$$i_C = I_M \sin \omega t \qquad (3.39)$$

passes through a pure capacitor as indicated in Figure 3.37 and we can deduce the voltage across the capacitor.
    We can calculate the impedance of this capacitor by obtaining the voltage across the capacitor as

$$v_L = \frac{1}{C} \int i\, dt = \frac{1}{C} \int I_M \sin \omega t = -\frac{I_M}{\omega C} \cos \omega t = \frac{I_M}{\omega C} \sin(\omega t - 90) \qquad (3.40)$$

Note that in a capacitor the phase of the voltage is equal to −90°. This indicates that the voltage lags the current by 90° or the current leads the voltage by 90°. Figure 3.38 shows the voltage and current in a pure capacitor in the time domain.
    Then phasor voltage for this capacitor can be written as

$$V_C = \frac{I_M}{\omega C} \angle 90 \qquad (3.41)$$

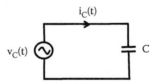

FIGURE 3.37   Ac circuit with a pure capacitor.

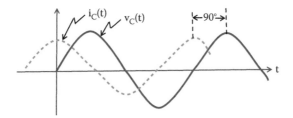

FIGURE 3.38 Voltage and current in a pure capacitor in the time domain.

and from the expression for the current in the capacitor in the time domain indicated by Equation 3.37 we can deduce the capacitor current phasor as

$$I_C = I_M \angle 0 \tag{3.42}$$

With the phasor voltage and phasor current we can calculate the impedance of this capacitor as

$$Z_C = \frac{V_C}{I_C} = \frac{(I_M / \omega C) \angle -90}{I_M \angle 0} = \frac{1}{\omega C} \angle -90 \tag{3.43}$$

The impedance of a capacitor can be written by complex numbers in polar form as above or in Cartesian form as below:

$$Z_C = -j \frac{1}{\omega C} \tag{3.44}$$

Using phasor and impedances the solution of ac circuits can be treated in the same way as dc circuits. Once in the frequency domain we can use Kirchhoff's laws to calculate voltages and currents in circuits. The only difference is that phasors and impedances are complex numbers.

Impedances can be expressed in rectangular, polar, and exponential forms and they can be added, subtracted, multiplied, and divided in the same way as complex numbers. If you are unfamiliar with manipulation of complex number it is advised that you revise this topic.

Using properties of complex numbers, the impedance of a capacitor indicated by Equation 3.44 can also be written as

$$Z_C = \frac{1}{jwC} \tag{3.45}$$

3.3.2.4 Impedance of a Resistance

The resistance is frequency independent and the voltage and current passing through a resistor are in phase. Then the ratio of the voltage to current in a resistor is the impedance

$$Z_R = R \tag{3.46}$$

TABLE 3.2

Impedances of R, L, and C

Element	Impedance
Z_R	R
Z_L	$j\omega L$
Z_C	$1/j\omega C$ or $-j(1/\omega C)$

Impedances combinations of series and parallel are combined in the same way as resistors.

In summary, Table 3.2 shows the impedance of resistors, inductors, and capacitors.

EXAMPLE 3.15

For the combination of R–L series circuit shown in Figure 3.39, determine the total impedance Z_{ab}, at a frequency of 50 Hz. Use accuracy up to two decimal places. At 50 Hz the angular frequency is $\omega = 2\pi f = 2 \cdot \pi \cdot 50 = 314$ rad/s. The impedance of the inductor is $Z_L = j\omega L = j(314 \times 10 \times 10^{-3}) = j3.14 [\Omega]$
Therefore,

$$Z_{ab} = Z_R + Z_L = R + j\omega L = 10 + j(314 \times 10 \times 10^{-3}) = 10 + j3.14 [\Omega]$$

3.3.2.5 Reactance X

In dc current Ohm's law provides the ratio of voltage to current and it is a measure of how the component opposes the flow of current, this is the resistance. In inductors and capacitors its opposition to circulation of current is termed as its reactance. Reactance is represented by the symbol X and is the magnitude of the impedance, that is, the value of the impedance excluding the phase, then

$$\text{Reactance of an inductor} = X_L = \omega L [\Omega] \tag{3.47}$$

$$\text{Reactance of a capacitor} = X_C = \frac{1}{\omega C} [\Omega] \tag{3.48}$$

Since reactance represents the ratio of voltage to current it has units of ohms. Therefore, the impedances can be expressed in term of their reactance as

$$Z_L = jX_L [\Omega] \tag{3.49}$$

FIGURE 3.39 R–L series circuit for Example 3.15.

$$Z_C = -jX_C [\Omega] \qquad (3.50)$$

3.3.2.6   Polar–Cartesian Forms

In some cases it is easier to manipulate complex numbers if they are in polar form, especially when multiplying and dividing numbers. On the other hand, it is easier to add or subtract complex numbers in Cartesian form.

3.3.2.7   Cartesian to Polar Form

Let us take a general complex number in Cartesian form, Z_1 and transform it to polar form:

$$Z_1 = x + jy$$

The magnitude can be calculated as

$$\text{Magnitude} = +\sqrt{x^2 + y^2}$$

and the phase can be calculated as

$$\text{Phase} = \angle \tan^{-1}\left(\frac{y}{x}\right)$$

Then, Z_1 can be written in polar form as

$$Z = x + jy = \sqrt{x^2 + y^2} \angle \tan^{-1}\left(\frac{y}{x}\right) \qquad (3.51)$$

3.3.2.8   Polar to Cartesian

Let us take a general complex number written in polar form, Z_2 and transform it to Cartesian form:

$$Z_2 = r\angle\varphi$$

We can calculate the real and imaginary part as

$$\text{Real part} = r\cos\varphi$$

$$\text{Imaginary part} = r\sin\varphi$$

and

$$Z_2 = r\angle\varphi = r\cos\varphi + j\sin\varphi \qquad (3.52)$$

3.3.2.9 Phasor Diagrams

The phasor contains the magnitude and phase of voltages and currents. For an ac circuit we can use phasor diagrams to represent magnitude and phase of voltages and currents within a single diagram. Usually a phasor diagram for a circuit will contain all voltages and currents in the circuit of very different values, so then they are usually represented in a graph using different scales for voltages and currents.

EXAMPLE 3.16

The phasor $V_a = 10\angle 45°$ can be represented in a phasor diagram as indicated in Figure 3.40.

EXAMPLE 3.17

For the RL series circuit shown in Figure 3.41 determine the impedance of the circuit, the current in this circuit, the circuit phase angle between voltage and current, and the voltage across the 12 Ω resistor. Calculations can be approximate to two decimals.

SOLUTION

The circuit impedance is given by

$$Z = R + j2\pi fL = 12 + j2\pi 50 \times 15.92 \times 10^{-3} = 12 + j5$$

or written in polar form

$$Z = \sqrt{12^2 + 5^2} \angle \tan^{-1}(5/12) = 13\angle 22.62°$$

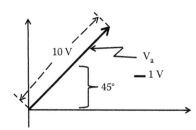

FIGURE 3.40 Phasor diagram representation of $V_a = 10\angle 45°$.

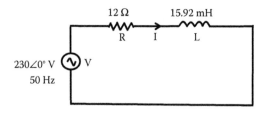

FIGURE 3.41 RL series circuit diagram for Example 3.17.

The impedance has a magnitude of 13 Ω and an angle of 22.62°.
The current in the circuit is given by

$$I = \frac{V}{Z} = \frac{230\angle 0°}{13\angle 22.62°} = 17.69\angle -22.62° \, A$$

and the phase angle between the voltage and the current is equal to 22.62°.
The voltage across the 12 Ω resistor is given by

$$V_R = I \times R = 17.69\angle -22.62° \times 12 = 212.28\angle -22.62° \, V$$

EXAMPLE 3.18

For the RC series circuit shown in Figure 3.42 determine the impedance of the
circuit, the phasor current in this circuit, the phasor voltage across the resistor, the
phasor voltage across the capacitor, and sketch a phasor diagram for this circuit.
Calculations can be approximate to two decimals.

SOLUTION

The circuit impedance is given by

$$Z = R + \frac{1}{j2\pi fC} = 10 - j\frac{1}{2\pi 50 \times 127.3 \times 10^{-6}} = 10 - j25$$

or written in polar form

$$Z = \sqrt{10^2 + 25^2} \angle \tan^{-1}\left(\frac{25}{10}\right) = 26.93\angle -68.20°$$

The impedance has a magnitude of 26.93 Ω and an angle of −68.2°.
The current in the circuit is given by

$$I = \frac{V}{Z} = \frac{230\angle 30°}{26.93\angle -68.2°} = 8.54\angle 98.2° \, A$$

and the phase angle between the voltage and the current is equal to 98.2°.

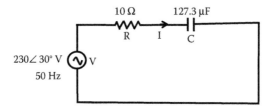

FIGURE 3.42 RC series circuit diagram for Example 3.18.

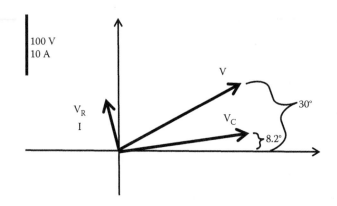

FIGURE 3.43   Sketch phasor diagram for Example 3.18.

The voltage across the 10 Ω resistor is given by

$$V_R = I \times R = 8.54 \angle 98.2° \times 10 = 85.4 \angle 98.2° \, V$$

The voltage across the capacitor is given by

$$V_C = I \times Z_C = 8.54 \angle 98.2° \times 25 \angle -90 = 213.5 \angle 8.2°$$

V, V_R, V_C, and I are plotted in Figure 3.43.

EXAMPLE 3.19

An inductor, L, is connected in series with a 24 Ω resistor and this combination is connected across an ac voltage source of 50 Hz frequency. If the total magnitude of the impedance is 26 Ω, calculate the value of the inductor and phase between the voltage and current of the total impedance.

SOLUTION

The total impedance, Z_T, is given by

$$Z_T = R + j2\pi fL = 24 + j2\pi 50 \times L$$

The magnitude of the total impedance is

$$|Z_T| = \sqrt{R^2 + (2\pi 50 L)^2}$$

or

$$2\pi 50 L = \sqrt{|Z_T|^2 - R^2} = \sqrt{26^2 - 24^2} = 10$$

Then,

$$L = \frac{10}{2\pi50} = 15.92 \text{ mH}$$

As the impedance is the ratio of the voltage to the current as $Z_T = V/I$, then the phase angle between the voltage and the current is given by the phase of the impedance as

$$\text{Phase angle of } Z_T = \arctan\left(\frac{2\pi fL}{R}\right) = \arctan\left(\frac{10}{24}\right) = 22.62°$$

EXAMPLE 3.20

Determine the expression for the current in the time domain i(t) for the RLC circuit shown in Figure 3.44 using the phasor domain. The voltage source is $v(t) = 230 \sin(2\pi50\,t + 30°)$ V. Calculation can be approximate to two decimal digits.

SOLUTION

By calculating the impedances and the phasor voltage for the source, we can obtain an equivalent circuit in the phasor domain.

From the expression of the voltage source we can obtain the source phasor voltage as

$$V = 230\angle30° \text{ V}$$

The impedance of the inductance is

$$Z_L = j2\pi fL = j2 \times 3.1416 \times 50 \times 318.3 \times 10^{-3} \,\Omega = j100\,\Omega$$

The impedance of the capacitance is

$$Z_C = \frac{1}{j2\pi fC} = \frac{1}{j2 \times 3.1416 \times 50 \times 127.3 \times 10^{-6}}\,\Omega = \frac{25}{j}\,\Omega = -j25\,\Omega$$

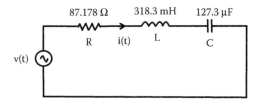

FIGURE 3.44   RLC series circuit in the time domain for Example 3.20.

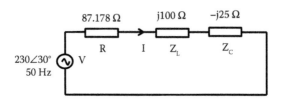

FIGURE 3.45   RLC series circuit in the phasor domain for Example 3.20.

With the impedances for the passive components and phasor for the active component we can create an equivalent circuit in the phasor domain as indicated in Figure 3.45.

We can add all impedance in series to one equivalent impedance Z_T as

$$Z_T = R + Z_L + Z_C = 87.178 + j100 - j25 = 87.178 + j75 \ \Omega$$

That can be written in polar form as

$$\text{Magnitude of } Z_T = |Z_T| = \sqrt{87.178^2 + 75^2} = 115 \ \Omega$$

$$\text{Phase of } Z_T = \angle Z_T = \arctan\left(\frac{75}{87.178}\right) = 40.71°$$

Then,

$$Z_T = 87.178 + j75 \ \Omega = 115\angle 40.71° \ \Omega$$

The current can be calculated as

$$I = \frac{V}{Z_T} = \frac{230\angle 30°}{115\angle 40.71°} = 2\angle -10.71 A$$

The current has a magnitude of 2 A and lags behind the voltage by 10.71°.

Since we know the frequency of the system and the sine shape of the waveform we can transfer from the phasor domain to the time domain. The current in the time domain will be

$$i(t) = 2\sin(2\pi 50 t - 10.71°) A$$

With the current phasor we can calculate the phasor voltages on the other components, then

$$V_L = I \cdot Z_L = (2\angle -10.71°) \cdot (100\angle 90°) = 200\angle 79.29° V$$

$$V_C = I \cdot Z_C = (2\angle -10.71°) \cdot (25\angle -90°) = 50\angle 100.71° V$$

$$V_R = I \cdot R = (2\angle -10.71°) \cdot (87.178) = 174.356\angle -10.71° \, V$$

These phasor voltages can be plotted to produce a phasor diagram. The magnitude of voltages are $V_L = 200$ V, $V_C = 50$ V, and $V_R = 174.356$ V.

EXAMPLE 3.21

For the circuit shown in Figure 3.46, calculate the current I_s.
   The impedance of the inductance is

$$Z_L = j2\pi fL = j2 \times 3.1416 \times 50 \times 15.92 \times 10^{-3} \, \Omega = j5 \, \Omega$$

The impedance of the capacitance is

$$Z_C = \frac{1}{j2\pi fC} = \frac{1}{j2 \times 3.1416 \times 50 \times 127.3 \times 10^{-6}} \, \Omega = \frac{25}{j} \, \Omega = -j25 \, \Omega$$

Now we can add Z_C and Z_L in parallel as Z_p

$$Z_p = \frac{Z_C \times Z_L}{Z_C + Z_L} = \frac{j5(-j25)}{j5 - j25} \, \Omega = \frac{125}{-j20} \, \Omega = +j6.25 \, \Omega$$

and the total impedance, Z_T, seen by the source is

$$Z_T = R + Z_p = 10 + j6.25 \, \Omega = 11.79\angle 32°$$

and I_s will be

$$I_s = \frac{V_s}{Z_T} = \frac{230\angle 0}{11.79\angle 32°} = 19.51\angle -32° \, A$$

The current has a magnitude of 19.51 A and lags the voltage by 32°.

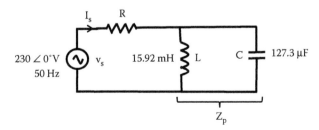

FIGURE 3.46   Circuit diagram for Example 3.21.

3.4   KEY POINTS

- In a circuit at any specific time the sum of currents flowing into a node is equal to the current going out of the same node (KCL).
- In a circuit at any specific time the sum of the pds around any loop in a circuit is equal to zero (KVL).
- Mesh current method offers a systematic approach to obtain a set of equations to determine voltages and currents in a circuit.
- Alternating waveforms vary with time; the most important waveform in an electrical system is the sinusoidal waveform.
- The frequency of a periodic waveform f is equal to the reciprocal of its period T.
- The magnitude of an alternating waveform can be described by its *root mean square (RMS)* value, its *peak* value, or its *peak-to-peak* value.
- An ac voltage signal can be represented by a sinusoidal waveform given by $v = E_p \sin(\omega t + \varphi)$, where E_p is the peak voltage, ω is the angular frequency, and φ is the phase of the waveform at $t = 0$.
- Waves of the same frequency can have phase differences.
- In an ac circuit the voltage across a pure resistor is *in phase* with its current.
- In an ac circuit the voltage across a pure inductor has a phase that *leads* its current by 90°.
- In an ac circuit the voltage across a pure capacitor has a phase that *lags* its current by 90°.
- The reactance of an inductor is $X_L = \omega L$, and the reactance of a capacitor is $X_C = 1/\omega C$.
- Phasor diagrams can be used to represent both the magnitude and the phase of quantities of the same frequency.
- The relationship between the current and the voltage within a circuit containing reactive components is described by its impedance Z.
- Complex numbers may be expressed in *rectangular form* $(x + jy)$ or in *polar form* $(r\angle\theta)$.
- Impedances can be used with ac circuits in the same way as resistances are used in dc circuits.

4 Diodes

4.1 INTRODUCTION

Any material can be classified according to its conductivity; there are conductor, insulator, and semiconductor materials. Semiconductor materials, as the name indicates, are materials in which conductivity is a great deal less than a conductor but is also substantially larger than an insulator. In a semiconductor material, the conductivity is due to the movement of negative and positive charges. If a semiconductor material has conductivity due mainly to positive charge, it is called a p-type semiconductor material. If the majority carrier charges are negative electrons, it is called an n-type semiconductor material. An n-type material in intimate contact with a p-type material will create a p–n junction. The p–n junction does have the useful property of a higher conductivity of current in one direction than in the other. When a p–n junction is formed a device known as a diode is created.

In this chapter, we look at the characteristics of semiconductor materials and investigate the behavior of semiconductor diodes. Several applications of semiconductor diodes are considered, including their use as rectifiers within power supplies. Some special purpose diodes are also described. This chapter not only provides information on the use of diodes, but also sets the scene for later material on transistors.

4.2 SEMICONDUCTOR MATERIAL

In pure semiconductor material, known as intrinsic semiconductor, thermal vibrations might weaken atoms bonds generating free electrons to move. As these free electrons move they leave *holes* that accept electrons from close atoms, therefore these holes also move. The mobility of a hole is less than an electron; physicists have given physical characteristics to holes so they can be considered as particles contributing to conductivity of semiconductor material. Electrons are negative charge carriers and holes are positive charge carriers. There are many materials that behave as semiconductors; silicon (Si) being the most commonly used in creating electronic devices. Figure 4.1 shows a representation of a structure of pure silicon semiconductor.

Figure 4.1a shows the structure of pure silicon with no external energy added to it, the electron bonds are shared among atoms. Figure 4.1b represents the atomic structure of silicon once enough energy is applied to break some bonds consequently leaving some holes and free electrons to move.

4.2.1 CONDUCTIVITY AND ENERGY BANDS IN SEMICONDUCTORS

In the initial studies on semiconductor materials, researchers published results in conductivity that have large discrepancies on the supposed same material under similar conditions. The influence on impurities in semiconductor conductivity was not

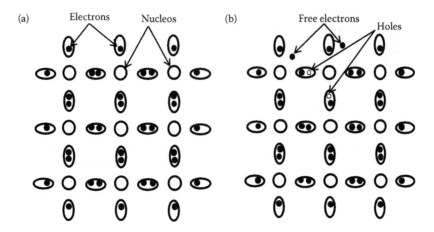

FIGURE 4.1 (a) Atomic structure of silicon. (b) Effect of external energy added to silicon.

well understood or controlled. In the 1920s, the introduction of the energy band role in explaining conductivity in semiconductors helped to develop the physics of semi-conductors to explain the behavior of semiconductor devices and clarify the large influence of impurities in the conductivity of semiconductor materials.

In a material the outer valence shell of an atom corresponds to energy levels form-ing a band. In equilibrium, valence electrons are restricted to this band—the *valence band*. If external energy is applied to the material some electrons might acquire enough energy to escape this valence band to another level of energy where they can move freely. This new energy level is one in which electrons can move to form a band: the *conduction band*. In terms of conductivity, materials can be classified according to the external amount of energy required for electrons to jump from the valence band to the conduction band. Between these two bands there is an energy gap that represents the energy necessary to induce a jump of electrons from the valence band to the conduction band. Figure 4.2 shows a representation of insulator, conductor, and semiconductor according to their energy band structure.

Insulators have a very large energy gap (E_g) and electrons normally do not jump into the conduction band, with the exception of an extreme case such as breakdown. In conductors the valence and conduction bands overlap, and then there are electrons free to move in the conduction band. For the semiconductor material, the energy gap is not as large as an insulator and a relative small amount of energy will cause electrons to jump into the conduction band.

In terms of conductibility in semiconductors, when an electron jumps from the valence band to the conduction band it leaves its space in the valence band. Another electron can move into the space left in the valence band. The movement of charge (electrons) in the valence band contributes to the current as well as the movement of electrons in the conduction band. Now the space left by an electron in the valence band, which is the lack of one electron, is known as a *hole*. So in semiconductor material we have contributions to current from electrons in the conduction band and holes in the valence band.

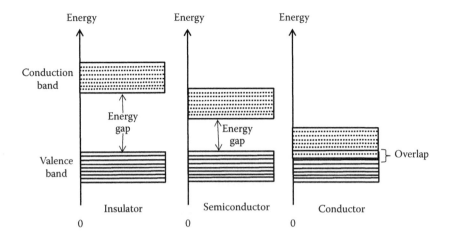

FIGURE 4.2 Energy band diagrams for insulator, semiconductor, and conductor.

4.2.2 DOPING

Adding external impurities to a semiconductor is known as doping a semiconductor. It was found that the addition of small amounts of impurities drastically changes its conductivity. Materials with impurities that produce an excess of *electrons,* in the conduction band, are n-type semiconductors. Materials with impurities that produce an excess of *hole, in the valence band*, are p-type semiconductors. Both n-type and p-type materials have much greater conductivity than pure semiconductors. The dominant charge carriers in a doped semiconductor are called *majority charge carriers*; the other type is *minority charge* carriers. For example, in an n-type semiconductor, electrons are the majority carriers and holes are the minority carriers.

In terms of the energy band diagram theory of semiconductors, the addition of impurities will result in an extra level of energy in the energy gap. Figure 4.3 shows the energy band diagrams of a semiconductor with impurities added.

In the case of an n-type material (adding *donor* impurities), electrons from the added impurity appear as a donor energy level (E_d) close to the conduction band. A small amount of energy is required to make these electrons jump into the conduction band and current is mainly due to electrons.

In the case of a p-type material (adding *acceptor* impurities) holes from the added impurity (i.e. empty level, ready to accept electrons) appear in an acceptor energy level (E_a) close to the valence band. A small amount of energy is required to make electrons jump from the valence band into this acceptor level and current is mainly due to a hole in the valence band.

In the energy band diagram, the Fermi level (E_F) represents the probability of finding an electron in the conduction band (or a hole in the valence band). In a pure semiconductor, any number of electrons jumping from the valence band to the conduction band will leave the same number of holes in the valence band. Then the probability of finding an electron in the conduction band is the same as finding a hole in the conduction band. Hence, E_F is in the middle of the gap for an intrinsic

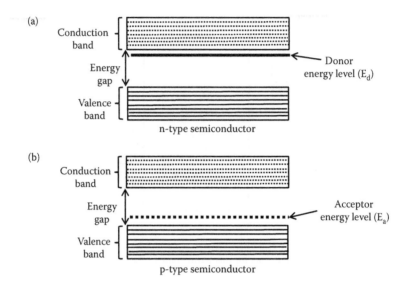

FIGURE 4.3 Energy band diagrams of a semiconductor with impurities added: (a) n-type semiconductor, (b) p-type semiconductor.

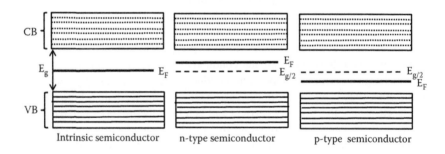

FIGURE 4.4 Energy band structure for an intrinsic, n-type, and p-type semiconductor.

semiconductor. The probability of finding electrons in the conduction band rather than a hole in the valence band is larger in an n-type semiconductor and E_F then is above the middle of the gap. The band energy structure of a semiconductor will provide information on the type of material and amount of impurity added. Figure 4.4 shows energy band structure for an intrinsic, n-type, and p-type semiconductor.

It is possible to deduce an expression to indicate the number of impurities added to an intrinsic semiconductor. This expression will be proportional to the different between Fermi level (E_F) and the intrinsic level ($E_i = E_{g/2}$).

4.3 P–N JUNCTION

When p-type and n-type materials are joined, a *p–n junction* is created. In practice, the junction is created together in the fabrication process. In the p–n junction the

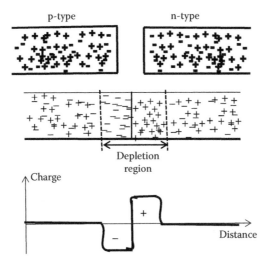

FIGURE 4.5   p–n Junction.

majority charge carriers on each side diffuse across the junction where they combine with the charge carriers of the opposite polarity. Around the junction there are then few free charge carriers creating a region known as the *depletion layer*, as shown in Figure 4.5.

The diffusion of positive charge in one direction and negative charge in the other produces a charge imbalance and creates a *potential barrier* (also known as the built-in voltage or internal voltage of a diode) across the junction. The potential barrier opposes the flow of *majority* charge carriers and only a small number have enough energy to overcome it. The concentration of positive and negative charges around this junction will generate a small *diffusion current*. The potential barrier formed facili-tates the flow of *minority* carriers and they will be swept over to produce a small *drift current*. In a p–n junction without external energy applied to it, these two currents, the diffusion and drift currents, must cancel each other and the net current is zero.

If an external energy is applied to the p–n junction the potential barrier will be affected. If the external energy is applied in the form of a source, the polarity of the source will produce a *forward bias* current or a *reverse bias* current depending on the polarity of the applied voltage.

4.4   DIODE CURRENT–VOLTAGE CHARACTERISTICS I–V

4.4.1   FORWARD BIAS

Forward bias happens when the p-type side is made *positive* with respect to the n-type side. In this case, the height of the barrier is reduced further. Majority charge carriers have sufficient energy to surmount this effect. The diffusion current there-fore increases while the drift current remains the same. There is thus a net large cur-rent flow across the junction, which increases exponentially with the applied voltage.

4.4.2 REVERSE BIAS

Reverse bias happens when the p-type side is made *negative* with respect to the n-type side. In this case, the height of the barrier is increased and the number of majority charge carriers that have sufficient energy to surmount it rapidly decreases. The diffusion current therefore vanishes, while the drift current remains the same. Thus, the only current is a small approximately constant drift current.

A diode's performance in a circuit depends on its I–V characteristic. The current–voltage characteristic is determined by the transport of charge carriers through the depletion region. At *forward bias* the external voltage opposes the potential barrier resulting in substantial electric current through the p–n junction. For silicon diodes, the potential barrier is approximately between 0.5 and 0.7 V and for germanium diodes around 0.3 V. The I–V characteristic in forward bias shows an exponential increase in the current at about 0.7 V. At *reversed bias* the external voltage has the same polarity as the potential barrier. Then the barrier is increased resulting in significant reduction of the electric current through the p–n junction. The I–V characteristic in reversed bias shows a very small almost constant reverse current mainly due to minority carriers known as the *saturation current*. Figure 4.6 shows the forward and reverse current in a diode.

At very large reverse bias, beyond the maximum reverse voltage indicated by the manufacture known as *peak inverse voltage* (PIV), a process called reverse breakdown occurs. This causes a large increase in reverse current that can usually damage the diode.

The current flowing through a diode can be related to the applied voltage by the following expression:

$$I = I_s \left[e^{qV/kT} - 1 \right] \tag{4.1}$$

where
  I is current through the diode
  q the electronic charge = 1.6×10^{-19} C
  T is the absolute temperature given in degrees Kelvin

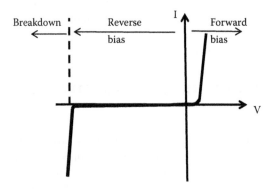

FIGURE 4.6 I–V diode characteristics.

k = Boltzmann's constant = 1.38×10^{-23} J/k
V applied voltage
I_s reverse saturation current

At normal room temperature, q/kT has a value of approximately 40 V^{-1}, therefore if a reverse bias V is applied the exponential term is approximately zero and I is given by

$$I \approx I_s(0-1) = -I_s \qquad (4.2)$$

where I_s is known as the saturation current (under reverse bias).

Similarly, if the forward bias applied (V) is greater than 0.1 V, the current I can be approximate to

$$I \approx I_s[e^{qV/kT}] = I_s[e^{40V}] \qquad (4.3)$$

and the current increases exponentially.

4.5   DIFFERENT TYPES OF DIODES

4.5.1   SEMICONDUCTOR DIODES

The diode is a p–n junction packaged as an electronic component; the symbol to represent a diode and some packaged samples are shown in Figure 4.7. The diode is a two-terminal device. One terminal is known as the anode (A) and the other the cathode (K). Thin bar in the package indicates the cathode.

Common p–n diodes, which operate as described above, are usually made of doped silicon or in the case of light-emitting diode (LED), they are made from some compound semiconductors. Diodes are fabricated with different characteristics to satisfy certain applications, below are some commonly used diodes.

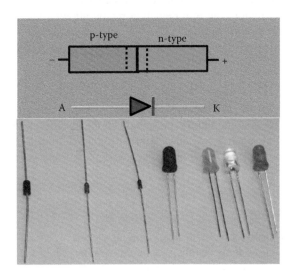

FIGURE 4.7   Typical diode symbol and package.

4.5.2 ZENER DIODES

Zener diodes are made to conduct under reversed bias. Zener diodes are fabricated to work in the breakdown region under a special tunneling breakdown called Zener breakdown. This breakdown occurs at a specifically defined value of voltage, as this breakdown voltage is maintained as the reverse current increases, this diode can be used as a voltage reference. The breakdown voltage is kept constant even if the current changes without damaging the diode.

4.5.3 AVALANCHE DIODES

These diodes behave in the same way as the Zener diodes, but the breakdown mechanism is different, known as the avalanche breakdown. The electric field across the p–n junction is large enough to accelerate the electrons within the depletion layer; these electrons collide with atoms and can free other electrons, producing an avalanche. Avalanche occurs in diodes with more lightly doped materials.

4.5.4 LIGHT-EMITTING DIODES

Light-emitting diodes (LEDs) are produced using some special semiconductor materials produced as a compound of elements. In these, diodes carriers that cross the junction emit light (photon) when they recombine. Doping these semiconductors with different types of impurities can produce photons with different wavelengths. The different wavelengths will provide different colours in an LED. It is possible to produce LED between the ultraviolet and infrared range.

4.5.5 TUNNEL DIODES

These diodes are produced with highly doped semiconductor materials creating a very thin depletion region; the thin depletion region allows some sort of quantum tunneling of charge carriers producing an I–V characteristic with a negative resistance region. They are also known as Esaki diodes.

4.5.6 GUNN DIODES

Gunn diodes are usually fabricated on compound semiconductor materials such as gallium arsenide (GaAs) and indium phosphide (InP). They have an I–V characteristic with a region of negative resistance (similar to tunnel diodes). They are usually used in applications to generate high-frequency signals in the order of microwave wavelength range.

4.5.7 PELTIER DIODES

These diodes can absorb energies as heat changing its current–voltage characteristics so that they can be used to detect changes in temperature. They can be used as thermoelectric sensors. They can also emit energies as heat.

4.5.8 PHOTODIODES

Semiconductors produce charges when light energy is added. Photodiodes are fabricated in semiconductor material most sensitive to light. They also contain a thin layer of undoped material to increase their sensitivity to light. They are packaged in such way that its p–n junction is exposed to light. They can be used in photometry and optical communication.

4.5.9 SOLAR CELL

Solar cell uses p–n junction reversed bias to absorb light energy and convert this energy into electrical energy. They are widely used as alternative way of generating electrical energy.

4.5.10 SCHOTTKY DIODES

Schottky diodes are fabricated as a junction between metal and semiconductor. Depending on the work function of the metal and the n or p type of semiconductor a device can be created that behaves as a diode and not as a simple contact. As the junction is between a metal and a semiconductor the charges can move fast when polarities are changed in this diode. In addition, the junction capacitance of these diodes is small, allowing a high switching speed required for high-frequency devices. It is also possible to fabricate Schottky diodes using amorphous semiconductor material. Although they present the diode behavior their capacitance and conductance differ from the Schottky diodes fabricated using crystalline semiconductors (Fernandez-Canque et al. 1983).

There are other devices that have the name of diode in it, but they are part of a combination of diodes and other components in integrated circuit (IC) form such as varactor diodes, super-barrier diodes, and photodiode.

4.6 DIODE APPLICATIONS

Diodes can be applied to produce different functions such as rectification, limiting voltages, radio frequency detection, protection, clamping, LED, and photodetectors.

4.6.1 RECTIFICATION

Due to a characteristic of diode to block current under reverse bias, one common application of diodes is to transform ac current into dc current, which is known as rectification.

Consider a sine wave ac voltage applied to a circuit containing a resistor as shown in Figure 4.8. The signal alternates between positive half-cycle and negative half-cycle, providing a forward and reverse bias to the diode. The output will indicate large conduction when only the diode is forward biased as shown in Figure 4.8.

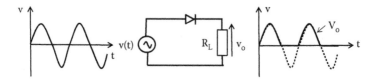

FIGURE 4.8   One diode rectifier circuit.

4.6.2   HALF-WAVE RECTIFIERS

As shown in Figure 4.8, the diode conducts when forward biased; this happens in the half-cycle of the ac input signal. The voltage drop across the diode will be about 0.7 V for a silicon diode. When the half-cycle reversing the voltage across the diode is applied, the diode does not conduct and the current is very small. Figure 4.9 shows a more detailed signal for the output voltage after the effect of the diode. The diode has changed the voltage from an alternating voltage to a voltage that does not alternate. In the example shown in Figure 4.9 we have only positive voltage.

For this half-wave rectified signal it is possible to deduce the root mean square (RMS) and mean value as

$$E_{mean} = \frac{E_P}{\pi}[V] \tag{4.4}$$

$$E_{RMS} = \frac{E_P}{2}[V] \tag{4.5}$$

If a smoothing capacitor, also known as reservoir capacitor, is connected across the load, the output voltage changes as indicated in Figure 4.10. As the input voltage increases from 0 to the maximum input voltage, E_P, the capacitor is charged and follows the input voltage. As the input voltage decreases after E_P the capacitor discharges. Due to its transient response the discharge takes time and it cannot follow the input voltage. The discharging of the capacitor will produce an output that

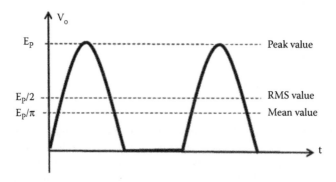

FIGURE 4.9   Half-wave rectified signal.

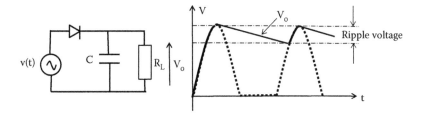

FIGURE 4.10   Half-wave rectifier with capacitor smoothing.

reduces gradually, and it does not fully discharge within the time of a period when
the input voltage is increasing again in the next cycle. The result output is a signal
that is close to a dc signal with a small ripple voltage at the top, as indicated in
Figure 4.10.

4.6.3   Full-Wave Rectifiers

It is possible to connect more than one diode to rectify both half-cycles of an ac signal. One such circuit is shown in Figure 4.11 using a center-tapped transformer. The
transformer can be used to change the ac voltage to a desired value. The diodes are
connected in such a way that in half-cycle one diode is forward bias and the other
is reverse bias. In the other half-cycle both diodes reverse polarities producing the
output signal shown in Figure 4.11.

For this full-wave rectified signal it possible to determine the RMS and mean
value. They are

$$E_{mean} = \frac{E_P}{0.5\pi}[V] \tag{4.6}$$

$$E_{RMS} = \frac{E_P}{\sqrt{2}}[V] \tag{4.7}$$

4.6.4   Single-Phase Bridge Rectifier Circuit

Another way of producing a full-wave rectification is to use four diodes connected as
shown in Figure 4.12; this is known as a bridge rectifier.

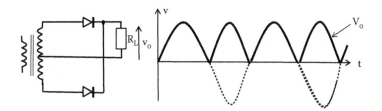

FIGURE 4.11   Transformer center tapped full-wave rectifier circuit.

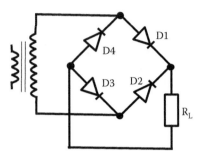

FIGURE 4.12   Full-wave bridge rectifier circuit.

A transformer is needed only if the output voltage is different from the source. In this circuit, for each half-cycle two diodes conduct. Figure 4.13 illustrates how the bridge rectifies the voltage.

For the positive half-cycle, diodes D1 and D3 are forward biased and diodes D2 and D4 are reverse biased; the current circulates as indicated in Figure 4.13b. For the negative half-cycle diodes D2 and D4 are forward biased and diodes D1 and D3 are reversed biased and the conductance is as indicated in Figure 4.13c. If a smoothing capacitor is connected across the load (R_L), then the output voltage changes as indicated in Figure 4.14. As the input voltage increases from 0 to E_p the capacitor is charged and follows the input voltage. As the voltage decreases after E_p the capacitor maintains the output voltage and reduces the ripple voltage as indicated in Figure 4.14.

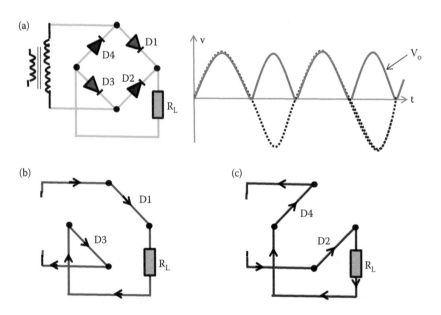

FIGURE 4.13   Illustration of conduction in a bridge rectifier: (a) bridge rectifier, (b) forward bias, and (c) reverse bias.

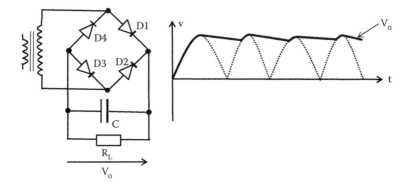

FIGURE 4.14   Full-wave rectifier with a smoothing capacitor.

TABLE 4.1
Example of Rectifier Diodes

Diode	Maximum Current (A)	Maximum Reverse Voltage (V)
1N4001	1	50
1N4002	1	100
1N4007	1	1000
1N5401	3	100
1N5408	3	1000

Rectifier diodes are usually made from silicon with a forward voltage drop of around 0.7 V. Table 4.1 shows some popular rectifier diodes with their maximum current and maximum reverse voltage.

EXAMPLE 4.1

Figure 4.15 shows a circuit of a diode with a saturation current of 5×10^{-11} A at 300°K. If the voltage source supplies a current of 100×10^{-3} A, calculate the voltage across the diode and the value of the variable resistance Rv. Values can be accurate to two decimal places.

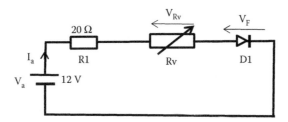

FIGURE 4.15   Circuit diagram for Example 4.1.

SOLUTION

At 300°K the expression q/KT has an approximate value of 1/0.025 V. The current passing through the diode, at 300°K, is 100 mA, then

$$I_a = 100 \times 10^{-3} = 5 \times 10^{-11} \left[e^{V_F/0.025} - 1 \right] [A]$$

or

$$e^{V_F/0.025} = \frac{100 \times 10^{-3}}{5 \times 10^{-11}} + 1 = 20 \times 10^8 + 1$$

and

$$V_F = 21.4 \times 0.025 = 0.54\,V$$

The voltage across the forward biased diode is 0.54 V. Applying KVL, the voltage across Rv can be calculated as

$$12 = 20 \times 100 \times 10^{-3} + V_{Rv} + 0.54$$

Given

$$V_{Rv} = 9.46\ V$$

and the value of Rv can be calculated using Ohm's law as

$$Rv = \frac{V_{Rv}}{I_a} = \frac{9.46}{100 \times 10^{-3}} = 94.6\ \Omega$$

EXAMPLE 4.2

For the circuit shown in Figure 4.16, the voltage source is given by $v_s = 50 \sin 2\pi 50t$ and D_1 is a silicon p–n junction. Sketch the waveform of the voltage across the resistor R_L and calculate the peak value, average value, and period of V_L. Values can be accurate to two decimal places.

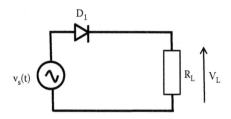

FIGURE 4.16 Circuit diagram for Example 4.2.

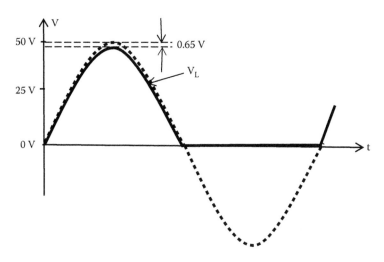

FIGURE 4.17 Sketch waveform of voltage across R_L in Example 4.2.

SOLUTION

As the diode D_1 consists of a silicon p–n junction, we can assume that under forward bias the voltage across D_1 is 0.65 V. Then, under forward bias the voltage across R_L will be reduced by 0.65 V. Figure 4.17 shows a sketch of V_L.

The peak value of V_L is

$$V_{Lpeak} = 50 - 0.65 = 49.35 \text{ V}$$

and the average value of V_L is

$$V_{Lav} = \frac{V_{Lpeak}}{\pi} = \frac{49.35}{\pi} = 15.71 \text{ V}$$

The waveform of V_L has the same frequency as the source voltage $f = 50$ Hz, then the period of V_L is

$$T = \frac{1}{f} = \frac{1}{50} = 0.020 \text{ s}$$

EXAMPLE 4.3

Two silicon diodes D_a and D_b are connected as indicated in Figure 4.18. Determine the value of V_o for (1) $V_a = V_b = 0$ V, (2) $V_a = 0$ V, $V_b = 10$ V, (3) $V_a = 10$ V, $V_b = 0$ V, and (4) $V_a = V_b = 10$ V.

SOLUTION

As D_a and D_b are silicon diodes, we can assume a forward bias voltage drop of 0.65 V.

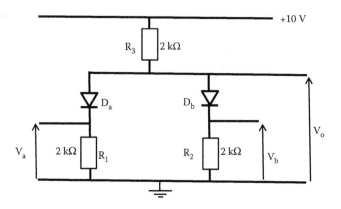

FIGURE 4.18   Circuit diagram for Example 4.3.

1. In this case R_1 and R_2 are short-circuited by $V_a = V_b = 0$ V and D_a and D_b are forward biased. Then

$$V_o = 0.65 \text{ V}$$

2. The pd across D_b is zero as V_a is the same as the bias voltage of 10 V. R_2 is short circuited and D_a is forward biased. Then

$$V_o = 0.65 \text{ V}$$

3. This is the same situation as (2) example as V_a is 10 V and $V_b = 0$ V. Then

$$V_o = 0.65 \text{ V}$$

4. The pd across D_a and D_b is zero as the voltage V_a and V_b is the same as the bias voltage. Then by KVL

$$V_o = V_b + V_{D_b} = 10 + 0 = 10 \text{ V}$$

or

$$V_o = V_a + V_{D_a} = 10 + 0 = 10 \text{ V}$$

4.6.5 DIODE AS VOLTAGE LIMITER

The circuit diagram of Figure 4.19 shows a circuit known as *positive limiter*. In this circuit the diode and the resistor are connected in parallel and they have the same voltage across.

When the signal is a positive voltage of more than 0.7 V, the diode D1 is forward biased and conducts all the current bypassing the resistor R_L. When the signal is a negative voltage or less than 0.7 V, the diode is reverse biased and does not conduct; then the current is circulated through R_L. Figure 4.19 shows the input and output signal for this positive limiter circuit.

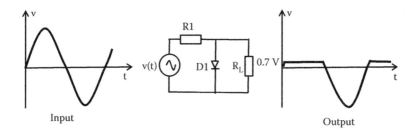

FIGURE 4.19   Circuit of a positive limiter.

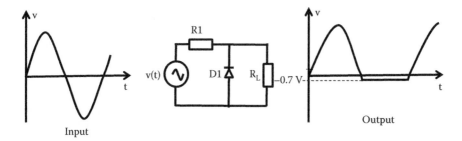

FIGURE 4.20   Circuit of a negative limiter.

If the diode is connected as indicated in Figure 4.20 it will create a *negative limiter.*

4.6.6   VOLTAGE DOUBLER

Figure 4.21 shows a circuit of a voltage doubler. First capacitor C_1 charges to the pick input voltage and keeps this voltage until the input voltage is increasing again adding the two voltages. The capacitor C_2 charges to nearly twice the peak input voltage.

Higher voltages, but low current, can be obtained by adding more combinations of the RC stages.

Diodes can also be used to protect transistors or integrated circuits (ICs) from the brief high voltage produced in an electronic system.

An LED can be connected in the vicinity of photodiode (or phototransistor) from another circuit creating what is known as an opto-isolator. The photodiode will switch on or off depending on the LED light. This combination electronically isolates two different circuits.

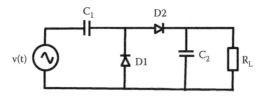

FIGURE 4.21   Circuit of a voltage doubler.

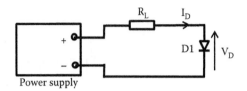

FIGURE 4.22 Circuit to test a diode.

4.7 TESTING DIODES

It is possible to perform a parametric test to check whether the diode meets the manufacturer's specifications. As semiconductor diodes are reasonably reliable, a functional test to establish whether a diode works is usually sufficient.

Measuring the diode's resistance, if the reading indicates infinity in one way and a small value the other way, then the diode is working. The diode is not working if the reading is 0 Ω in both ways (short circuited); or if the reading is infinity both ways (open circuit).

The circuit shown in Figure 4.22 can be used to test a diode. The level of currents in the forward and reverse bias can be measured to determine whether the diode is working or not. In the case of an LED the diode should light up with the right polarity. LEDs have a larger voltage drop across (1.5–2.5 V) than ordinary diodes.

If you want to test a diode connected to a circuit, you should disconnect the diode before testing to avoid the effect of other components in the circuit.

4.8 KEY POINTS

- The introduction of small amounts of impurities dramatically changes the properties of diodes.
- Doping of semiconductors with appropriate impurities will produce n-type or p-type materials.
- The use of band energy diagrams can explain conduction in semiconductors.
- A diode can be created by a junction between n-type and p-type semiconductor materials.
- Diodes allow current to flow in one direction (forward bias) but not the other (reverse bias).
- Forward bias happens when the p-type side is made *positive* with respect to the n-type side.
- Reverse bias happens when the p-type side is made *negative* with respect to the n-type side.
- Diodes are used in a range of applications, including rectification, demodulation, and signal clamping.

REFERENCE

Fernandez-Canque, H. L, Thompson, M. J., Allison, J. 1983. The capacitance of R.F. sputtered a-Si:H Shottky barrier diodes. *Journal of Applied Physics*, 54, 7025.

5 Bipolar Junction Transistor

5.1 INTRODUCTION

Transistor is one of the fundamental components of electronics. Advancements made in modern electronics are due to the invention and improvement of transistors. Developments in transistor technologies have enabled massive advancements in telecommunications, computing, and media technology. The colossal explosion of digital electronics in new wireless media technology and computing is due to advances in semiconductor technology to create transistors in huge quantities and diminutive sizes that switch very fast. Fabrication of fast transistors has allowed the dominance of digital electronics over analog electronics in communication and media technology. Before bipolar junction transistor (BJT) the device used to manipulate analog electrical signal was the electronic valve or tube. The valve has many problems when used in electronic circuits. They are very unreliable: their characteristics vary considerably from nominal values, they are bulky, require two different circuits (one for the amplifier behavior and another for the heating of the cathode filament), life expectancy is short, and they require large amounts of energy dissipation. The BJT solves or reduces all these problems as it is fabricated using solid-state material.

The BJT was the first transistor invented, the field effect transistor followed with certain advantages in digital applications. In this chapter, we look at properties of BJT, bipolar transistor operation and bipolar transistor characteristics; Chapter 6 will look at the field effect transistor.

5.2 BIPOLAR JUNCTION TRANSISTOR

The BJT transistor consists of two p–n junctions, in the same semiconductor crystal. The device is fabricated in such a way that almost all minority carriers injected from one junction (the *emitter*) into the region common to both junctions (the *base*) diffuses to the second junction (the *collector*); rather than being lost by recombination.

The two p–n junctions can have a p-type semiconductor in common creating an NPN transistor or a common n-type material to create a PNP transistor as indicated in Figure 5.1.

The region that serves as the source for the first junction to inject minority carriers into the center region is called the *emitter* (E). The center region into which the minority carriers are swept through to the second junction is called the *base* (B) and the region that collects the injected minority carriers is called the *collector* (C). Typically, the emitter–base junction is forward biased and the collector–base junction is reverse biased. Under these conditions, current from the emitter flows across the base to the collector. The collector current is carried by holes in the PNP

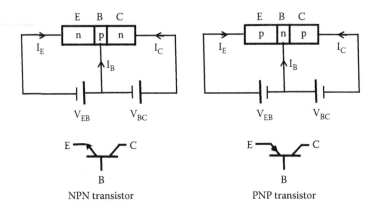

FIGURE 5.1   NPN and PNP BJT transistors configurations.

transistor and by electrons in the NPN transistor. Electrons have higher mobility that makes the NPN transistor preferable for fast response.

Let us consider an NPN transistor. The band structure of an unbiased NPN structure is shown in Figure 5.2, where W_b is the width of the base region.

As the two junctions are not biased (no external voltage applied) the Fermi level is continuous through the transistor to maintain its equilibrium; potential barriers are established at the emitter–base and base–collector junctions that prevent diffusions of majority carriers across the junctions.

With the normal active-mode bias voltage applied, that is the emitter–base junction forward biased and base–collector junction reversed biased; the band structure under biasing is modified as shown in Figure 5.3, where V_o is the potential barrier and W_b is the width of the base region.

The potential barrier between E and B is reduced due to the forward bias to this junction as $(V_o - V_{BE})$. In this junction an electron current, I_{en}, is injected from E into B region and a hole current, I_{hp}, in the reverse direction. Efficient transistor

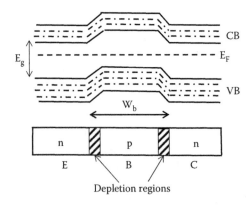

FIGURE 5.2   Band diagram structure of an unbiased NPN transistor.

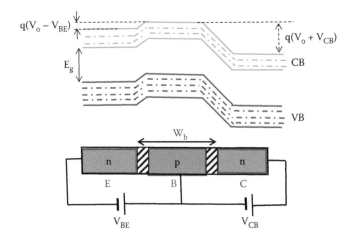

FIGURE 5.3 Band diagram structure of an active-mode biased NPN transistor.

operation is achieved when most of the current across this junction is carried by carriers originating in the emitter; that is, when the electron current I_{en} is much larger than the current due to holes I_{hp}. This is achieved by doping the emitter to a higher degree than the base.

In the collector–base junction, the potential barrier height is increased to (V_o + V_{CB}) due to the reverse bias V_{CB} applied. The result is that there is no majority carrier diffusion and the only current flow across this junction is due to minority carriers. A very small saturation current, the collector leakage current, flows across the C–B junction, which has contributions from minority carriers in the base and collector regions.

For efficient transistor operation it is necessary to ensure that electrons passing through the base do not recombine. This can be achieved with a very narrow base width, W_b. Some recombination will occur reducing the current gain of a transistor.

Then the minority carrier concentration in the base region is as indicated in Figure 5.4.

In an efficient transistor with a highly doped emitter and narrow base there is little recombination in the base region. Nearly all the electrons injected from the

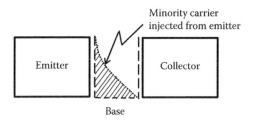

FIGURE 5.4 Minority carrier concentration in the base in transistor biased in active mode.

emitter defuse across the base and are swept across the base–collector junction into the collector. As the emitter is more heavily doped than the base the majority of the current across the emitter–base junction is transported by electrons. The collector current, I_C, is then only slightly less than the emitter current I_E.

For a PNP transistor the analysis is the same, but the polarities of the junctions will be reversed and the carrier involved will be holes.

5.3 BJT CHARACTERISTICS

5.3.1 TRANSISTOR CONFIGURATIONS

Bipolar transistors are three terminal devices. The transistor can use different terminals as input and output and one of the terminals can be common to input and output. Figure 5.5 shows the emitter, collector, and base common configurations.

The common emitter configuration is a configuration in which the emitter terminal is common to the input and output. We can consider the input and output characteristics of the transistor under this configuration.

5.3.2 INPUT CHARACTERISTICS

Measuring the input voltage between the emitter and base, V_{BE}, and the input current through the base, I_B, we can obtain the input characteristics of the device. Figure 5.6 shows a typical form input characteristics of a common emitter configuration transistor.

The input characteristics of a transistor are similar to the form of a forward-biased p–n junction.

5.3.3 OUTPUT CHARACTERISTICS

Measuring the output voltage between the collector and emitter, V_{CE}, and the output current through the collector, I_C, when the base current is kept constant, we can obtain the output characteristics of the common emitter transistor. Figure 5.7 shows the output characteristics of a transistor.

Near the origin there is a region where the current–voltage relationship is similar to the behavior of a resistor; this region is sometimes called the ohmic region or the

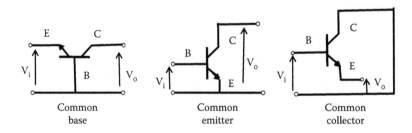

Common base   Common emitter   Common collector

FIGURE 5.5 Transistor configurations.

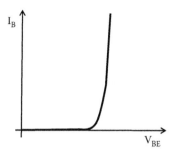

FIGURE 5.6 Common emitter transistor input characteristics.

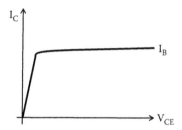

FIGURE 5.7 Transistor output characteristics at one base current.

resistive region. After this region the current is almost constant. It is usual to present the output characteristics including a family of curves at different values of base currents (I_B). Figure 5.8 shows the output characteristics of a transistor at various values of base currents.

With V_{CE} very low, the collector is not efficient at collecting carriers from the emitter passing through the base region. After V_{CE} exceeds a particular value known as the "knee" voltage, the collector current I_C does not depend largely on the collector–emitter voltage. It depends on the base current I_B. Even with the base current, $I_B = 0$, a small current flows into the collector, which is the leakage current, I_{CEo}. For a small base current a considerably larger collector current, I_C, flows, which gives a

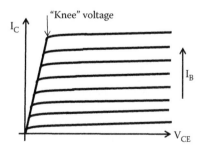

FIGURE 5.8 Transistor output characteristics at various base currents.

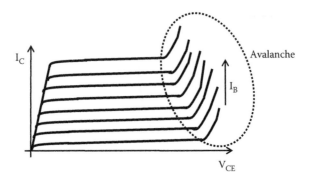

FIGURE 5.9   BJT output characteristic showing an avalanche region.

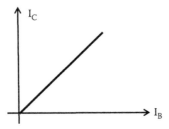

FIGURE 5.10   Relationship between the collector current and the base current in a BJT.

current gain much greater than unity. As the voltage between the collector and emitter increases the reversed-biased voltage between collector and base increases as well. This can reach a point, like any diode, where avalanche effect occurs. Figure 5.9 shows the output characteristics including an avalanche region.

The avalanche region should be avoided in a transistor. A good transistor is one with a low leakage current, high V_{CE} before avalanche, evenly spaced lines on the output characteristic and a low knee voltage.

Relationship between the collector current and the base current in a bipolar transistor is approximately linear. The magnitude of collector current is generally many times that of the base current, the device provides current gain. Figure 5.10 shows a typical relationship between the collector current and the base current in a BJT.

5.3.4   DATA FOR A TYPICAL NPN TRANSISTOR

There are four main sections noticeable in a manufacturer's data. They are

- The safe maximum ratings allowed for the device
- The normal operating voltages, currents, and gains
- Some parameters for a model of the transistor
- Typical graphical characteristics

5.3.5 RATING AND SELECTION OF OPERATING POINT

When a transistor is being used, a region is selected where the transistor will work. Within this region, the transistor is biased to obtain a particular value of collector–emitter voltage and a collector current within its output characteristics. This is known as the *operating point*, Q, sometimes called the working or quiescent point of the device. Q should be in a "safe" area of its characteristic and in the majority of cases of analog system; it should be in an area where the output characteristics are equally spaced. Chapter 6 will detail how the operation point is selected and how it can be designed in various circuits.

The manufacturer's data for a transistor usually provides

- Maximum collector current, I_{Cmax}
- Maximum collector voltage, V_{CEmax}
- Maximum power dissipation, P_{max}

The electrical power input to the transistor must be less or equal to P_{max} for safe operation. In a common emitter transistor usually the base current is small compared to the collector current and the voltage across base and emitter is small compared to the voltage across the collector–base junction. The input power can be approximated to

$$\text{Input power} \approx V_{CE}I_C \tag{5.1}$$

The product of V_{CE} and I_C must not exceed the maximum power dissipation. Both V_{CE} and I_C cannot be maximum at the same time. If V_{CE} is maximum, I_C can be calculated as

$$I_C = \frac{P_{max}}{V_{CE}} \tag{5.2}$$

If I_C is maximum, V_{CE} can be calculated by rearranging Equation 5.2 as follows:

$$V_{CE} = \frac{P_{max}}{I_C} \tag{5.3}$$

For a given transistor, a maximum power dissipation curve can be plotted on the transistor output characteristic output curves as shown in Figure 5.11.

This clearly limits the usable part of the output characteristic to the region avoiding the lines of

A: Maximum safe collector current
B: Maximum safe collector voltage
C: Maximum power dissipation

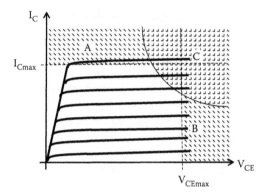

FIGURE 5.11 Maximum limits of transistor usage.

EXAMPLE 5.1

A transistor's characteristics provided by the manufacturer include a maximum power, $P_{max} = 800$ mW, $V_{CEmax} = 30$ V, and an $I_{Cmax} = 300$ mA. If the transistor is required to work at maximum collector–emitter voltage, calculate the upper value of current that can be used. If the transistor is required to work at maximum collector current, calculate the upper limit of voltage between collector and emitter set as a limit.

SOLUTION

The maximum collector current and maximum V_{CE} cannot be used at the same time. If the maximum collector–emitter voltage is used the collector current limit is

$$I_{Climit} = \frac{P_{max}}{V_{CEmax}} = \frac{800 \times 10^{-3}}{30} = 26.67 \text{ mA}$$

which is smaller than I_{Cmax} specified by the manufacturer.

The maximum collector current and maximum V_{CE} cannot be used at the same time. If the maximum collector current is used then collector–emitter voltage limit is

$$V_{CElimit} = \frac{P_{max}}{I_{Cmax}} = \frac{800 \times 10^{-3}}{300 \times 10^{-3}} = 2.67 \text{ V}$$

which is smaller than V_{CEmax} specified by the manufacturer.

5.4 GAIN PARAMETERS OF BJT: RELATIONSHIP OF α AND β PARAMETERS

5.4.1 COMMON BASE CONNECTION

Figure 5.12 shows a transistor connected as common base configuration, in this configuration the input current is the emitter current and the output current is the collector current.

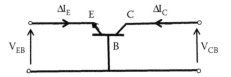

FIGURE 5.12 Common base configuration. Current gain.

If the emitter current is changed by increment ΔI_E, there will be a corresponding incremental change in collector current ΔI_C. The relative change is described in terms of a gain parameter, α, which is defined by

$$\alpha = \frac{\text{Change in collector current}}{\text{Change in emitter current}} = \frac{\Delta I_C}{\Delta I_E}\bigg|_{V_{CB}=\text{CONSTANT}} \tag{5.4}$$

where α is the common base current gain.

In an efficient transistor almost all electrons emitted at the emitter are collected at the collector. Then in a common base configuration a variation in the emitter current will produce approximately the same variation in the collector current. Therefore,

$$\Delta I_C \approx \Delta I_E \tag{5.5}$$

and α is only slightly less than unity, typical values being in the range 0.900–0.999; α will be dependent on the number of electrons injected from the emitter into the base, and the proportion of these electrons that diffuses across the base without recombination, to the collector.

5.4.2 COMMON EMITTER CONFIGURATION

The common base configuration then produces no current gain as α is almost 1; but it is more frequent for a transistor to be operated in the common emitter configuration as shown in Figure 5.13.

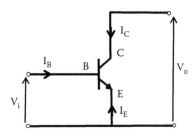

FIGURE 5.13 Common emitter configuration. Current gain.

A change in base current ΔI_B produces a corresponding change in collector current, ΔI_C, and the common emitter gain β is then defined by

$$\beta = \frac{\text{Change in collector current}}{\text{Change in base current}} = \frac{\Delta I_C}{\Delta I_B}\bigg|_{V_{CE}=\text{CONSTANT}} \tag{5.6}$$

Now for the transistor common emitter configuration the input current is I_B and the output current is I_C.

The relationship for the current in the transistor can be written as

$$I_B + I_C + I_E = 0 \tag{5.7}$$

and using the relationship for α we can write an expression for the collector current taking into account the direction of currents as

$$I_E = -\frac{I_C}{\alpha} \tag{5.8}$$

Then, Equation 5.7 becomes

$$I_B + I_C - \frac{I_C}{\alpha} = 0$$

Rearranging this equation we obtain

$$I_C \frac{\alpha - 1}{\alpha} + I_B = 0$$

and the current gain for a common emitter transistor can be expressed in terms of α as

$$\beta = \frac{I_C}{I_B} = \frac{\alpha}{1-\alpha} \tag{5.9}$$

Since α is slightly less than unity then β can be large, typical values being in the range 10–1000.

EXAMPLE 5.2

A BJT has a base current I_B = 12 mA and a collector current, I_C = 1.8 A. Determine the value of β, α, and the emitter current I_E.

SOLUTION

$$\beta = \frac{I_C}{I_B} = \frac{1.8}{12 \times 10^{-3}} = 150$$

and

$$\beta = \frac{\alpha}{1-\alpha}$$

or

$$\alpha = \frac{\beta}{1+\beta} = \frac{150}{1+150} = 0.9934$$

Also,

$$I_E = \frac{I_C}{\alpha} = \frac{1.8}{0.9934} = 1.812 \text{ A}$$

I_E has a value approximate to I_C.

EXAMPLE 5.3

A 10 V source supplies current to the base of a transistor through a base resistor, R_B as shown in Figure 5.14. If the voltage between the base and emitter, V_{BE}, is 0.65 V and $\alpha = 0.97$, calculate the base current I_B, β, collector current, and the emitter current.

SOLUTION

By KVL we can write

$$E_s = V_{R_B} + V_{BE} = 10 = V_{R_B} + 0.65$$

and

$$V_{R_B} = 9.35 \text{ V}$$

and the base current is

$$I_B = \frac{V_{R_B}}{R_B} = \frac{9.35}{80 \times 10^3} = 116.875 \text{ μA}$$

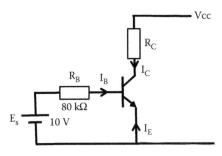

FIGURE 5.14   Circuit diagram for Example 5.3.

The value of β is

$$\beta = \frac{\alpha}{1-\alpha} = \frac{0.97}{1-0.97} = 32.33$$

The collector current is

$$I_C = \alpha I_B = 32.33 \times 116.875 \times 10^{-6} = 3.78\,mA$$

and the emitter current is

$$I_E = \frac{I_C}{\alpha} = \frac{3.78\,mA}{0.97} = 3.897\,mA$$

5.5   TESTING TRANSISTORS

As with diodes, a transistor can fail to work. A few tests with an ohmmeter can show whether a transistor works. A transistor, for the testing, can be regarded as two back-to-back diodes. Figure 5.15 shows a transistor as two back-to-back diodes.

Manufacturers' data will indicate whether the transistor is an NPN or a PNP. The data sheet will also indicate where B, C, and E are, and then just test the B–C junction and the B–E junction as if they were standard diodes. If a p–n junction is a defective diode, then the transistor is faulty. Also, check the resistance from C to E using a higher ohms scale. If the transistor is working, you should get an open-circuit reading from collector to emitter. To check a transistor connected in a network, first disconnect the transistor.

5.6   EFFICIENT BJT AS AMPLIFICATION DEVICE

This section includes deduction of current gain in a BJT from the semiconductor physics point of view. It will indicate how certain parameters affect an efficient transistor. It is not intended to deduce the different equations that fit the theory of

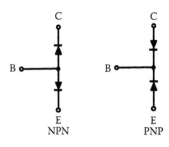

FIGURE 5.15   BJT represented as two back-to-back diodes.

semiconductor devices, but we can use these equations to determine how they affect the behavior of a transistor. In that sense some equations will be used without being deduced. This will reduce the mathematical content to facilitate the understanding of the electronics processes in a transistor.

It is convenient to express the current gain, α, in common base configuration in terms of the efficiency of the emitter as injector of carriers, efficiency of the base transfer of carriers, and efficiency of the collector as collector of carriers.

Then α can be written as the product of these three components: η (emitter injection efficiency), χ (base transport factor), and τ (collector of carrier efficiency) as

$$\alpha = \eta \cdot \chi \cdot \tau \tag{5.10}$$

Considering an NPN transistor η, the emitter injection efficiency, is the ratio of the electron current injected into the base from the emitter to the total emitter–base junction current; χ, the base transport factor, is the ratio of the electron current at the collector junction to that at the emitter junction, and τ is the collector efficiency to collect electrons. The efficiency of a collector in the modern fabrication technologies of BJT as collector of carriers is very good and can be taken as 100% efficient; that is, all carriers arriving at the collector will be collected.

$$\text{Collector efficiency} = \tau = 1 \tag{5.11}$$

Current transport is an NPN transistor and is shown schematically in Figure 5.16, where I_{en} is the electron current from E to B, I_{hn} is the hole current from C to B, I_{hp} is the hole current from B to E, I_{ep} is the electron current from B to C and I_B is the base current.

An efficient transistor will have a value of α close to 1. In an NPN transistor we would like to have all electrons emitted at the emitter be collected at the collector, but in a bipolar transistor the emitter current will have a hole component, and this will contribute to make α being less than 1.

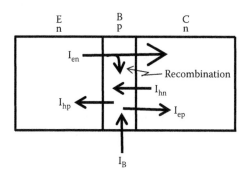

FIGURE 5.16 Schematic of current in a transistor.

5.6.1 Emitter Injection Efficiency η

The emitter injection efficient for an NPN transistor can be expressed as the percentage of electrons compared to the total number of carriers, then

$$\eta = \frac{\text{Electrons current}}{\text{Electrons current} + \text{holes current}} = \frac{I_{en}}{I_{en} + I_{hp}} = \frac{I_{EBn}}{I_{EBn} + I_{EBp}} \qquad (5.12)$$

where I_{hp} and I_{en} are the hole and electron components of the current crossing the emitter–base junction; I_{EBn} is the current due to electrons in the emitter–base junction; and I_{EBp} is the current due to holes in the emitter–base junction.

The contribution from holes to the emitter current, I_{hp} is made very small by doping the n-region much more heavily than the p-region, that is

$$I_{en} \gg I_{hp}$$

Now the equation for a current in a p–n junction can be deduced by obtaining an expression for the density of carrier electrons and holes and adding the two. The hole contribution, I_{hp}, can be represented by the following expression (Sze 1969):

$$I_{hp} = \frac{qD_pp_nA}{L_p}\left[\exp\left(\frac{qV_{EB}}{kT}\right) - 1\right] \qquad (5.13)$$

where q is the charge of an electron, D_p the hole diffusion coefficient, P_n the number of holes in the emitter, A being the effective cross-sectional area of the base, and L_p the diffusion length of holes in the emitter.

The electron contribution to the current in the emitter–base junction, I_{en}, may be obtained in a similar way as Equation 5.13 for an n–p junction, but in a transistor the p side (base) of this n–p junction is very narrow. We can use the equation for electrons in an n–p junction with L_e replaced by W_b; in the emitter–base junction the width of the base, W_b is shorter than the diffusion length of electrons, L_e. Then, electron components of the current crossing the emitter–base junction can be expressed as (Sze 1969)

$$I_{en} = \frac{qD_nn_pA}{W_b}\left[\exp\left(\frac{qV_{EB}}{kT}\right) - 1\right] \qquad (5.14)$$

where D_n is the electron diffusion coefficient, n_p the number of electrons in the emitter, A being the effective cross-sectional area of the base, and W_b the width of the base.

And the emitter injection efficiency can be obtained by replacing I_{hp} and I_{en} from Equations 5.13 and 5.14 into Equation 5.12 and the *emitter injection efficiency* η becomes

$$\eta = \frac{I_{en}}{I_{en} + I_{hp}} = \frac{qD_n n_p A/W_b[\exp(qV_{EB}/kT) - 1]}{qD_n n_p A/W_b[\exp(qV_{EB}/kT) - 1] + qD_p p_n A/L_p[\exp(qV_{EB}/kT) - 1]}$$

or

$$\eta = \frac{D_n n_p/W_b}{D_n n_p/W_b + D_p P_n/L_p} = \frac{1}{1 + D_p P_n W_b/L_p D_n n_p} \tag{5.15}$$

Now using other equations in semiconductor physics that relate electrons and holes and their mobility

$$n_p p_p = n_n p_n \tag{5.16}$$

and also

$$\frac{D_p}{D_n} = \frac{\mu_p}{\mu_n} \tag{5.17}$$

where μ_n is the electron mobility and μ_p is the hole mobility.

We can write η as function of the conductivity of the p-region σ_p and in the n-region σ_n, as

$$\eta = \frac{1}{1 + (\sigma_p W_b/\sigma_n L_p)} \tag{5.18}$$

where $\sigma_p = D_p P_n = qP_n\mu_p$ and $\sigma_n = qn_p\mu_n$ are the conductivities of the p and n regions, respectively.

In the n–p junction between emitter and base the conductivity of the emitter is larger than the conductivity of base then $\sigma_n \gg \sigma_p$ and we can write η as the following approximation:

$$\eta \approx 1 - \frac{\sigma_p W_b}{\sigma_n L_p} \tag{5.19}$$

and η will be close to 1 if the second term of Equation 5.19 is small. Equation 5.19 corroborates that an efficient emitter injection ratio is achieved by making the base region narrow and also by making emitter n-region heavily doped with respect to the base.

5.6.2 BASE TRANSPORT FACTOR χ

The base contributes to the efficiency of a transistor if the recombination in this region is small. Ideally no recombination is desired, so all electron will diffuse across the base to the collector.

The base transport factor χ is the ratio of electron injected from the emitter to electrons that reach the collector. It can be expressed as

$$\chi = \frac{I_e(\text{at collector junction})}{I_e(\text{at emitter junction})} = \frac{I_e(x = W_b)}{I_e(x = 0)} \tag{5.20}$$

where x is the distance from the emitter junction, taken as the origin zero the emitter–base junction, then the distance from the emitter–base junction to the base–collector junction will be the width of the base, W_b.

The collector current due to electron current can be found by applying the continuity equation for excess electrons in the base region, Δn. The excess of electrons injected in the base, Δn, as a function of distance x is given by (Sze 1969)

$$\Delta n(x) = \frac{n_{po} - n_p}{1 - \exp\{2W_b/L_e\}} \left[\exp\frac{x}{L_e} \exp\left(\frac{2W_b}{L_e}\right) \exp\left(\frac{-x}{L_e}\right) \right] \tag{5.21}$$

Now, the electron diffusion current density, J(x), in terms of the electron density as function of distance x can be written as

$$J(x) = qD_n \frac{d\,\Delta n(x)}{dx} \tag{5.22}$$

Performing the differentiation of $\Delta n(x)$ with respect to x we can obtain the current density as a function of distance x. Then, including the particular boundary values of $x = 0$ at the emitter–base junction and $x = W_b$ at the base–collector junction we can obtain the base transport factor χ as

$$\chi = \frac{J_e(\text{at } x = W_b)}{J_e(\text{at } x = 0)} = \frac{d\Delta n(x)/dx\big|_{x=W_b}}{d\Delta n(x)/dx\big|_{x=0}} = \frac{2\exp W_b/L_e}{1 + \exp 2W_b/L_e} \tag{5.23}$$

Now since the electron diffusion length L_e is much less than the base width W_b, Equation 5.23 can be approximated to

$$\chi = \left[1 + \frac{1}{2}\left(\frac{W_b}{L_b}\right)^2 \right]^{-1} \approx 1 - \frac{1}{2}\left(\frac{W_b}{L_e}\right)^2 \tag{5.24}$$

Thus, as before for high efficiency of the base χ large L_e and narrow W_b is required.

The total current gain α in terms of semiconductor physic parameters can be approximated to

$$\alpha = \eta\chi \approx \left[1 - \frac{\sigma_p W_b}{\sigma_n L_p} \right]\left[1 - \frac{1}{2}\left(\frac{W_b}{L_e}\right)^2 \right] \tag{5.25}$$

Equation 5.25 corroborates that for an efficient transistor a good emitter is needed. This can be achieved by doping the emitter with large number of impurities compared to the number of impurities of the base and making the base as narrow as possible.

5.6.3 PUNCH-THROUGH

There is a limitation on how thin you can make the base. The base cannot be made too thin because when voltage reverse bias is applied to the base–collector junction, the depletion region extends and can connect to the depletion junction of the emitter–base junction. This phenomenon is known as *punch-through* resulting in direct conduction and the device stop working as a transistor.

5.7 KEY POINTS

- The BJT was the first transistor invented to replace the electronic valve.
- BJT consists of three layers of solid-state semiconductor material either NPN or PNP.
- Bipolar transistors are one of the most important electronic components and an understanding of their operation and use is vital for anyone working in analog electronics.
- BJT is used in a wide range of applications of both analog and digital circuits.
- Electrons have higher mobility making the NPN transistor favored when used as a faster switch.
- In a normal active-mode transistor, the emitter–base is forward biased and base–collector reverse biased.
- In a BJT the common base current gain α is close to unity and the current gain for a common emitter transistor is usually large.
- An efficient BJT will require the efficiency of the emitter, efficiency of the base, and efficiency of the collector to be close to unity.

REFERENCE

Sze, S. M. 1969. *Physic of Semiconductor Devices.* Wiley Interscience, USA. ISBN 10-0-471-14323–5.

6 Field Effect Transistors

6.1 INTRODUCTION

Once the BJT was successfully developed, semiconductor technology underwent rapid advances to fabricate thin layers of materials with better doping control and small geometries in order to produce a new type of transistor: the field effect transistor (FET). This new type of transistor behavior is similar to a BJT, but it is based on the control of carriers using an electric field. The introduction of FETs has a large impact on the systems using ICs. This is due to the FET physical size, easy fabrication, better isolation of devices, and packing characteristics. New technologies allow fabricating very small FETs in very large ICs (VLSI—very-large-scale integration). This chapter highlights the similarities between these two forms of transistors (FETs and BJTs) as well as point out the differences. This allows readers to transfer knowledge of BJT devices and circuits while appreciating the essential differences between these two forms of components. Transistors are used in a similar way in analog as well as in digital circuits; the main difference is the region where the transistor works. In the case of digital circuits, the transistor is usually used as a switch between the saturated and cut-off regions of a transistor.

6.2 FABRICATION OF FET

FETs are the simplest form of transistor in terms of fabrication. They are commonly used in analog and digital applications; they have high input impedance and small physical size. They can be used in circuits with a low power consumption combined with its small size that suits VLSI well.

FETs are four-terminal devices:

- Drain (D)
- Source (S)
- Gate (G)
- Substrate

In many applications, the substrate is connected to the source, then for the analysis it is usually taken as three-terminal devices, D, S, and G.

Figure 6.1 illustrates the basic structure of an FET including notation used for labeling voltages and currents. MOSFET (metal–oxide–semiconductor field effect transistor) with an n-channel is shown in Figure 6.1a. In this FET the starting material is a p-type material that forms the substrate, and then impurities are introduced to create two regions of n-type material that form the source and drain. Between the source and the drain, a layer of insulation material usually silicon dioxide (SiO_2) is grown. On top of this insulating material a conducting material forms the gate. With this structure it is possible to generate a thin layer of n-type semiconductor between

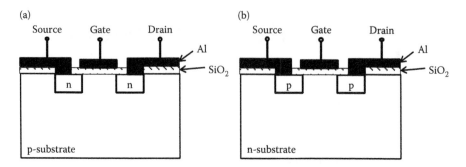

FIGURE 6.1   Structure of a FET: (a) n-channel MOS, (b) p-channel MOS.

source and drain, known as the channel, by applying a voltage in the gate. This structure creates an n-channel MOS transistor. If the starting material in the substrate is an n-type material, it is possible to create a p-channel MOS as shown in Figure 6.1b.

6.3   DIFFERENT TYPES OF FET

According to the different structures and types of semiconductor material there are different types of FETs. The two basic forms are

- Insulated-gate FETs
- Junction gate FETs

6.3.1   INSULATED-GATE FETS

The FET shown in Figure 6.1 is known as insulated-gate FETs, because the gate is insulated. Such devices are called IGFETs or sometimes MOSFETs or sometimes just MOS. ICs using these devices are usually described as using MOS technology. There is a further division within the MOSFET in that they can have a physical channel created under fabrication, known as depletion MOSFET, Figure 6.2b, or the enhancement FET where the channel is induced by applying a voltage to the gate, Figure 6.2a.

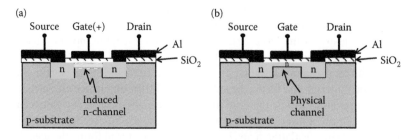

FIGURE 6.2   (a) Enhancement MOSFET, (b) depletion MOSFET.

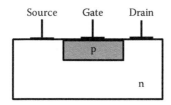

FIGURE 6.3   n-channel JFET.

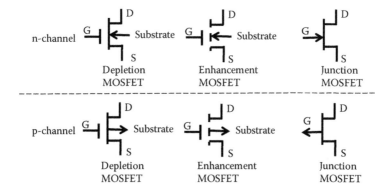

FIGURE 6.4   Circuit symbols for different types of FET transistors.

6.3.2   JUNCTION GATE FETs

In this type of transistor there is no insulating material to insulate the gate. The junction gate FET is sometimes known as a *JUGFET* or more commonly as *JFET*. Figure 6.3 shows an n-channel JFET.

In this transistor the insulated gate of a MOSFET is replaced with a reverse-biased p–n junction. The gate junction is always reverse biased. No current flows into the gate and it acts as if it were insulated.

6.3.3   FET CIRCUIT SYMBOLS

Both forms of MOSFET and JFET are available as either n-channel or p-channel devices. Figure 6.4 shows a summary of the different MOS and JFET.

6.4   MOS FET

6.4.1   THE MOS IGFET

Figure 6.5 below shows a typical structure of an IGFET (insulated-gate FET). The metal gate terminal is entirely insulated from the substrate semiconductor by a thin insulating layer of silicon dioxide, SiO_2. The MOS transistor shown in Figure 6.5 is an n-channel type and the conducting carriers are electrons. As electrons have a better mobility than holes, this transistor has better operating performance than

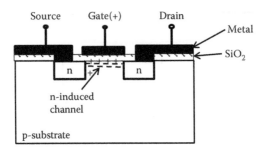

FIGURE 6.5   Insulated-gate field effect transistor.

a p-channel type. The operating speed and gain of an MOS transistor are directly related to its carrier mobility, so n-channel devices are usually favored.

6.4.2 MOS Operation

Without a voltage applied to the gate, there is no current between the drain and source. Source and drain form p–n junctions with the substrate, voltage of any polarity between drain and source will reverse bias one of these p–n junctions and no current will circulate between source and drain. In order to have a current between source and drain a channel of n-type material is needed between the source and drain. This channel is produced by applying a positive voltage at the gate as follows.

A positive gate potential will induce an electrostatically negative charge, similar to a capacitor, on the substrate. Then a positive voltage at the gate will repel the holes (positively charged) and attract electrons (negatively charged) in the region below the gate within the p-type substrate. This action changes the charge concentration of a thin layer under the gate increasing the number of electrons in this region, the minority carrier in the p-type substrate. If a larger positive voltage is applied to the gate, then at the surface, the concentration of minority carrier (electrons) will eventually become larger than the density of majority carrier (holes), and the channel is created. It is usually referred to as this region being depleted (of holes). Once the channel is created by this electric field produced by the voltage applied to the gate, there can be conduction between the drain and source and this device operates as a transistor.

The substrate may be either p-type or n-type corresponding to n- or p-channel transistors, respectively.

The operation of an MOS can also be explained in terms of the energy band structure of materials. The gate region is essentially a capacitor; in this region the electric field is created by applying a voltage to the gate. The gate is a conducting material, the oxide layer SiO_2 is an insulator and the semiconductor is a conducting material (of higher resistance than metals), so it is a capacitor structure. The name of this device comes from this structure: metal (M), oxide insulator (O), and semiconductor (S). The MOS structure is basically a capacitor in which one plate is a semiconductor and the other is the metal layer.

Let us concentrate on the gate region of an n-channel transistor, where the field effect for transistors takes place. With zero voltage applied the transistor is in

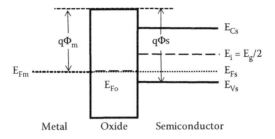

FIGURE 6.6   Band energy of an MOS structure at equilibrium.

equilibrium. Figure 6.6 shows the band diagram for the gate region at equilibrium ($V = 0$), where E_i is the energy level at the middle of the gap, that is, $E_F = E_i = E_g/2$ for an intrinsic material. E_{Fm} is the metal Fermi level; E_{Fs} the semiconductor Fermi level; E_{Fo} the oxide Fermi level; E_{Cs} the bottom of the conduction band; and E_{Vs} is the top of the valence band of the semiconductor.

Now let see what appends when an external energy is applied, that is, a voltage to the gate. Then the system is not in equilibrium and electron and holes will be affected by the electric field created by the voltage applied to the gate. We have two cases: with a negative voltage to the gate an *accumulation* of holes will happen and with a positive voltage to the gate *depletion* of holes is created.

6.4.3   Accumulation

If we apply a *negative voltage,* V, to the gate, the capacitor behavior indicates that we are in fact depositing negative charges on the metal of the gate. Then the same amount of positive charges will appear on the other plate (at the surface of the semiconductor material). The applied negative voltage lowers the electrostatic potential of the metal relative to the semiconductor, the electron energies are raised in the metal relative to the semiconductor. As a result, the Fermi level for the metal E_{Fm} is raised from the equilibrium position by the external energy applied qV. The raising E_{Fm} up in energy relative to E_{Fs} causes the oxide conduction band to tilt, as indicated in Figure 6.7.

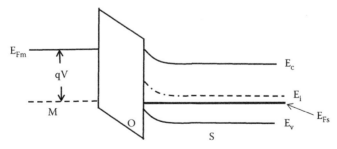

Negative applied voltage to gate: accumulation

FIGURE 6.7   Band energy of an MOS structure under accumulation.

As a consequence, the conduction and valence bands bend as shown in Figure 6.7. The intrinsic level that indicates the same concentration of electrons and holes bend as well, as indicated in Figure 6.7. There is no current conduction through the insulator of this MOS structure, and then the Fermi level E_{Fs} within the semiconductor does not bend but remains flat as indicated in Figure 6.7.

Now if we examine at the Fermi level in the p-type semiconductor near the interface between the oxide and semiconductor we see that E_{Fs} is closer to the valence band; this indicates that in this thin region there is a larger hole concentration (accumulation) compared with the concentration at the bulk (away from the oxide–semiconductor interface). Therefore, in this region, negative voltage applied to the gate has created a thin layer where the concentration of holes has increased, this is called *accumulation*. This has been achieved by creating an electric field. The name of this type of transistor comes from this field effect.

6.4.4 DEPLETION

Now, if we apply a *positive voltage* V from the metal to the semiconductor (gate voltage) this lowers the metal Fermi level by qV relative to its equilibrium position as shown in Figure 6.8.

This positive voltage deposits positive charge on the metal producing an equivalent net negative charge at the surface of the semiconductor. This negative charge in p-type material increases the number of electrons resulting in the *depletion* of holes at the region nearer the surface. The energy bands bend upward in this case, and E_{Fs} within the semiconductor does not bend but remains flat as indicated in Figure 6.8. Now if we examine the region at the Fermi level in the p-type semiconductor near the interface between the oxide and semiconductor we see that E_{Fs} is further away from the valence band; this indicates that in this thin region there is a larger electron concentration (depletion of holes) compared with the concentration at the bulk (away from the oxide–semiconductor interface). Thus, in this region, positive voltage applied to the gate has created a thin layer where the concentration of electrons has increased, this is called *depletion*. The energy bands bend downward in this case as indicated in Figure 6.8.

Now if we increase the positive voltage at the gate, the energy bands at the semiconductor surface bend downward more strongly until a region at the interphase of

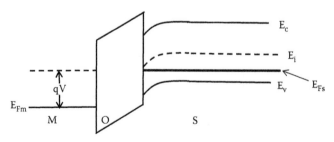

Positive applied voltage to gate: depletion

FIGURE 6.8   Band energy of an MOS structure at depletion.

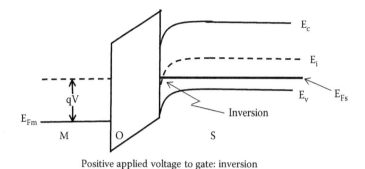

Positive applied voltage to gate: inversion

FIGURE 6.9   Band energy of an MOS structure at inversion.

oxide–semiconductor is changed from a majority of holes to a majority of electrons. Then an *inversion* has occurred. This thin layer is called the channel and the semiconductor has been inverted from p-type to n-type by the application of this electric, as indicated in Figure 6.9.

At the surface region (the channel), the Fermi level is above the intrinsic level and therefore the channel is n-type material; at the bulk (away from the surface region), the Fermi level is below the intrinsic level and hence is p-type.

The channel has conduction properties typical of n-type material and this layer is formed not by doping, but instead by *inversion* of the originally p-type semiconductor due to external applied voltage.

6.4.5   Strong Inversion

The positive voltage applied to the gate can create a channel that can disappear with a small variation of voltage, and then the channel is weak. In order to use an FET we need the channel to exist even if small variations of voltages occur. Then we need to apply a voltage level that produces a *strong inversion* in the semiconductor material. An accepted condition for *strong inversion* is that the channel should be as strongly n-type as the substrate is p-type. In other words we need to apply a positive voltage at the gate in such a way that produces a band bending where the difference of energy between the Fermi level and intrinsic level in the channel layer and in bulk region is the same. Figure 6.10 shows the band energy diagram for the boundary between the oxide material and the p-type semiconductor. The band bending happens in the semiconductor at the interface with the oxide insulator; the channel region is created where the Fermi level is above the intrinsic level (n-type). At the bulk of the semiconductor (away from the channel), the Fermi level is below the intrinsic level (p-type). Figure 6.10 shows the band structure for an n-channel transistor under strong inversion.

The energy difference due to the band bending and the intrinsic energy level E_i as a function of the distance from the surface of the oxide, can be represented by the expression $q\Phi(x)$. At the interface between oxide and semiconductor x has a value of 0. Then, the energy at $x = 0$ is $q\Phi(x = 0) = q\Phi_s$, where Φ_s is the potential voltage

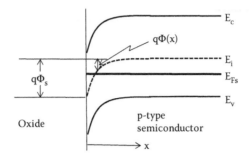

FIGURE 6.10 Band energy of an MOS structure at strong inversion.

that produces the band bending. A condition for strong inversion is that $q\Phi_s$ is equal to twice the difference of energy between E_i and E_{Fs}, that is,

$$q\Phi_s = 2(E_i - E_{Fs}) \tag{6.1}$$

6.4.6 THRESHOLD VOLTAGE, V_T

The external voltage applied to the gate V_G at which the surface just becomes strongly inverted is called the threshold voltage V_T. Below the threshold voltage an NMOS transistor is off and no current flows, while above the threshold the inversion channel is established and the device conducts. The threshold voltage at the gate, V_T, is the voltage value at which the channel current, I_D, conducts (above the leakage levels). V_T is an important parameter.

The channel then allows a current between the source and drain. If the channel is not created at a gate voltage less than V_T then there is no conduction between source and drain. When the channel is created we can have conduction through the channel between sources and drain when a voltage is applied.

6.5 JFET OPERATION

The JFET operation is based on the variation of the depletion region with a voltage applied to a p–n junction. As we saw in Chapter 4, at the junction of a p–n junction a depletion layer is formed. In a JFET, this depletion layer is modulated by a reverse voltage to control the conductivity between source and drain. Figure 6.11 shows the structure of an n-channel JFET. Commonly, the JFET and other semiconductor components are fabricated using planar technology so all terminals are in one plane. In Figure 6.11 the gate is split into two to provide an easier explanation of how this transistor works, in practice there is only one gate. As shown in Figure 6.11, the depletion layer exists even if no voltage is applied. We take the source voltage as reference then V_{GS} is written as V_G and V_{DS} is written as V_D for notation. Figure 6.11 shows the case when $V_G = V_{GS} = 0$ and $V_D = V_{DS} = 0$.

In this case, the depletion region exists even if no drain voltage is applied.

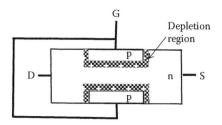

FIGURE 6.11   JFET structure under zero bias applied $V_G = V_{GS} = 0$ and $V_D = V_{DS} = 0$.

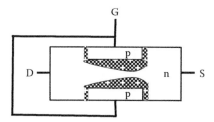

FIGURE 6.12   JFET structure under $V_G = 0$, $V_{D1} \neq 0$ small, $I_{D1} \neq 0$ small.

Now if a small voltage is applied between the source and the drain, V_{D1}, the thickness of the depletion layer will increase taking the shape indicated in Figure 6.12. The irregular shape of the depletion region is because the reverse voltage between the source and the gate on the source side is larger. Figure 6.12 shows the depletion region change when a small voltage is applied between source and drain. The drain current will then be different from zero, but small.

Further increase of the drain voltage to V_{D2} ($V_{D2} > V_{D1}$) causes the channel to narrow and its resistance to increase. The thickness of the depletion layer will increase taking the shape indicated in Figure 6.13. In this case we have $V_G = V_{GS} = 0$, V_{D2} larger than V_{D1}, then the new drain current I_{D2} will be larger than I_{D1}.

Further increase of drain voltage to $V_{D3} = V_{DP}$ ($V_{DP} > V_{D2} > V_{D1}$) causes the channel to narrow, the depletion layers that extend across the channel almost close at the drain end. This point is called *pinch-off*. The thickness of the depletion layer will increase taking the shape indicated in Figure 6.14. In this case we have $V_G = V_{GS} = 0$, V_{DP} larger than V_{D2}, then the new drain current I_{DP} will be larger than I_{D2}.

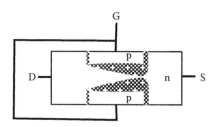

FIGURE 6.13   JFET structure under $V_G = 0$, $V_{D2} > V_{D1}$, $I_{D2} > I_{D1}$.

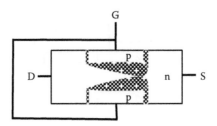

FIGURE 6.14   JFET structure under $V_G = 0$, $V_{D3} = V_{DP}$ ($V_{DP} > V_{D2} > V_{D1}$).

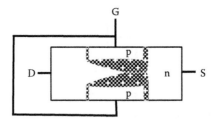

FIGURE 6.15   JFET structure under $V_G = 0$, $V_{D4} > V_{DP}$, $I_{D4} = I_{DP}$.

Further increase of drain voltage, larger than V_{DP}, causes the channel to narrow to the point of closure. Further increase of the drain voltage will not increase the drain current, so for any increase of V_D above V_{DP} the current will remain constant. The thickness of the depletion layer will increase taking the shape indicated in Figure 6.15. In this case we have $V_G = V_{GS} = 0$, V_{D4} larger than V_{DP}, then the new drain current I_{D4} will remain constant at I_{DP}.

As the drain voltage is increased from V_{D1} to V_{D3} the drain current increases gradually similar to the behavior of a resistance. This region before pinch-off is known as the ohmic region or resistive region. At pinch-off the current stop increasing the channels is strangled and the current stays constant. This region is known as the saturation region of the I–V curve.

Figure 6.16 shows a plot of the drain current against the drain voltage.

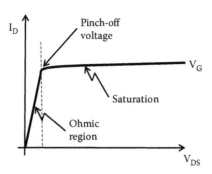

FIGURE 6.16   JFET output I–V characteristics.

Note that in a JFET the channel is created at fabrication, so even with a zero voltage at the gate it is possible to have a current between drain and source. The I–V curve illustrated above takes $V_{GS} = 0$. By changing the gate voltage a similar curve can be obtained, creating a familiar set of I–V curves of transistor behavior.

6.6   STATIC CHARACTERISTICS OF FET

MOSFETs and JFETs operate in different ways, but their characteristics are similar.

6.6.1   INPUT CHARACTERISTICS

In both type of transistors, MOSFETs and JFETs, the gate is effectively insulated from the remainder of the device.

6.6.2   OUTPUT CHARACTERISTICS

In n-channel devices usually the drain is more positive than the source. Figure 6.17 shows the output characteristics of an n-channel FET.

6.6.3   TRANSFER CHARACTERISTICS

The transfer characteristics of all FET are similar in shape but with a different offset. Figure 6.18 shows the transfer characteristics for different forms of FET.

The transfer characteristics are not a linear response, but over a small region might be approximate to a linear response.

6.7   CURRENT–VOLTAGE CHARACTERISTICS

The deduction of an expression of the drain current as a function of the drain voltage leads to a complex equation because it has to take into account the physical changes of the different region within the transistor as voltage is applied as well as

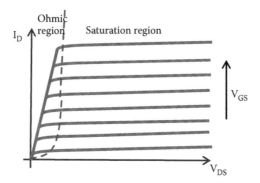

FIGURE 6.17   FET output characteristics.

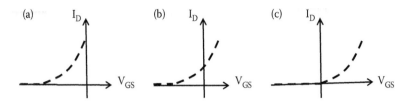

FIGURE 6.18 Transfer characteristics for (a) JFET, (b) depletion MOSFET, and (c) enhancement MOSFET.

the variation of carriers. An accurate approximation to the behavior of the drain current with voltage can include mathematical terms of various orders of magnitude. An equation of high complexity for the drain current is not always used. Usually a simpler approximate expression is commonly used and I_D can be approximated to

$$I_D \approx \frac{W}{L} \mu_n C_{ox} \left[(V_G - V_T)V_D - \frac{V_D^2}{2} \right] \tag{6.2}$$

For

$$0 < V_D < (V_G - V_T)$$

where W is the width of the gate, L the length of the gate, μ_n the mobility of electrons, C_{ox} the capacitance due to the insulator at the gate, V_G the voltage at the gate with respect to the source, V_T the threshold voltage, and V_D is the voltage at the drain with respect to the source.

Then the drain current depends on the geometrical dimension of the gate as well as voltage applied. The ratio W/L is referred to as the MOS transistor *aspect ratio*. The expression of Equation 6.2 provides the drain current for the drain voltage between 0 and $(V_G - V_T)$ the pinch-off voltage.

Pinch-off voltage V_{DP} is

$$V_{DP} = (V_G - V_T) \tag{6.3}$$

As V_D increases above the pinch-off voltage the drain current remains almost constant. For V_D above pinch-off, the current flow is said to be saturated.

6.7.1 SATURATION

At drain voltage above pinch-off, the drain current remains almost constant from the values reached at pinch-off, I_{DSS}. An approximated expression for the drain current in the saturation region can be obtained assuming the voltage at saturation, V_{Dsat},

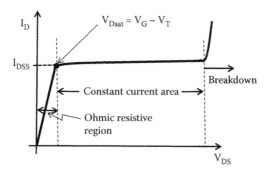

FIGURE 6.19 FET current–voltage characteristics.

remain constant at $(V_G - V_T)$ for I_D in the saturation region. Therefore, replacing $(V_G - V_T)$ by V_{Dsat} in Equation 6.2, the drain current I_D becomes

$$I_{Dsat} \approx \frac{W}{L}\mu_n C_{ox}\left[V_{Dsat} V_{Dsat} - \frac{V_{Dsat}^2}{2}\right]$$

or

$$I_{Dsat} = \frac{W}{L}\mu_n C_{ox}\left[\frac{V_{Dsat}^2}{2}\right] \tag{6.4}$$

$$\text{For} \quad V_D \geq (V_{GS} - V_T)$$

In practice there is a slight increase in current with increasing V_D above pinch-off, as a result of the reduction in the channel length.

Figure 6.19 shows the drain current against voltage indicating the two regions—ohmic and saturation regions—of the I–V curve.

If the gate voltage is varied then it is possible to create a family of I–V curves for a FET as shown in Figure 6.17.

EXAMPLE 6.1

The N-MOSFET of Figure 6.20 has a threshold voltage $V_T = 1$ V and $\mu C_{ox} W/L = 0.3$ mA/V^2. Calculate the values of the voltage between the source and drain, V_{DS}, and the drain current I_D.

SOLUTION

$$I_D = \frac{W}{L}\mu_n C_{ox}\left[(V_G - V_T)V_D - \frac{V_D^2}{2}\right] = 0.3\left[(5-1)V_D - \frac{V_D^2}{2}\right] \text{mA}$$

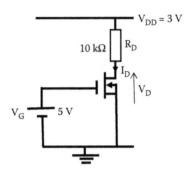

FIGURE 6.20   Circuit diagram for Example 6.1.

Also from KVL

$$V_{DD} = R_D I_D + V_D$$

or

$$I_D = \frac{V_{DD} - V_D}{R_D} = \frac{3 - V_D}{10 \times 10^3}$$

And solving the two simultaneous equations, we can obtain the drain voltage V_D as

$$V_D = 0.237 \text{ V}$$

Then the drain current will be given by

$$I_D = \frac{V_{DD} - V_D}{R_D} = \frac{3 - 0.237}{10 \times 10^3} = 0.2763 \text{ mA}$$

6.8   KEY POINTS

- FETs are widely used in both analog and digital applications.
- FET is characterized by very high input impedance and a small physical size. They are less noisy and can be used to produce circuits with very low-power consumption.
- FET operation depends upon the flow of majority carriers only. It is therefore a unipolar device.
- FETs are simpler to fabricate and occupy less space in IC form.
- There are two basic forms of FETs, namely the MOSFET and the JFET. Although MOSFET and the JFET operate in rather different ways, many of their characteristics are similar.

- All FETs are voltage-controlled devices in which the voltage on the gate controls the current that flows between the drain and the source.
- The characteristics of depletion MOSFETs, enhancement MOSFETs, and JFETs are similar, except that they require different bias voltages.
- FETs can be used not only to produce amplifiers, but also in a range of other applications such as voltage-controlled attenuators, analog switches, and logic gates.

7 Bipolar Junction Transistor Biasing

7.1 INTRODUCTION

A transistor can be used as a current and voltage amplifier device. We have seen from the output characteristics that the transistor has different regions of operation. To make the transistor operate in a desired region, according to the application required, we need to establish the voltages and currents where the transistor will operate. This can be achieved by connecting external components to the transistor to select the region and point of operation. This is what we call biasing a transistor. The biasing can be achieved by using dc voltage sources or current sources; these sources provide the energy required to amplify a signal.

7.2 LOAD LINE

The output characteristics of a transistor used in common emitter configuration shows the variation of the voltage between the collector and emitter, V_{CE}, versus the collector current I_C for different values of base current I_B as seen in Figure 7.1. We can make the transistor operate at different configurations of V_{CE} and I_C. We can determine and impose the values of I_C and V_{CE} in a transistor by applying dc sources and resistors to determine the value of I_C and V_{CE} required. The value of I_C and V_{CE} selected is known as the operating point or the quiescent point Q of a transistor. It is usually the case that we have a fixed voltage source. Different voltage values can then be achieved by adding external passive components. Figure 7.1 shows an example of a transistor with operation point $V_{CE} = 5$ V and $I_C = 1$ mA.

Usually, external passive components are resistors that can be changed to obtain different values of operating point as the voltage source is fixed. Let us consider a simple example to bias a transistor using a single source V_{cc} and a resistor at the collector R_C as seen in Figure 7.2.

Applying KVL to the loop formed by the source V_{cc}, the resistor R_C and the voltage V_{CE} we obtain the following equation:

$$V_{CC} = I_C R_C + V_{CE} \qquad (7.1)$$

In this equation the two terms, V_{CE} and I_C, are parameters of the output characteristic of the transistor and they select the Q point. V_{CE} and I_C can be varied and this gives a linear relationship creating what is called the *load line*. It is called the load line because R_C can be the load in this circuit and a variation of this load will produce different values of I_C and V_{CE}. Figure 7.3 shows an example of a load line imposed on the output characteristic of a transistor.

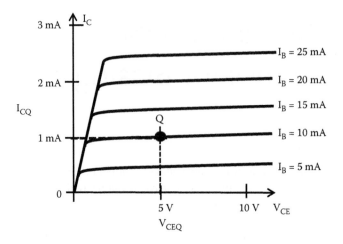

FIGURE 7.1   Example of operating point, Q, in a transistor in common emitter configuration.

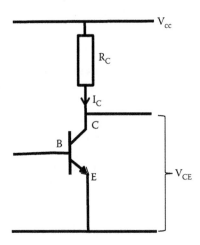

FIGURE 7.2   Example of biasing a transistor with a source and a resistor.

From Equation 7.1 it is possible to determine the two points D and E where this line intersects the V_{CE} and I_C axes. They are given by

$$\text{At } I_C = 0 \quad \text{then } V_{cc} = V_{CE} \text{ point D}$$

and

$$\text{At } V_{CE} = 0 \quad \text{then } I_C = V_{cc}/R_C \text{ point E}$$

The line joining D to E defines the only combinations of V_{CE} and I_C that are possible by a value of R_C once V_{cc} has been selected. It is desired to select the operating

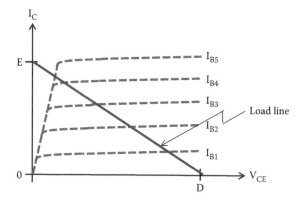

FIGURE 7.3 Example of a load line.

point within a load line and avoid regions where the transistor does not operate safely. In analog electronics it is usually desired to select an appropriate Q within the liner region to obtain an output voltage without distortion.

EXAMPLE 7.1

A BJT has a collector resistance $R_C = 10\ k\Omega$ as indicated in Figure 7.4. Determine two points that will allow you to draw the load line for this circuit.

SOLUTION

We can draw the load line superimposed on the output characteristics of the transistor.

One point (P_1) can be determined when the collector current $I_C = 0$, then the potential difference across R_C will be zero and the 12 V from the source will be applied between the collector and emitter. Then at $I_C = 0$, $V_{CE} = 12$ V. This gives us a point P_1 in the load line as P_1 (0 A, 12 V).

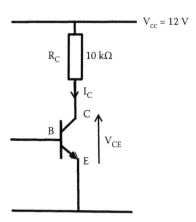

FIGURE 7.4 Example for load line with $R_C = 10\ k\Omega$.

Another point in this load line (P_2) can be determined when the voltage between collector and emitter is equal to zero, and then the 12 V from the source will be applied across R_C. Then at $V_{CE} = 0$, I_C is equal to

$$I_C = \frac{V_{R_C}}{R_C} = \frac{12}{10 \times 10^3} = 1.2 \text{ mA}$$

and the second point will be P_2 (1.2 mA, 0 V). The straight line for the load line can be drawn with these two points.

7.2.1 Cut Off

When the base current is equal to zero, the transistor is in what is known as the cut-off region. Figure 7.5 shows a cut-off region in the output characteristics of a transistor.

Under cut-off conditions there is a very small amount of collector leakage current, I_{ceo}, due mainly to thermally produced carriers. In cut off, the base–emitter and the base–collector junctions are reverse biased.

7.2.2 Saturation

In the output characteristics of a transistor, if the voltage between collector and emitter is increased then the collector current increases, but there is a value of V_{CE} when the I_C current saturates, that is, remains almost constant. This happens for different values of I_B creating what is known as the *saturation region*. At saturation, the B–C junction becomes forward biased and I_C can increase no further. Figure 7.6 shows the saturation region in the output characteristics of a transistor.

At saturation, the relation $I_C = \beta I_B$ is no longer valid. The V_{CEsat} for a transistor is usually only a few tenths of a volt for Si transistors.

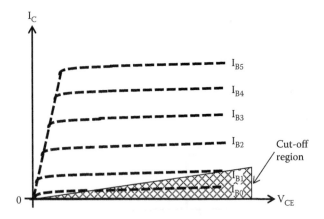

FIGURE 7.5 Output characteristics. Cut-off region in a BJT.

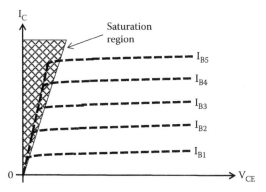

FIGURE 7.6   Saturation region in a BJT.

It is usually the case that in a circuit that works with analog signals the cut-off region and the saturation region are avoided (in digital systems the transistor it is made to work between these two regions).

7.3   BIASING A BJT

Biasing can be achieved using current sources; this is usually the case when biasing transistors in IC forms. Biasing can also be achieved using external passive components and sources. There are some forms of biasing that are commonly used in designing analog circuits. In this section, we look at the most popular circuits to bias BJT transistors.

One factor to consider when biasing a transistor using components "off the shelf" is that transistors are mass produced and manufacturers cannot guarantee a particular value of β in a transistor. In the data provided by the manufacturer, the nominal value, the minimum, and the maximum values are indicated. This means that the value of parameter β can be within this range of values. Then β varies from transistor to transistor and β also varies in a particular transistor with V_{CE}, I_C, and also with variation of temperature. The disadvantage of this is that a variation in β causes both I_C and V_{CE} to change, thus changing the Q point of the transistor, making the biased transistor β-dependent.

7.3.1   FIXED BIAS

Fixed bias biasing is a simple and economic way of biasing a transistor. It consists of one source and two external resistors as indicated in Figure 7.7.

Considering the voltages around the closed circuit that include V_{cc}, voltage across R_B and V_{BE} and applying KVL we obtain the following equation:

$$V_{CC} = I_B R_B + V_{BE}$$

Rearranging this equation we have

$$I_B = \frac{V_{CC} - V_{BE}}{R_B}$$

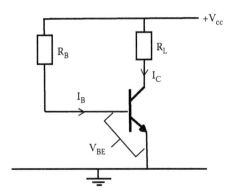

FIGURE 7.7 Fixed bias circuit.

Replacing the value of I_B in terms of I_C using the expression for $\beta = I_C/I_B$ leading to $I_B = I_C/\beta$ then

$$\frac{I_C}{\beta} = \frac{V_{CC} - V_{BE}}{R_B}$$

and

$$R_B = \frac{V_{CC} - V_{BE}}{I_C}\beta \tag{7.2}$$

Applying KVL around the loop to include voltages V_{CC}, V_{CE}, and voltage across R_L, we have

$$V_{CC} = I_C R_L + V_{CE} \tag{7.3}$$

With these two expressions we can design this fixed bias biasing. With this type of biasing the variation of β has a large effect on the operating point. This is illustrated by the following example.

EXAMPLE 7.2

Design a fixed biasing of a transistor with $\beta = 100$. The operating point for this transistor is given by $I_C = 5$ mA and $V_{CE} = 5$ V. The collector resistance, R_L, is 1 kΩ.
   Once the biasing is designed we can assume that β changes and calculate the operation point for $\beta = 50$ and 150.

SOLUTION

The fixed bias biasing circuit to be used is indicated in Figure 7.8.
   Calculate V_{cc} from Equation 7.3

$$V_{CC} = I_C R_L + V_{CE} = 5 \times 10^{-3} \times 1 \times 10^3 + 5 = 10 \text{ V}$$

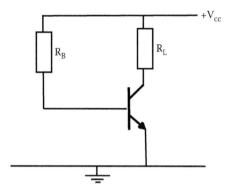

FIGURE 7.8 Fixed bias circuit for numerical Example 7.2.

Calculate R_B.

When the silicon transistor B–E junction is forward biased, it is like a diode and has a forward voltage drop between about 0.5 and 0.7 V. Use $V_{BE} = 0.65$ V without losing much accuracy, then

$$R_B = \frac{V_{CC} - V_{BE}}{I_C}\beta = \frac{10 - 0.65}{5 \times 10^{-3}} \times 100 = 187 \text{ k}\Omega$$

187 kΩ is not normally manufactured. Manufacturers produce components of certain specified values. These components are known as the *nominal preferred value* (npv). In this example, the nearest npv that exist commercially is 180 Ω. Then the circuit designed is shown in Figure 7.9.

Now let us change β to 50 and 150 and calculate the new operating points (I_C, V_{CE}). Repeating the calculation above with the designed values and new β we obtain Table 7.1.

From Table 7.1 we notice that the problem with this biasing is the large variation of the operating point as β changes.

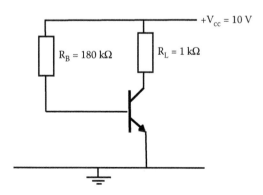

FIGURE 7.9 Designed fixed bias circuit.

TABLE 7.1
Fixed Bias with Different Values of β

β	I_c (mA)	V_{CE} (V)
100	5	5
150	7.5	2.5
50	2.5	7.5

EXAMPLE 7.3

The operating point for the circuit shown in Figure 7.10 has an operating point given by $V_{CE} = 7$ V and $I_c = 7$ mA. In order to achieve this operating point, determine the value of resistors R_B and R_C. The transistor has $V_{BE} = 0.65$ V and $β = 120$.

SOLUTION

Using KVL we can obtain the following equation:

$$V_{CC} = I_C R_C + V_{CE} = 7 \times 10^{-3} \times R_C + 7 = 16 \text{ V}$$

Then,

$$R_C = \frac{16 - 7}{7 \times 10^{-3}} = 1285.71 \Omega$$

The value of I_B can be calculated as

$$I_B = \frac{I_C}{β} = \frac{7 \times 10^{-3}}{120} = 5.83 \times 10^{-5} \text{ A}$$

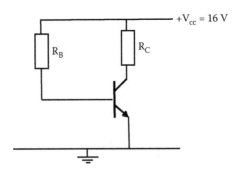

FIGURE 7.10   Fixed bias amplifier circuit for Example 7.3.

and KVL will provide the relationship to calculate R_B as

$$V_{CC} = I_B R_B + V_{BE} = 5.83 \times 10^{-5} \times R_B + 0.65 = 16\,V$$

and

$$R_B = \frac{16 - 0.65}{5.83 \times 10^{-5}} = 263293.31\,\Omega$$

7.3.2 Auto Bias

Auto bias is also known as self-bias, potential divider bias, or emitter bias. The auto bias circuit consists of four resistors and a voltage source as shown in Figure 7.11. This circuit is more stable and most frequently used than the fixed biasing circuit. The necessary steady voltage at the base is provided by the potential divider formed by resistors R_1 and R_2. It uses more resistors but it provides a more stable operating point.

The analysis of this circuit is made easier if we reduce R_1, R_2, and V_{cc} to obtain an equivalent circuit using the Thevenin theorem. The resulting circuit is shown in Figure 7.12.

Applying KVL to the right-hand side (RHS) loop we can write

$$V_{CC} = I_C R_C + V_{CE} + I_E R_E \tag{7.4}$$

Also applying KVL to the left-hand side (LHS) loop we can write

$$\frac{R_2 V_{CC}}{R_1 + R_2} = \frac{R_1 R_2}{R_1 + R_2} I_B + V_{BE} + I_E R_E$$

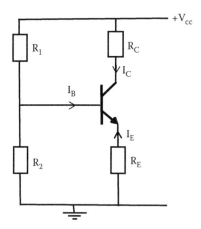

FIGURE 7.11   Auto bias circuit.

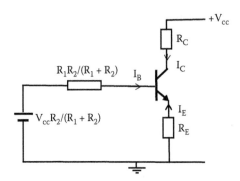

FIGURE 7.12 Equivalent circuit of an auto bias circuit.

Rearranging

$$\frac{R_2 V_{CC}}{R_1 + R_2} - V_{BE} - I_E R_E = \frac{R_1 R_2}{R_1 + R_2} I_B$$

Now the collector current is almost the same value as the emitter current:

$$I_E \approx I_C$$

and

$$I_C = \beta I_B$$

Replacing I_B and I_C in the equation above we have

$$\frac{R_2 V_{CC}}{R_1 + R_2} - V_{BE} - I_C R_E = \left(\frac{R_1 R_2}{R_1 + R_2} \right) \frac{I_C}{\beta}$$

or

$$\frac{R_2 V_{CC}}{R_1 + R_2} - V_{BE} = I_C \left\{ R_E + \left(\frac{R_1 R_2}{R_1 + R_2} \right) \frac{1}{\beta} \right\}$$

and

$$I_C = \frac{R_2 V_{CC}/(R_1 + R_2) - V_{BE}}{R_E + R_1 R_2/[(R_1 + R_2)\beta]} \tag{7.5}$$

The collector current can be independent of β if R_E is larger than $R_1 R_2/[(R_1 + R_2)\beta]$ to provide a stable Q. This is the same as to consider I_C larger than I_B.

EXAMPLE 7.4

An auto bias biasing circuit was designed as shown in Figure 7.13 for an operating point Q (I_C, V_{CE}). Other values for this circuit are $R_1 = 20\ k\Omega$, $R_2 = 10\ k\Omega$, $R_C = 2\ k\Omega$, and $R_E = 1\ k\Omega$. Calculate the operation point V_{CE} and I_C for a transistor with $\beta = 100$. Also calculate the variation of Q for changes at $\beta = 50$ and 150.

SOLUTION

We can assume that the transistor was fabricated in silicon then $V_{BE} = 0.65$ V.

Replacing the values of the resistors, β and source we can obtain I_C and V_{CE} required by operating point Q. For $\beta = 100$ we have

$$I_C = \frac{R_2V_{CC}/(R_1+R_2)-V_{BE}}{R_E+R_1R_2/(R_1+R_2)\beta} = \frac{10k \times 15/(20k+10k)-0.65}{1k+20k \times 10k/(20k+10k)100} = 4.078\ mA$$

From $V_{CC} = I_C R_C + V_{CE} + I_E R_E$ the value of V_{CE} can be calculated as

$$V_{CE} = V_{CC} - I_E R_E - I_C R_C = 15 - 4.078 \times 10^{-3} \times 1 \times 10^3 - 4.125 \times 10^{-3} \times 2 \times 10^3 = 2.672\ V$$

Table 7.2 shows the calculated Q values for the different values of β.

We can see that the variation of I_C and V_{CE} is small, given a stable operating point.

EXAMPLE 7.5

For the circuit shown in Figure 7.14, determine the values of I_B, I_C, α, I_E, V_{CE}, and V_{BC}. The transistor has a value of $\beta = 225$ and a $V_{BE} = 0.65$ V.

SOLUTION

The base current I_B is the same current passing through the resistance R_B, then the base current is

$$I_B = \frac{E_B - V_{BE}}{R_B} = \frac{3 - 0.65}{8.6 \times 10^3} = 273.23\ \mu A$$

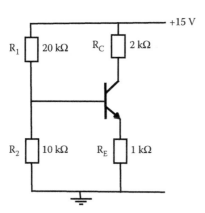

FIGURE 7.13   Auto bias circuit diagram for numerical Example 7.4.

TABLE 7.2

Calculated Q Values for Different Values of β

β	I_c (mA)	V_{CE} (V)
100	4.078	2.672
150	4.214	2.358
50	3.9	3.3

Note: Auto bias circuit.

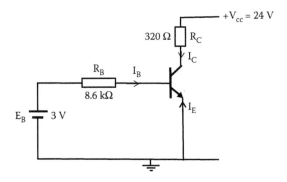

FIGURE 7.14 Circuit diagram for Example 7.5.

and

$$I_C = \beta I_B = 225 \times 273.23 \times 10^{-6} = 61.48 \, mA$$

$$\alpha = \frac{\beta}{\beta+1} = \frac{225}{225+1} = 0.9956$$

$$I_E = \frac{I_C}{\alpha} = \frac{61.48[mA]}{0.9956} = 61.27 \, mA$$

By KVL

$$V_{CE} = V_{CC} - I_C R_C = 24 - 61.48 \times 10^{-3} \times 320 = 4.326 \, V$$

and

$$V_{BC} = V_{CE} - V_{BE} = 4.326 \, V - 0.65 = 3.676 \, V$$

7.3.3 COLLECTOR-FEEDBACK BIAS

The collector-feedback bias is also economical to build with two resistors and a voltage source as shown in Figure 7.15. In this biasing circuit if β increases, which can be due to temperature change, then the collector current increases and the voltage across the resistor R_C increases, but the voltage across the resistor R_B decreases to keep the voltage V_{cc} constant. As a consequence of the reduction of voltage across R_B the base current is reduced and I_C reduces. The result is an I_C stable and Q stable.

The advantage of the negative feedback connection provides a stable Q point by reducing the effect of variations in β and is simple in terms of components required.

Using Kirchhoff laws, equations for I_C and V_{CE} required to determine the operation point can be deduced. Applying KVL to the close circuit that include R_C, R_B, V_{BE}, and V_{cc} we can write

$$V_{CC} = I_C R_C + I_B R_B + V_{BE} \tag{7.6}$$

Rearranging an expression for the base current can be written as

$$I_B = \frac{V_{CC} - I_C R_C - V_{BE}}{R_B} \tag{7.7}$$

But $I_B = I_C/\beta$ and Equation 7.7 becomes

$$\frac{I_C R_B}{\beta} + I_C R_C = V_{CC} - V_{BE} \tag{7.8}$$

and finally we can write an expression for I_C as

$$I_C = \frac{V_{cc} - V_{BE}}{(R_B/\beta) + R_C} \tag{7.9}$$

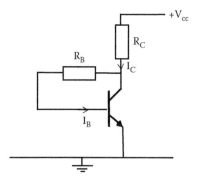

FIGURE 7.15 Collector-feedback bias circuit.

Applying KVL to the close circuit that include R_C, V_{CE}, and V_{cc} we can write

$$V_{CC} = I_C R_C + V_{CE} \qquad (7.10)$$

or

$$V_{CE} = V_{CC} - I_C R_C$$

Equations 7.9 and 7.10 provide expressions to calculate the operating point Q.

EXAMPLE 7.6

Calculate Q for the circuit shown in Figure 7.16 for β 100, 150, and 50

SOLUTION

For $\beta = 100$
  From Equation 7.9

$$I_C = \frac{V_{CC} - V_{BE}}{(R_B/\beta) + R_C}$$

$$I_C = \frac{10 - 0.65}{(100\,\text{K}/100) + 10\,\text{K}} = 0.85\,\text{mA}$$

$$V_{CE} = V_{CC} - I_C R_C = 10 - 0.85 \times 10^{-3} \times 10 \times 10^3 = 1.5\,\text{V}$$

As in the case of auto bias the variation of I_C and V_{CE} is small for this circuit providing a stable operating point (Table 7.3).

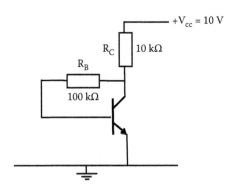

FIGURE 7.16   Circuit for collector-feedback bias example.

TABLE 7.3

Q Collector-Feedback Bias for Different Values of β

β	I_c (mA)	V_{CE} (V)
100	0.850	1.50
150	0.8812	1.1875
50	0.7833	2.16

7.3.4 Two Sources

Using two sources is more commonly used in IC, where the two sources allow the use of more complicated biasing circuit such as creating current sources and current mirrors as load. Two sources also provide the possibility to design flexible differential amplifiers. Two sources will be used in more detail in Chapters 18 and 19.

In Figure 7.17 an example of biasing a transistor with two sources is shown.

As before, we can use Kirchhoff's laws to obtain relationships for voltages and currents. These relationships will lead to equations to determine the operating point of the transistor. These relationships will help to select the external components to fix the required operating point.

Applying KVL to the close circuit that include R_C, V_{CE}, and V_{cc} we can write

$$V_{CC} = I_C R_C + V_{CE}$$

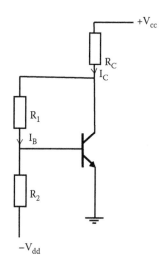

FIGURE 7.17 Two-source biasing.

Also from close circuit including components R_C, R_1, V_{cc}, and the voltage between base and emitter V_{BE} by KVL we can write

$$V_{CC} = I_C R_C + R_1 I_B + V_{BE} \tag{7.11}$$

Also from close circuit including components R_2, V_{dd}, and the voltage between base and emitter V_{BE} by KVL we can write

$$R_2 I_B = V_{dd} + V_{BE} \tag{7.12}$$

The above relationships will help to select the external components to fix the required operating point.

7.4 KEY POINTS

- For a fix resistor at the collector, V_{CE} and I_C can be varied and this gives a linear relationship creating what is called the *load line*.
- The operating point, Q, establishes the voltage between collector and emitter and the collector current where the transistor will operate. This can be achieved by connecting external components to the transistor to select the point of operation.
- Making the transistor operate at the Q point is biasing a transistor. This makes the transistor work in a desired region, according to the application required.
- The current gain parameter of a transistor β varies from transistor to transistor and β also varies in a particular transistor with V_{CE}, I_C, and temperature. One of the aims of biasing is to use a circuit that reduces or eliminates the influence of any variation of β.
- Fixed bias is very economical as it requires only two resistors and dc source. Auto bias makes the operation more stable as well as collector feedback bias. Using two sources for biasing increases the flexibility of the circuit, but the two sources increase the requirements in an application.

8 Modeling Transistors

8.1 INTRODUCTION

The property of any two-port or four-terminal network may be specified by the relationship between the currents and voltages at its terminals. Figure 8.1 shows a general four-terminal network with its input and output voltages and currents, where i_1 is the input current, v_1 the input voltage, v_2 the output voltage, and i_2 the output current.

There are six ways in which two quantities may be chosen from four. This provides six possible pairs of equations and six sets of parameters to characterize this network. These sets of parameters are chosen according to how well they describe the network to be analyzed in terms of the function of the network. For example, there are G parameters that provide a good characterization of communication networks, Y parameters to analyze transfer functions, A parameters that provide a good way of analyzing transmission systems. The h parameters and T parameters are usually used to analyze analog systems containing transistors. In this chapter, we will describe some of the models used based on these parameters. In Chapter 9 these models will be used to analyze transistors as part of amplifiers.

8.2 HYBRID h PARAMETERS

Taking i_1 and v_2 as independent variables and v_1 and i_2 as dependent variables we can write the following set of equations for the two-port network:

$$v_1 = h_{11}i_1 + h_{12}v_2 \qquad (8.1)$$

$$i_2 = h_{21}i_1 + h_{22}v_2 \qquad (8.2)$$

This creates the "hybrid" or h parameters. They are termed as hybrid because they are not all alike dimensionally. The h parameters used to model a transistor are labeled as follows:

$$h_{11} = h_i \quad h_{21} = h_f \quad h_{12} = h_r \quad h_{22} = h_o$$

These parameters are normally specified on a manufacturer's data sheet. The four basic h parameters are as follows:

h_i = input resistance (Ω)
h_r = voltage feedback ratio (dimensionless)
h_f = forward current gain (dimensionless)
h_o = output conductance (Siemen = 1/Ohms)

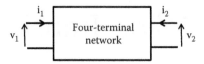

FIGURE 8.1   Block diagram for a general two-port network.

8.2.1   h Parameters Common Emitter Configuration

An extra letter is added for transistors used in different configuration. For example, in a common emitter configuration the h parameters are as follows: h_{ie}, h_{re}, h_{fe}, h_{oe}, and h_{fe} (= β).

Figure 8.2 shows a transistor as common emitter configuration.

For the common emitter configuration shown in Figure 8.2, an equivalent circuit in terms of its h parameters is shown in Figure 8.3.

Similarly, equivalent circuits and equations can be obtained for the other configurations, common collector, and common base. They will have the same equivalent circuit but the actual values for the h parameters will vary.

The h parameters are usually given by the manufacturers; they can also be obtained from the static characteristics of transistors. For example, for a common emitter transistor the h_{ie} can be obtained as

$$h_{ie} = \frac{\delta V_{BE}}{\delta i_B}\bigg|_{V_{CE}\text{constant}} \tag{8.3}$$

where $\delta V_{BE}/\delta i_B$ denotes partial derivative of V_{BE} with respect to i_B. Similarly, other h parameters can be obtained in the same way

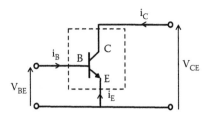

FIGURE 8.2   Circuit diagram for a transistor connected in common emitter configuration.

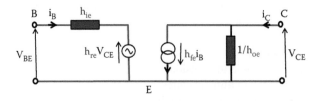

FIGURE 8.3   Common emitter configuration equivalent circuit using h parameters.

$$h_{re} = \frac{\delta V_{BE}}{\delta V_{CB}}\bigg|_{i_B \text{constant}} \tag{8.4}$$

$$h_{fe} = \frac{\delta i_C}{\delta i_B}\bigg|_{V_{CE}\text{constant}} \tag{8.5}$$

$$h_{oe} = \frac{\delta i_C}{\delta V_{CE}}\bigg|_{i_B \text{constant}} \tag{8.6}$$

Experimentally they can be obtained using the experimental setup shown in Figure 8.4.

Once the input and output characteristics of a transistor are obtained then the h parameters can be deduced from the curves as indicated in Figure 8.5.

Then, h_{ie} can be obtained as

$$h_{ie} = \frac{\Delta V_{BE}}{\Delta i_B}\bigg|_{V_{CE}\text{constant}}$$

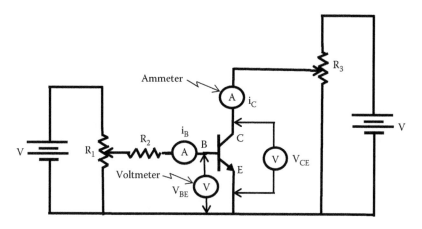

FIGURE 8.4   Experimental setup to obtain the h parameters of a BJT.

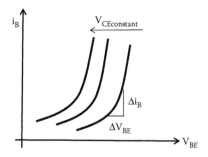

FIGURE 8.5   Input or base characteristic for BJT common emitter.

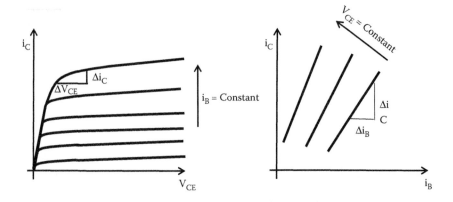

FIGURE 8.6   Output or collector characteristic for a BJT common emitter.

and from the output characteristics of a BJT the h_{oe} and h_{fe} can be obtained. Figure 8.6 shows the output characteristics for a BJT common emitter configuration.

Then, h_{oe} can be obtained as

$$h_{oe} = \frac{\Delta i_C}{\Delta V_{CE}}\bigg|_{i_{B\text{constant}}} \quad \text{and} \quad h_{fe} = \frac{i_C}{\Delta I_B}\bigg|_{V_{CE\text{constant}}}$$

8.3   ADMITTANCE Y PARAMETERS

The Y parameters are useful for analyzing a circuit at high frequency. It is used in telecommunications and also to analyze response behavior of amplifiers at high frequency. Figure 8.7 shows a general equivalent circuit using the Y parameters.

These parameters have the unit of admittance *Siemens* (S). The set of equations that define the Y parameters are

$$I_{in} = Y_i V_{in} + Y_r V_o \tag{8.7}$$

$$I_o = Y_f V_{in} + Y_o V_o \tag{8.8}$$

where
  Y_i is short circuit input admittance (S)
  Y_f forward transfer admittance (S)

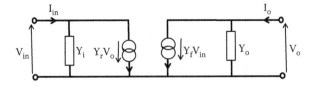

FIGURE 8.7   General Y parameters equivalent circuit for a transistor.

Y_r short reverse transfer admittance (S)
Y_o short circuit output admittance (S)

Y_r is usually relatively small and can in many cases be neglected.

8.4 GENERAL THREE-PARAMETER MODEL

In some cases it is useful to use a general three-parameter model to analyze an amplifier. Figure 8.8 shows an equivalent circuit using the three-parameter model, where R_o is the output resistance, R_{in} the input resistance, and A_v is the open circuit voltage gain.

A_v is the gain when the circuit is open (load connected equal to infinitive) then the output current is equal to zero and

$$A_V = \frac{v_O}{v_i} \tag{8.9}$$

The input and output resistance can be measured to create this model.

8.5 T-EQUIVALENT TWO-PARAMETER MODEL

8.5.1 AC-Simple T-Parameters Transistor Model

This is a very simple model for a transistor that holds well for IC planar technology transistors and small currents. Figure 8.9 shows the T-equivalent circuit for the common emitter configuration and how the parameters relate to the structure of the transistor. The resistor r_e represents the forward-biased base–emitter junction resistance and the current source represents the current gain of the transistor.

The emitter–base p–n junction results in a diode relationship expressed as

$$I = I_s(e^{qV/kT} - 1) \tag{8.10}$$

where I_s is the saturation current, V the junction voltage, k the Boltzmann constant, T the temperature in Kelvin, and q the electronic charge as indicated in Chapter 4.

This equation can be written in terms of "thermal voltage," V_T.

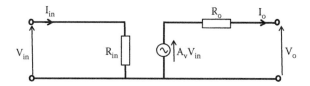

FIGURE 8.8 General equivalent circuit three parameters for a transistor.

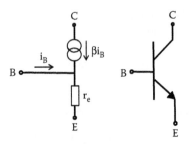

FIGURE 8.9   Simplify T-model equivalent circuit for a transistor.

$$V_T = \frac{kT}{q} \tag{8.11}$$

V_T has the dimension of a voltage and it depends on the temperature. At a temperature of 300°K, usually taken as room temperature, then V_T is approximately 25 mV. Then, Equation 8.10 can be written in terms of V_T

$$I = I_s(e^{V/V_T} - 1) \tag{8.12}$$

8.5.2   EMITTER RESISTANCE, r_e (SMALL SIGNAL)

The emitter–base junction is forward biased and presents a resistance that varies with dc. This resistance can be obtained as the first derivative of the current in the forward direction with respect to the voltage applied and is given by

$$\frac{1}{r_e} = \frac{dI}{dV} = \frac{d}{dV}[I_s(e^{V/V_T} - 1)] \tag{8.13}$$

This gives

$$\frac{1}{r_e} = \frac{1}{V_T} I_s e^{V/V_T} \tag{8.14}$$

or

$$\frac{1}{r_e} \approx \frac{I}{V_T} \tag{8.15}$$

Then, r_e is

$$r_e = \frac{V_T}{I} \tag{8.16}$$

and at room temperature r_e can be written as

$$r_e = \frac{25 \, mV}{I} \, \Omega \tag{8.17}$$

The resistor r_e represents the forward-biased base–emitter junction, which is obtained from the slope of the I_E against V_{BE} plot. For a transistor, the r_e parameter can be calculated as

$$r_e = \frac{25 \, mV}{I_E}$$

as before.

Thus, the T equivalent circuit reduces to two components, the resistance of the forward-biased base–emitter junction r_e and a current generator βi_b to represent the collector current, which is generated when base current flows.

8.6   MUTUAL CONDUCTANCE MODEL

For small signals at low frequency and modest collector loads we can ignore h_{oe} and h_{re} in the h parameters. Then, the equivalent circuit is reduced to a two-parameter model as shown in Figure 8.10.

The mutual conductance parameter, g_m, is defined as

$$Mutual \; conductance = g_m = \left.\frac{\delta I_C}{\delta V_{BE}}\right|_{V_{CE}=constant} \tag{8.18}$$

This is the reciprocal of the emitter resistance parameter, r_e, then

$$g_m = \frac{1}{r_e} = \frac{h_{fe}}{h_{ie}} = \frac{eI_C}{kT} \tag{8.19}$$

g_m at room temperature, 300°K, can be expressed as

$$g_m = \frac{I_C}{25 \times 10^{-3}} = 40 I_C \tag{8.20}$$

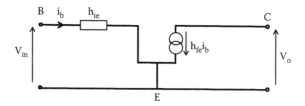

FIGURE 8.10   Equivalent circuit for two h parameters model.

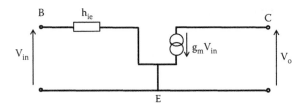

FIGURE 8.11 Equivalent circuit for the mutual conductance model.

Now, from circuit shown in Figure 8.10 we can write

$$h_{fe}i_b = h_{fe}\frac{V_{in}}{h_{ie}} = g_m V_{in} \qquad (8.21)$$

and the mutual conductance model can be created based on the two h parameter as shown in Figure 8.11, where g_m is the mutual conductance of the transistor and h_{ie} is the h parameter representing the input resistance.

8.7   KEY POINTS

- Different transistor models may be specified by the relationships between the currents and voltages at its terminals.
- These sets of parameters to model a transistor are chosen according to how well they describe the network to be analyzed.
- G parameters provide a good characterization of communication networks.
- Y parameters can be used to analyze transfer functions.
- A parameters provide a good way of analyzing transmission systems.
- The h parameters and T parameters are usually used to analyze analog systems containing transistors at small signal.
- Manufacturers provide some value of parameters to model a transistor.
- Parameters for a model can be obtained from experimental result of the characteristics of a transistor.

9 Small-Signal Analysis of an Amplifier under Different Models

9.1 INTRODUCTION

Transistors are usually used to amplify ac signals. The biasing of a transistor uses dc voltages and currents to determine the operating point, Q. The amplification of ac signal is achieved by superimposing the ac signal on the biasing dc currents. It means that the operating point, Q, will change along with the ac variations of the input signal.

A capacitor plays an important function in blocking a dc current and allowing ac signals to pass through it. The impedance of a capacitor is inversely proportional to the frequency. As we saw in Chapter 3, the reactance of a capacitor can be expressed as

$$X_C = \frac{1}{2\pi\, fC}$$

For low frequencies the reactance will be large, in the limit at a frequency equal to zero (dc) the reactance will be infinity, that is, open circuit to dc signal. On the other hand if the frequency is high, the reactance will be small, in the limit at a frequency equal to infinitive, the reactance will be zero, that is, a short circuit to ac signal. Then capacitors are used to separate the dc biasing of a transistor and the amplification of an ac signal.

In this chapter, we will see models of transistor that are investigated in Chapter 8 as part of an amplifier and deduce the characteristic of these amplifiers.

9.2 ANALYSIS OF TRANSISTOR AMPLIFIERS USING h PARAMETERS

In general, an amplifier can consist of a transistor and external components to bias this transistor. Coupling capacitors can couple the input signal to be amplified and the output signal to the load. At the frequency range of operation of an amplifier, the impedance of the coupling capacitors can be taken as negligible. Then, a general equivalent ac circuit for an amplifier, using h parameter, can be represented as in Figure 9.1. In later sections of this chapter, an analysis of the different biasing techniques is presented.

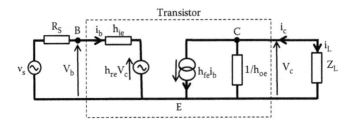

FIGURE 9.1 Basic CE amplifier equivalent circuit using h parameters.

Figure 9.1 shows a basic common emitter (CE) amplifier circuit using an h parameters model for the transistor.

In this equivalent, circuit R_S is the resistance of the voltage source and Z_L represents the load impedance.

9.2.1 CURRENT GAIN, G_i

The current gain is given as the ratio of the output current to the input current, from Figure 9.1 we can write the current gain as

$$G_i = \frac{\text{Output current}}{\text{Input current}} = \frac{i_L}{i_b} = -\frac{i_c}{i_b} \tag{9.1}$$

Applying KCL at node C we can write

$$i_c = h_{fe}i_b + h_{oe}V_c \tag{9.2}$$

and

$$V_c = -i_c Z_L \tag{9.3}$$

Hence, replacing V_c from Equation 9.3 into Equation 9.2 we have

$$i_c = h_{fe}i_b - h_{oe}i_c Z_L$$

or rearranging

$$i_c(1 + h_{oe}Z_L) = h_{fe}i_b$$

and

$$\frac{i_c}{i_b} = \frac{h_{fe}}{1 + h_{oe}Z_L}$$

and finally, the current gain can be expressed as

$$G_i = \frac{i_L}{i_b} = -\frac{i_c}{i_b} = -\frac{h_{fe}}{1+h_{oe}Z_L} \tag{9.4}$$

Now h_{oe} has a value of the order of 10^{-5} S. It is not uncommon to have $h_{oe}Z_L$, very small, that can be neglected, then

$$G_i \approx -h_{fe} \tag{9.5}$$

That is why the parameter h_{fe} provides the current gain.

9.2.2 INPUT IMPEDANCE, Z_i

The input impedance is the ratio of the input voltage to the input current, from Figure 9.1 we can write Z_i as

$$Z_i = \frac{\text{Input voltage}}{\text{Input current}} = \frac{V_b}{i_b} \tag{9.6}$$

But applying KVL at the input of circuit shown in Figure 9.1 we can write

$$V_b = h_{ie}i_b + h_{re}V_c \tag{9.7}$$

But from Equation 9.1 we can write i_c in terms of the current gain G_i as

$$-i_c = G_i i_b$$

Replacing i_c into Equation 9.3 the voltage at the collector V_c becomes

$$V_c = -i_c Z_L = G_i i_b Z_L$$

Hence, replacing V_c into Equation 9.7 we have

$$V_b = h_{ie}i_b + h_{re}G_i i_b Z_L = i_b(h_{ie} + h_{re}G_i Z_L)$$

and finally, the input impedance Z_i can be expressed as

$$Z_i = \frac{V_b}{i_b} = h_{ie} + h_{re}G_i Z_L = h_{ie} - \frac{h_{re}h_{fe}Z_L}{1+h_{oe}Z_L} \tag{9.8}$$

Note that the input impedance is a function of load impedance Z_L. Typical values of h parameters will give a value for $h_{re}h_{fe}Z_L/(1 + h_{oe}Z_L)$ much smaller than h_{ie}, provided that Z_L in not very large, then the input impedance can be approximated to

$$Z_i \approx h_{ie} \tag{9.9}$$

9.2.3 VOLTAGE GAIN, G_v

The voltage gain is given as the ratio of the output voltage to the input voltage, from Figure 9.1 we can write the gain as

$$G_v = \frac{\text{Output voltage}}{\text{Input voltage}} = \frac{V_c}{V_b} \tag{9.10}$$

The output voltage can be expressed as

$$V_c = i_L Z_L = -i_c Z_L = G_i i_b Z_L$$

and the input voltage can be expressed as

$$V_b = Z_i i_b$$

and replacing the expression of V_c and V_b in terms of G_i and Z_i into Equation 9.10 we can obtain the voltage gain G_v as

$$G_v = \frac{V_c}{V_b} = \frac{G_i Z_L i_b}{Z_i i_b} = \frac{G_i Z_L}{Z_i} = -\frac{h_{fe}Z_L}{(1 + h_{oe}Z_L)(h_{ie} - (h_{re}h_{fe}Z_L/1 + h_{oe}Z_L))} \tag{9.11}$$

Using the previous approximations $G_i \approx -h_{fe}$ and $Z_i \approx h_{ie}$ we can obtain an approximation for G_v as

$$G_v = -\frac{h_{fe}Z_L}{h_{ie}} \tag{9.12}$$

9.2.4 OUTPUT IMPEDANCE, Z_o

The output impedance, Z_o, can be calculated as a combination of the impedances at the output circuit of Figure 9.1, that is, RHS circuit of Figure 9.1. The output impedance at point C will be the parallel combination of current source's impedance and $1/h_{oe}$. The impedance of the current source, Z_{source}, can be obtained as the voltage across the source, V_c, divided by the current passing through it, $h_{fe}i_b$. Then,

$$Z_{source} = \frac{V_c}{h_{fe}i_b}$$

and adding the impedance, $1/h_o$ and Z_{source} in parallel we can obtain an expression of the reciprocal of the output impedance, $1/Z_o$ as

$$\frac{1}{Z_o} = \frac{1}{1/h_{oe}} + \frac{1}{V_c/h_{fe}i_b} = h_{oe} + \frac{h_{fe}i_b}{V_c} \tag{9.13}$$

Applying a short circuit at the input, to deduce the output impedance, we can calculate the current i_b with v_s short circuited as

$$i_b = -\frac{h_{re}V_c}{h_{ie} + R_S} \tag{9.14}$$

The negative sign is due to direction of current polarity of voltage. Replacing i_b from Equation 9.14 into Equation 9.13 we have

$$\frac{1}{Z_0} = h_{oe} - \frac{h_{fe}h_{re}}{h_{ie} + R_S}$$

The output impedance Z_o is the reciprocal of the above expression

$$Z_o = \frac{1}{h_{oe} - (h_{fe}h_{re}/h_{ie} + R_S)} \tag{9.15}$$

When the transistor is driven from a high-impedance source R_S, then Z_o can be approximated as

$$Z_o \approx \frac{1}{h_{oe}} \tag{9.16}$$

9.2.5 Power Gain, G_p

The power gain can be written as

$$G_P = G_v G_i \tag{9.17}$$

Using the expression for the current gain $G_i \approx -h_{fe}$ from Equation 9.5, and the voltage gain $G_v = -h_{fe}Z_L/h_{ie}$, Equation 9.12, an approximation of the power can be obtained as

$$G_P = G_v G_i = \frac{h_{fe}^2 Z_L}{h_{ie}} \tag{9.18}$$

The h parameters provide the main characteristics of a transistor. They are easy to measure and are quoted by manufacturers.

9.3 SMALL-SIGNAL PRACTICAL CE AMPLIFIER UNDER h-PARAMETER MODEL

For small signal the h-parameter h_{re} can almost always be neglected while to neglect the parameter h_{oe} will depend on the value of Z_L.

We will use the h-parameter model neglecting h_{re} to represent a CE transistor as shown in Figure 9.2. This model will be used in the following section to analyze amplifiers circuits, that is, transistor plus bias resistor plus coupling capacitors.

9.3.1 FIXED BIAS AMPLIFIER

Figure 9.3 shows a fixed bias amplifier including the biasing and capacitor C_{c1} to block the dc voltage at the input and C_{c2} blocking dc at the output. C_{c1} permits the ac signal trough and C_{c2} allows ac signal out, which is collected across the load resistor R_L.

The input ac signal voltage causes the base voltage to vary above and below its dc bias level. The variation in base current creates a big variation in the collector current since the current gain transistor is usually high. This variation of the collector current creates a variation of the voltage across the resistor R_c and then the voltage between collector–emitter, V_{CE}, also varies to keep the same V_{cc}.

Figure 9.4 shows an ac equivalent circuit for this amplifier using the h-parameter model for the transistor. It is assumed that at the working frequencies, medium

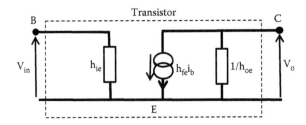

FIGURE 9.2 Practical h parameter model for a transistor in CE configuration.

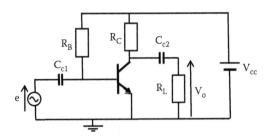

FIGURE 9.3 Circuit for fixed bias amplifier.

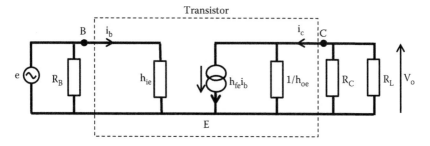

FIGURE 9.4   Fixed biased amplifier ac equivalent circuit at medium frequencies.

frequencies, the impedance for the capacitors is very small that can be neglected. Also note that the impedance of the dc source can be considered a short circuit due to its low impedance to ac.

Under medium frequencies, the impedance of coupling capacitor C_{c2} is assumed to be zero, then the resistor R_L will appear in the ac equivalent circuit as connected between collector and ground. The resistor at the collector R_C also appears to be connected between collector and ground as the impedance of the V_{cc} source under ac is considered zero. Then, in the ac equivalent circuit of this amplifier R_L and R_C appear in parallel between collector and emitter (ground). At the input the bias resistor at the base R_B (under ac medium frequencies equivalent circuit) will appear connected between base and emitter (ground).

The ac analysis of the fix bias amplifier is the same for an fixed bias amplifier circuit as the equivalent circuit will have identical form. The value of the components may be different but the two equivalent circuits can be reduced to have the same format. The analysis done for the auto bias amplifier will be valid for the fixed bias amplifier using the appropriate value of resistors for these amplifiers. In the next section, we will analyze the auto bias amplifier. This analysis is also valid for the fixed bias amplifier with the appropriate values of components.

9.3.2   Auto Bias Amplifier

Figure 9.5 shows an auto bias amplifier including the biasing and coupling capacitor C_{c1} and C_{c2}. The resistor R_E is required by the dc bias circuit but on ac R_E reduces the gain, then C_E is used to bypass R_E at ac.

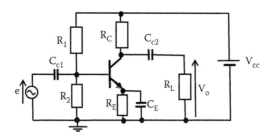

FIGURE 9.5   Auto biased amplifier circuit.

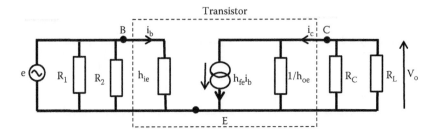

FIGURE 9.6 Auto biased amplifier ac equivalent circuit at medium frequencies.

Figure 9.6 shows the equivalent circuit for the auto bias amplifier using the h-parameter model for the transistor under ac. It is assumed, as before, that at medium frequencies, the impedance for the capacitors is very small and can be neglected. Note that the R_E is not included in the ac equivalent circuit as it is by-passed by the low impedance of capacitor C_E.

In this equivalent circuit, if the parallel resistors R_1 and R_2 are reduced to one resistor then the auto biased and the fixed biased ac equivalent circuits have the same circuit format. In both cases the bias resistor (R_B in fixed bias case or the parallel combination of R_1 and R_2 in the auto bias case) is always much larger than h_{ie}, therefore they can be neglected (if they are not neglected, they can be replaced by a resistor that is the combination in parallel of biasing resistors and h_{ie} at the input). Figure 9.7 shows the simplified version of the amplifier equivalent circuit, R'_C is the combined resistance of R_C and $1/h_{oe}$ in parallel as

$$\frac{1}{R'_C} = \frac{1}{R_C} + \frac{1}{1/h_{oe}} = \frac{1}{R_C} + h_{oe} = \frac{1 + R_C h_{oe}}{R_C}$$

So,

$$R'_C = \frac{R_C}{1 + R_C h_{oe}} \tag{9.19}$$

where R'_C is the combined resistance of R_C and $1/h_{oe}$ in parallel.

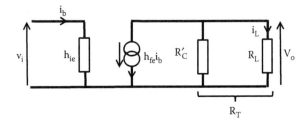

FIGURE 9.7 Simplified equivalent circuit auto bias amplifier h parameter.

This circuit can be simplified further by combining R'_C and R_L in parallel into one resistor R_T as

$$\frac{1}{R_T} = \frac{1}{R_L} + \frac{1}{R'_C} = \frac{1}{R_L} + \frac{1+R_C h_{oe}}{R_C} = \frac{R_C + R_L(1+R_C h_{oe})}{R_L R_C}$$

and

$$R_T = \frac{R_L R_C}{R_C + R_L(1+R_C h_{oe})} \tag{9.20}$$

9.3.2.1 Voltage Gain

In Figure 9.7, considering the input circuit, the input voltage v_i can be written as

$$v_i = h_{ie} i_b \tag{9.21}$$

and from the output circuit the output voltage v_o can be written as

$$v_o = -h_{fe} i_b R_T \tag{9.22}$$

and the voltage gain at medium frequency, G_v can be written as

$$G_v = \frac{v_o}{v_i} = -\frac{h_{fe}}{h_{ie}} R_T \tag{9.23}$$

The voltage gain at medium frequencies, G_v, can also be expressed in terms of h_{oe}, R_L, and R_C by replacing the expression for R_T.

9.3.2.2 Current Gain

In the ac equivalent circuit for the auto bias amplifiers shown in Figure 9.7, the output current i_L can be deduced using Kirchhoff's laws or the current divider rule at the output of this equivalent circuit as

$$i_L = -\frac{R'_C}{R'_C + R_L}(h_{fe} i_b) \tag{9.24}$$

The negative sign is due to the opposite direction of the currents. The current gain can be written as

$$G_i = \frac{i_L}{i_b} = -\frac{h_{fe} R'_C}{R'_C + R_L} \tag{9.25}$$

The current gain at medium frequencies, G_i, can also be expressed in terms of h_{oe} and R_C by replacing the expression for R_C'.

9.3.2.3   Power Gain

As described earlier, the power gain can be expressed as the product of the voltage and currents gains as

$$G_P = G_i G_v$$

or the power gain can be deduced from the ratio of the output power, $P_o = v_o^2/R_L$ to the input power $P_i = v_i^2/h_{ie}$ as

$$G_P = \frac{P_o}{P_i} = \frac{v_o^2/R_L}{v_i^2/h_{ie}} = G_v^2 \frac{h_{ie}}{R_L} \tag{9.26}$$

It is also possible to obtain an expression for the power gain in terms of the current gain.

EXAMPLE 9.1

For the amplifier shown in Figure 9.8, draw a mid-band frequency equivalent circuit using h parameters. Calculate the ac current, voltage, and power gains of the amplifier. The transistor has $h_{ie} = 2\ k\Omega$, $h_{fe} = 45$, $h_{oe} = 30\ \mu S$, and $h_{re} = 0$.

SOLUTION

At mid-band frequency the coupling capacitors can be assumed to have zero impedance. The ac mid-band equivalent circuit is shown in Figure 9.9.
The value of R_C' is the parallel combination of R_C and $1/h_{oe}$, then

$$R_C' = \frac{R_C}{1 + R_C h_{oe}} = \frac{1 \times 10^3}{1 + 1 \times 10^3 \times 30 \times 10^{-6}} = 970.87\ \Omega$$

and the current gain is

$$G_i = -\frac{h_{fe} R_C'}{R_C' + R_L} = -\frac{45 \times 970.87}{970.87 + 10 \times 10^3} = -3.98$$

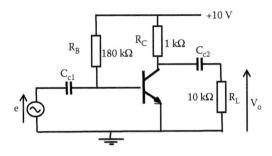

FIGURE 9.8   Fixed bias amplifier circuit for Example 9.1.

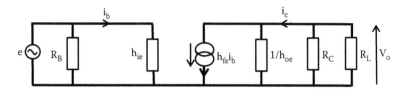

FIGURE 9.9   Equivalent circuit for amplifier of Example 9.1 using h parameters.

The voltage gain is given by

$$G_v = -\frac{h_{fe}R_T}{h_{ie}}$$

The value of R_T is the parallel combination of R_C' and R_L.

$$R_T = \frac{R_L R_C}{R_C + R_L(1 + R_C h_{oe})} = \frac{10 \times 10^3 \times 1 \times 10^3}{1 \times 10^3 + 10 \times 10^3 (1 + 1 \times 10^3 \times 30 \times 10^{-6})}$$

$$R_T = 884.96 \ \Omega$$

and the voltage gain G_v is

$$G_v = -\frac{h_{fe}R_T}{h_{ie}} = -\frac{45 \times 884.96}{2 \times 10^3} = -19.91$$

For a more accurate calculation, the influence of R_B can be included by replacing h_{ie} with the parallel combination of h_{ie} and R_B, which provide a value close to h_{ie}.

The power gain is the product of the voltage gain and the current gain as

$$G_P = G_i G_v = 3.98 \times 19.91 = 79.24$$

EXAMPLE 9.2

For the amplifier shown in Figure 9.10, draw a mid-band frequency equivalent circuit using h parameters and calculate the ac current gain, voltage gain, power gain, and the output voltage of the amplifier. The transistor has $h_{ie} = 2 \ k\Omega$, $h_{fe} = 45$, $h_{oe} = 30 \ \mu S$, and $h_{re} = 0$.

SOLUTION

At mid-band frequency, the coupling capacitors can be assumed to have zero impedance. The ac mid-band equivalent circuit is shown in Figure 9.11.

The value of R_C' is the parallel combination of R_C and $1/h_{oe}$, then

$$R_C' = \frac{R_C}{1 + R_C h_{oe}} = \frac{2 \times 10^3}{1 + 2 \times 10^3 \times 30 \times 10^{-6}} = 1886.79 \ \Omega$$

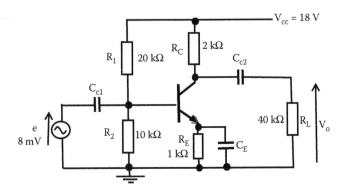

FIGURE 9.10 Auto bias amplifier circuit for Example 9.2.

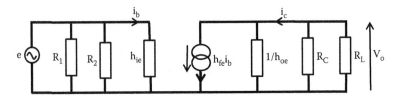

FIGURE 9.11 Equivalent circuit for amplifier of Example 9.2 using h parameters.

and the current gain is

$$G_i = -\frac{h_{fe}R'_C}{R'_C + R_L} = -\frac{45 \times 1886.79}{1886.79 + 40 \times 10^3} = -2.03$$

The voltage gain is given by

$$G_v = -\frac{h_{fe}R_T}{h_{ie}}$$

The value of R_T is the parallel combination of R'_C and R_L.

$$R_T = \frac{R_L R_C}{R_C + R_L(1 + R_C h_{oe})} = \frac{40 \times 10^3 \times 2 \times 10^3}{2 \times 10^3 + 40 \times 10^3(1 + 2 \times 10^3 \times 30 \times 10^{-6})}$$

$$R_T = 1801.80 \ \Omega$$

and the voltage gain, G_v is

$$G_v = -\frac{h_{fe}R_T}{h_{ie}} = -\frac{45 \times 1801.80}{2 \times 10^3} = -40.54$$

For a more accurate calculation, the influence of R_1 and R_2 can be included by replacing h_{ie} with the parallel combination of h_{ie}, R_1, and R_2 that provide a value close to h_{ie}.

The power gain is the product of the voltage gain and the current gain as given below:

$$G_P = G_I G_V = 2.03 \times 40.54 = 82.30$$

and the output voltage, V_o, can be obtained as

$$V_o = G_V \times e = 40.54 \times 8 \times 10^{-3} = 324.32 \, mV$$

9.4 ANALYSIS OF TRANSISTOR AMPLIFIERS USING Y PARAMETERS

Figure 9.12 shows an equivalent circuit for the auto bias amplifier shown in Figure 9.5, using the Y parameters under medium frequencies. As before the Y_{re} parameter has been neglected due to its small value and also biasing resistors R_1 and R_2 in parallel are much larger than Y_{ie} in parallel.

The collector resistance and the load resistances are represented by its admittance Y_C and Y_L, respectively. The three admittances in parallel, Y_{oe}, Y_C, and Y_L, can be reduced to one admittance, Y_T, by adding the three admittances in parallel as

$$Y_T = Y_{oe} + Y_C + Y_L$$

Figure 9.13 shows the equivalent circuit with Y_T.

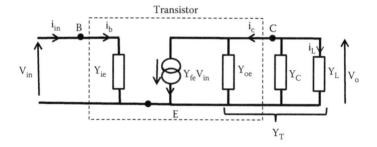

FIGURE 9.12 Equivalent circuit auto bias amplifier using Y parameters.

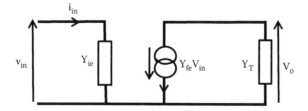

FIGURE 9.13 Simplified equivalent circuit auto bias amplifier Y parameters.

9.4.1 Voltage Gain, G_v

Then, from Figure 9.13 we can see that output voltage can be expressed in terms of the current passing through the total admittance Y_T as

$$V_o = -\frac{Y_{fe}V_{in}}{Y_T} \qquad (9.27)$$

The output voltage is in terms of the input voltage in Equation 9.27. Dividing Equation 9.27 by the input voltage, we can deduce an expression for the voltage gain as

$$G_v = \frac{V_o}{V_{in}} = \frac{-Y_{fe}V_{in}/Y_T}{V_{in}} = -\frac{Y_{fe}}{Y_T} \qquad (9.28)$$

Y_T includes the term of the admittance of collector and the load resistance.

9.4.2 Current Gain, G_i

From the circuit shown in Figure 9.12 the output current, i_L, can be written as

$$i_L = V_o Y_L \qquad (9.29)$$

and the input current, i_{in}, can be written as

$$i_{in} = V_{in} Y_{ie} \qquad (9.30)$$

and the current gain can be obtained as the ratio of the output current to the input current

$$G_i = \frac{i_L}{i_{in}} = \frac{V_o Y_L}{V_{in} Y_{ie}} = \frac{G_v Y_L}{Y_{ie}} \qquad (9.31)$$

9.4.3 Power Gain, G_p

As described earlier, the power gain can be expressed as the product of the voltage and current gains as given below:

$$G_p = G_i G_v$$

This can be written as a function of the voltage gain or the current gain

$$G_p = G_iG_v = G_i^2 \frac{Y_{ie}}{Y_L} \tag{9.32}$$

or as function of the voltage gain

$$G_p = G_iG_v = G_v^2 \frac{Y_L}{Y_{ie}} \tag{9.33}$$

EXAMPLE 9.3

For the amplifier shown in Figure 9.14, draw a mid-band frequency equivalent circuit using Y parameters. Calculate the ac voltage, current, and power gains of the amplifier. The transistor has $Y_{ie} = 5 \times 10^{-4}$ S, $Y_{fe} = 3$ mA/V, and $Y_{oe} = 30$ μS.

SOLUTION

At mid-band frequency the coupling capacitors can be assumed to have zero impedance. The ac mid-band equivalent circuit is shown in Figure 9.15.

The admittance Y_{oe}, Y_C, and Y_L can be reduced to one equivalent admittance, Y_T, as given below:

$$Y_T = Y_{oe} + Y_C + Y_L = 30 \times 10^{-6} + \frac{1}{8 \times 10^3} + \frac{1}{20 \times 10^3} = 2.05 \times 10^{-4} \text{ S}$$

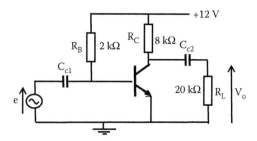

FIGURE 9.14 Fixed bias amplifier circuit for Example 9.3.

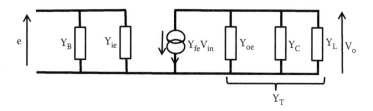

FIGURE 9.15 Equivalent circuit for amplifier of Example 9.3 using Y parameters.

and the voltage gain for this amplifier is given by

$$G_v = \frac{V_o}{V_{in}} = -\frac{Y_{fe}}{Y_T} = -\frac{3 \times 10^{-3}}{2.05 \times 10^{-4}} = -14.63$$

and the current gain is given by

$$G_i = \frac{G_v Y_L}{Y_{ie}} = \frac{14.63 \times 5 \times 10^{-5}}{5 \times 10^{-4}} = 1.463$$

For a more accurate calculation, the influence of R_B can be included by adding its admittance $1/R_B$ and Y_{ie} in parallel instead of Y_{ie}.

The power gain is the product of the voltage gain and the current gain as given below:

$$G_P = G_i G_v = 1.463 \times 14.63 = 21.40$$

9.5   ANALYSIS OF TRANSISTOR AMPLIFIERS USING THE GENERAL THREE-PARAMETER MODEL

The general equivalent model for an amplifier can be represented as the circuit shown in Figure 9.16. This circuit can be an equivalent circuit for a fixed bias or auto bias amplifier. In this equivalent circuit R_{in} can be the combination of the input and biasing resistors; and R_L can also include the resistor at the collector plus load resistor. In both cases we can arrive at a circuit as shown in Figure 9.16, where R_{in} is the input resistance of the amplifier, R_o the output resistance, and A_v is the open voltage gain of the amplifier.

9.5.1   VOLTAGE GAIN, G_v

The voltage gain can be obtained by examining the output circuit of Figure 9.16. The output circuit loop contains expression for the input and output voltages. Applying

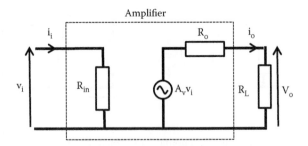

FIGURE 9.16   Equivalent circuit of an amplifier using the general three-parameter model.

KVL to the output loop formed by the voltage source $A_v v_i$, voltage across R_o, and voltage across R_L we can write

$$A_v v_i = i_o(R_o + R_L)$$

Rearranging this equation we can obtain an expression for the input voltage as

$$v_i = \frac{i_o(R_o + R_L)}{A_v}$$

Also, we can write an expression for the output voltage in terms of the output current i_o and the load resistor R_L as

$$v_o = i_o R_L$$

Then, the overall gain G_v is given as the ratio of the output voltage to the input voltage as given below:

$$G_v = \frac{v_o}{v_i} = \frac{i_o R_L}{i_o(R_o + R_L)/A_v}$$

and the voltage gain can be written as

$$G_v = \frac{A_v R_L}{(R_o + R_L)} \tag{9.34}$$

9.5.2 Current Gain, G_i

As before, KVL at the output closed circuit of Figure 9.16 gives

$$A_v v_i = i_o(R_o + R_L)$$

Then, the output current is

$$i_o = \frac{A_v v_i}{(R_o + R_L)}$$

and from the input network of Figure 9.16 we can write an expression for the input voltage as

$$v_{in} = i_i R_{in}$$

or

$$i_i = \frac{v_{in}}{R_{in}}$$

The ratio of the output current to the input current will provide the current gain as

$$G_i = \frac{i_o}{i_i} = \frac{A_v v_i / (R_o + R_L)}{v_i / R_{in}}$$

and finally, the current gain is

$$G_i = \frac{i_o}{i_i} = \frac{A_v R_{in}}{(R_o + R_L)} \tag{9.35}$$

Now if the R_L is equal to zero, that is, short circuit at the output, then we can obtain an expression for the *short circuit current gain*, A_i as

$$A_i = \frac{i_o}{i_i} = \frac{A_v R_{in}}{R_o} \tag{9.36}$$

or

$$A_v = \frac{A_i R_o}{R_{in}} \tag{9.37}$$

Replacing A_v in the G_i expression we obtain the current gain in terms of A_i as

$$G_i = \frac{A_i R_o}{(R_o + R_L)} \tag{9.38}$$

9.5.3 Power Gain, G_p

The power gain can be expressed as the product of the voltage and current gains as given below:

$$G_p = G_i G_v = \frac{A_v R_{in}}{(R_o + R_L)} \frac{A_v R_L}{(R_o + R_L)} = \frac{A_v^2 R_L R_{in}}{(R_o + R_L)^2} \tag{9.39}$$

Multiplying and dividing by R_L allows the power gain to be arranged as

$$G_p = \left[\frac{A_v R_{in}}{(R_o + R_L)} \right]^2 \frac{R_L}{R_{in}} = G_i^2 \frac{R_L}{R_{in}} \tag{9.40}$$

Then the power gain can be expressed in term of the voltage gain G_v as

$$G_p = G_v^2 \frac{R_{in}}{R_L}$$

(9.41)

EXAMPLE 9.4

The equivalent circuit of an amplifier using the general three-parameter model is shown in Figure 9.17. If the overall voltage gain, G_v is equal to 4500 and the overall current gain, G_i is equal to 9000, calculate the values of R_{in} and R_o for this amplifier.

SOLUTION

The equivalent circuit of Figure 9.17 indicate a value for the open voltage gain, A_v, equal to 5000. The overall voltage gain is

$$G_v = \frac{A_v R_L}{(R_o + R_L)} = \frac{5000 \times 500}{R_o + 500} = 4500$$

Solving this equation we find

$$R_o = 55.56 \ \Omega$$

Also, the overall current gain G_i is given by

$$G_i = \frac{A_v R_{in}}{(R_o + R_L)} = \frac{5000 \times R_{in}}{55.56 + 500} = 9000$$

and

$$R_{in} = 1 \ k\Omega$$

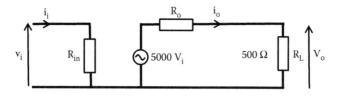

FIGURE 9.17 Amplifier equivalent circuit using a general three-parameter model for Example 9.4.

9.6 ANALYSIS OF TRANSISTOR AMPLIFIERS USING THE T-MODEL TWO-PARAMETERS

9.6.1 VOLTAGE GAIN, G_v

9.6.1.1 CE Amplifier Using T Parameters

The ac equivalent circuit for a CE amplifier using the T parameter model for the transistor is shown in Figure 9.18.

The biasing resistances R_1 and R_2 are in practice much larger than r_e and they can be omitted from the ac equivalent circuit as almost all input current will go to the base of the transistor. Then, the equivalent circuit for the amplifier can be simplified as shown in Figure 9.19.

This circuit is valid for fixed biasing as well as auto biasing. In the case of fixed bias R_B can be the combination in parallel of R_1 and R_2. Also R_c can be the collector resistor or it can be combined with a load resistor.

The input voltage is equal to the voltage across r_e. Applying KCL we can obtain the current flowing through r_e as the sum of currents i_b and βi_b, the input voltage v_i can be expressed as

$$v_i = (i_b + \beta i_b)r_e \tag{9.42}$$

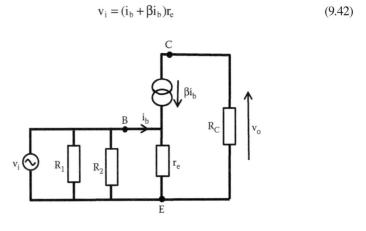

FIGURE 9.18 Ac equivalent circuit for an auto bias amplifier using the simplified T-model.

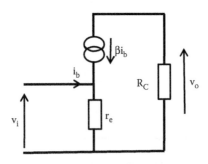

FIGURE 9.19 Simplified ac T-model equivalent circuit for an auto bias amplifier.

and the output voltage, v_o, is the voltage across R_C and it is given by

$$v_o = -\beta i_b R_C \qquad (9.43)$$

The negative sign follows the convention of the direction of current and voltage; then the gain voltage G_v is given, ratio of output voltage to input voltage, as

$$G_v = \frac{v_o}{v_i} = \frac{-\beta i_b R_C}{(i_b + \beta i_b)r_e} = \frac{-\beta R_C}{(1+\beta)r_e} \qquad (9.44)$$

In practice, β is much larger than 1 then the voltage gain can be approximated to

$$G_v \approx -\frac{R_C}{r_e} \qquad (9.45)$$

Therefore, the gain is dependent on the dc current as r_e depends on I_E.

9.6.2 Current Gain, G_i

The current gain G_i can be easily deduced as the input and output currents are indicated in the equivalent circuit.

Input current $i_i = i_b$ and output current $i_o = \beta i_b$, then the current gain is

$$G_i = \frac{i_o}{i_i} = \frac{i_c}{i_b} = \frac{\beta i_b}{i_b} = \beta \qquad (9.46)$$

9.6.3 Input Impedance, Z_i

The input impedance for a CE amplifier can be obtained from the equivalent circuit including the biasing resistances R_1 and R_2 shown in Figure 9.20.

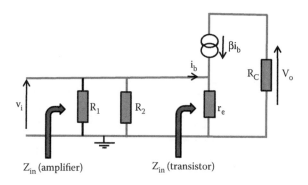

FIGURE 9.20 Auto bias amplifier equivalent circuit input impedance model.

From the equivalent circuit shown in Figure 9.20 the input voltage v_i can be obtained as

$$v_i = r_e(i_b + \beta i_b) = i_b r_e(1 + \beta)$$

The input impedance at the base of the transistor Z_{in} (transistor) is given by

$$Z_{in}(\text{transistor}) = \frac{v_i}{i_b}$$

Replacing the value of v_i into equation for Z_{in} (transistor) we can obtain the input impedance at the base of the transistor as

$$Z_{in}(\text{transistor}) = \frac{v_i}{i_b} = \frac{(1+\beta)r_e i_b}{i_b} = (1+\beta)r_e \qquad (9.47)$$

The amplifier input impedance of the total amplifier, Z_{in} (amplifier), includes R_1 and R_2 and can be calculated by adding the three impedances in parallel as follows:

$$\frac{1}{Z_{in}(\text{amplifier})} = \frac{1}{R_1} + \frac{1}{R_2} + \frac{1}{(1+\beta)r} \qquad (9.48)$$

The reciprocal of this expression will provide the total input impedance of the amplifier.

9.6.4 Output Impedance, Z_o

The output impedance for a CE amplifier can be deduced from the simplified T-parameter equivalent circuit shown in Figure 9.21.

The small-signal impedance of an ideal current generator is infinite so it behaves as an open circuit. To calculate Z_o (transistor) we have to include the impedance of

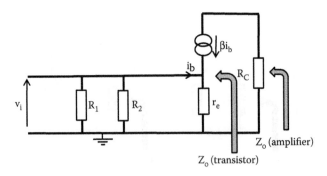

FIGURE 9.21 Auto bias amplifier equivalent circuit input impedance T model.

the current source βi_b, then ideally the impedance Z_o (transistor) is infinitive and behaves as an open circuit. This means that R_C is connected in parallel with an open circuit. Therefore, the output impedance Z_o (amplifier) is equal to R_C.

$$Z_o(\text{amplifier}) = R_C \tag{9.49}$$

9.6.4.1 Effect of a Load Resistance

The output from an amplifier would be connected to a load that can be the input resistance of the next stage. In the ac equivalent circuit this is represented by a resistance R_L that appears in parallel with the collector resistance R_C as indicated in Figure 9.22.

The voltage gain expression obtained in Equation 9.45 can be modified to replace the resistor R_c with the parallel combination of R_C and R_L as given below

$$G_v = \frac{R_C R_L / (R_C + R_L)}{r_e} = \frac{R_C R_L}{(R_C + R_L) r_e} \tag{9.50}$$

In the case of the current gain G_i, the expression given in Equation 9.46 has to be modified because, in this case, the output current is the current passing through the load, i_{RL}.

The current i_{RL} can be obtained using the current divider rule as follows:

$$i_{RL} = \frac{R_C}{R_C + R_L} \beta i_b \tag{9.51}$$

Then, the overall current gain is

$$G_i = \frac{i_{RL}}{i_b} = \frac{R_C \beta i_b / (R_C + R_L)}{i_b} = \frac{R_C \beta}{R_C + R_L} \tag{9.52}$$

9.6.4.2 Effect of the Source Resistance

In practice the amplifier will be connected to a source that is not ideal or the previous stage will have an output resistance. Figure 9.23 shows the amplifier including a source resistance R_S.

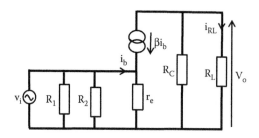

FIGURE 9.22 Ac auto bias amplifier equivalent circuit including load resistor R_L.

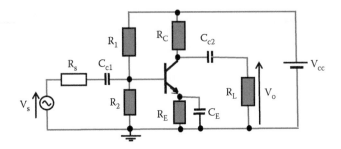

FIGURE 9.23 Ac auto bias amplifier equivalent circuit including source resistor R_S.

When voltage gain was calculated the series resistance R_S was assumed equal to zero. R_S connected in series will reduce the input voltage by the potential drop across R_S. To calculate the new gain including the source resistance we can represent the amplifier by its input impedance Z_{in} (amplifier) as expressed in Equation 9.48. This forms a potential divider as shown in Figure 9.24.

From the circuit shown in Figure 9.24 we can write an equation for v_{in} as

$$v_{in} = \left[\frac{Z_{in}(\text{amplifier})}{[R_s + Z_{in}(\text{amplifier})]} \right] v_s$$

but

$$G_v = \frac{v_o}{v_{in}} \Rightarrow \text{then} \Rightarrow v_{in} = \frac{v_o}{G_v}$$

Then, replacing the expression for v_{in} the voltage overall gain for the amplifier from the voltage source to the load, G_v(overall), is given by

$$G_v(\text{overall}) = \frac{v_o}{v_s} = \left[\frac{Z_{in}(\text{amplifier})}{(R_s + Z_{in}(\text{amplifier}))} \right] G_v \qquad (9.53)$$

From this equation we can see that to minimize the effect of R_s in the overall voltage gain it is desired to have an amplifier with large input resistance.

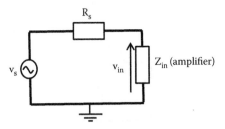

FIGURE 9.24 Potential divider to calculate the voltage gain with a source resistor.

9.6.5   POWER GAIN, G_p

The power gain can be expressed as the product of the voltage and current gains as given below:

$$G_p = G_i G_v$$

Excluding the effect of the load and source resistors the power gain can be approximated to

$$G_p = G_i G_v \approx \frac{\beta R_C}{r_e} \tag{9.54}$$

Therefore, the power gain is dependent on the dc current as r_e depends on I_E.

EXAMPLE 9.5

For the amplifier shown in Figure 9.25, draw a mid-band frequency equivalent circuit using the T-model two parameters and calculate the ac voltage gain, G_v of the amplifier. The transistor has $\beta = 200$ and a collector current $I_C = 8$ mA.

SOLUTION

The parameter r_e can be calculated as

$$r_e = \frac{KT/q}{I_E}$$

As indicated in Chapter 8, assuming room temperature $T = 300°K$ the parameter r_e can be approximated to

$$r_e \approx \frac{25 \text{ mV}}{I_E} = \frac{25 \times 10^{-3}}{8 \times 10^{-3}} = 3.125 \ \Omega$$

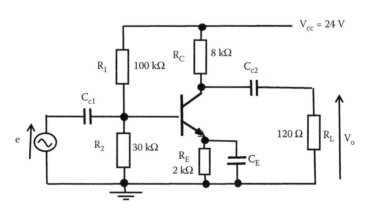

FIGURE 9.25   Auto bias amplifier circuit for Example 9.5.

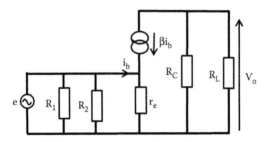

FIGURE 9.26 Equivalent circuit for amplifier of Example 9.5 using a T-model two parameters for the transistor.

At mid-band frequency the coupling capacitors can be assumed to have zero impedance. Then, the ac mid-band equivalent circuit, using the T-model two parameters for the transistor is shown in Figure 9.26.

The voltage gain, G_v, for the amplifier is given by

$$G_v \approx -\frac{R'_C}{r_e}$$

where R'_C in this case is the parallel combination of R_C and R_L.

Then,

$$R'_C = \frac{R_C R_L}{R_C + R_L} = \frac{8 \times 10^3 \times 120}{8 \times 10^3 + 120} = 118.23$$

and the voltage gain is

$$G_v = -\frac{R'_C}{r_e} = \frac{118.23}{3.125} = 37.83$$

9.7 KEY POINTS

- Amplification is a fundamental part of most electronic systems.
- Equivalent circuits allow us to model the behavior of an amplifier and simplify circuit analysis.
- There are various models that can predict the behavior of a transistor. Depending on the application, some models make it easier to analyze certain behaviors of amplifiers or any electronic system.
- The load resistance and the source resistance affect the gain in an amplifier.
- Power gain is the ratio of the power delivered to the load to that absorbed at the input.

10 Amplifiers Frequency Response

10.1 INTRODUCTION

The components in an amplifier contain capacitors and may contain inductors as well. Capacitors and inductors change their impedance as the frequency changes; this will affect the gain of an amplifier.

The response of an amplifier, containing transistors, to frequency change will in general behave as indicated in Figure 10.1. Voltage gains in modern electronic amplifiers are often very high; it is often convenient to express the gain in logarithmic expressions. This is usually done in decibels, dB. Also because the frequency response is indicated over a large range of frequencies the scale used to represent the frequency is given in a logarithmic scale as indicated in Figure 10.1.

The voltage gain at low and high frequencies is usually small compared to medium frequencies. There are two points where the power gain is reduced to half the gain at medium frequencies. These particular frequencies are the low-frequency cutoff, f_L and the high-frequency cutoff, f_H. Between f_L and f_H is the mid-frequency range. The bandwidth (BW) represents the range of frequencies between f_L and f_H providing the range of frequencies over which an amplifier can be used.

10.2 HALF-POWER GAIN, CONCEPT OF 3 dB

Decibel notation of a power gain is expressed as

$$\text{Power gain (dB)} = 10\log\left(\frac{P_o}{P_i}\right) \tag{10.1}$$

where P_o and P_i are the output and input power, respectively.

We can calculate the number of decibels when the power gain is reduced to half as

$$10\log\left(\frac{P_o}{P_i}\right) = 10\log\left(\frac{1}{2}\right) = -10\times0.3 = -3\,\text{dB} \tag{10.2}$$

Therefore, we can specify the cutoff frequency at frequency point where the gain is reduced by 3 dB. Now power gain is related to voltage gain as squared of voltage; it is possible to express decibels in term of voltages as

$$\text{Power gain (dB)} = 10\log\left(\frac{P_o}{P_i}\right) = 10\log\left(\frac{V_o^2/R}{V_i^2/R}\right) = 10\log\left(\frac{V_o}{V_i}\right)^2 = 20\log\left(\frac{V_o}{V_i}\right) \tag{10.3}$$

179

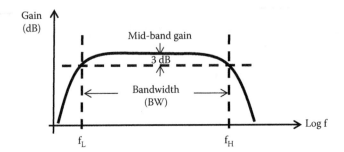

FIGURE 10.1 General frequency response of an amplifier.

Resistors at the output and input are taken as equal; this expression is used even if these resistors are different.

Then the decibel notation can be used in terms of voltages as

$$1\,\text{dB} = 20\log\left(\frac{V_o}{V_i}\right) \tag{10.4}$$

Then, f_L and f_H can be found at the frequency where the voltage gain is reduced by 3 dB from the mid-frequencies as indicated in Figure 10.1. The BW of an amplifier is the frequency region between f_L and f_H and is usually the frequency range where the amplifier is used.

The frequency cutoff can also be specified in terms of voltage gain. For example, if we consider the low-frequency cutoff we can find the ratio of voltage gain at low- and mid-frequency in terms of voltage gains. From absolute values the difference between medium and low frequency will be 3 dB at cutoff, and then we can write

$$20\log\left|\frac{Gv_{mf}}{Gv_{LF}}\right| = 10\log\left|\frac{Gv_{mf}}{Gv_{LF}}\right|^2 = 3$$

or

$$\log\left|\frac{Gv_{mf}}{Gv_{LF}}\right|^2 = 0.3$$

and

$$\left|\frac{Gv_{mf}}{Gv_{LF}}\right|^2 = 2$$

Inverting

$$\left|\frac{Gv_{LF}}{Gv_{mf}}\right|^2 = \frac{1}{2}$$

Finally,

$$\frac{Gv_{LF}}{Gv_{mf}} = \sqrt{\frac{1}{2}} = \frac{1}{\sqrt{2}} \tag{10.5}$$

Then, at cutoff the power gain has fallen by half or the voltage gain has dropped by a factor of $1/\sqrt{2}$.

EXAMPLE 10.1

The voltage gain of an amplifier is 60 dB. This amplifier is connected to a load of 10 kΩ. Calculate the output voltage and the power dissipated by the load when the input voltage is 5 mV.

SOLUTION

The voltage gain of the amplifier in decibels is

$$20\log\left(\frac{V_o}{V_i}\right) = 60 \text{ dB}$$

Then,

$$\log\left(\frac{V_o}{V_i}\right) = \frac{60}{20} = 3 \text{ dB}$$

or

$$\left(\frac{V_o}{V_i}\right) = 10^3 = 1000$$

and

$$V_o = 1000 \times V_i = 1000 \times 5 \times 10^{-3} = 5 \text{ V}$$

The power dissipated by the load is

$$P_o = \frac{V_o^2}{R_L} = \frac{5^2}{10 \times 10^3} = 2.5 \text{ mW}$$

10.3   CE AMPLIFIER AT LOW FREQUENCY

The coupling capacitors that are in series will reduce the gain at low frequency, because their impedance increases as the frequency decreases. This creates a higher voltage drop across the capacitor in series as a consequence of a loss in voltage gain.

10.3.1   Voltage Gain at Low Frequency

In particular, the output capacitor in series will reduce the output signal. Figure 10.2a shows a common emitter auto bias amplifier used in Chapter 9. Figure 10.2b shows an equivalent circuit for a CE amplifier including the output coupling capacitor connected to the load resistor R_L. Figure 10.2b can be used to obtain a voltage gain at low frequencies, Gv_{LF}, where R_C' is the combined resistance of R_C and the parameter $1/h_{oe}$ of the transistor in parallel.

Using the current divider rule the output current, i_L, can be obtained as

$$i_L = -\frac{R_C'}{R_C' + (R_L + 1/j\omega C_{c2})} h_{fe}i_b \tag{10.6}$$

and the output voltage, v_o, is equal to the current i_L multiplied by R_L and can be written as

$$v_O = R_L i_L = -R_L \frac{R_C'}{R_C' + R_L + 1/j\omega C_{c2}} h_{fe}i_b \tag{10.7}$$

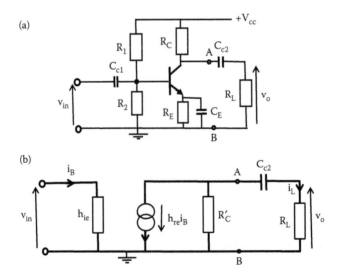

FIGURE 10.2   (a) CE amplifier and (b) an equivalent circuit of this amplifier at low frequency.

and v_{in} the input voltage is

$$v_{in} = i_b h_{ie} \qquad (10.8)$$

Then, with the input and output voltages we can obtain an expression for the voltage gain at low frequency Gv_{LF} as

$$Gv_{LF} = \frac{v_O}{v_{in}} = \frac{-R_L(R_C'/R_C' + R_L + 1/j\omega C_{c2})h_{fe}i_b}{i_b h_{ie}}$$

Dividing numerator and denominator by $(R_C' + R_L)$ and rearranging Gv_{LF} expression can be written as

$$Gv_{LF} = -\frac{h_{fe}}{h_{ie}}\left[\frac{R_C'R_L/(R_C' + R_L)}{(R_C' + R_L)/(R_C' + R_L) + 1/j\omega C_{c2}(R_C' + R_L)}\right]$$

But $R_C'R_L/(R_C' + R_L)$ is R_T, the parallel combination of R_C' and R_L, then the voltage gain at low frequency Gv_{LF} can be written as

$$Gv_{LF} = -\frac{h_{fe}}{h_{ie}}\left[\frac{R_T}{1 + 1/j\omega C_{c2}(R_C' + R_L)}\right] \qquad (10.9)$$

which can also be expressed in terms of the voltage gain at medium frequencies $(Gv_{mf} = -h_{fe}R_T/h_{ie})$ as

$$Gv_{LF} = \left[\frac{Gv_{mf}}{1 + (1/j\omega C_{c2}(R_C' + R_L))}\right] \qquad (10.10)$$

We can see from this expression that as the frequency increases the low-frequency voltage gain increases. Above the low frequency cutoff the voltage gain will reach the mid-frequency voltage gain.

10.3.2 Low-Frequency Cutoff

Rearranging Equation 10.10 we can obtain a ratio of the voltage gain at low frequency to the voltage gain at medium frequencies as

$$\frac{Gv_{LF}}{Gv_{mf}} = \left[\frac{1}{1 + (1/j\omega C_{c2}(R_C' + R_L))}\right]$$

and the low frequency cutoff, f_L, can be obtained when the magnitude of the voltage gain is reduced by a $1/\sqrt{2}$ factor; then,

$$\left|\frac{Gv_{LF}}{Gv_{mf}}\right| = \frac{1}{\sqrt{2}} = \left|\frac{1}{1 + (1/j\omega C_{c2}(R_C' + R_L))}\right| = \frac{1}{1^2 + \{(1/j\omega C_{c2}(R_C' + R_L))\}^2} \qquad (10.11)$$

The parallel bar denotes absolute value. From Equation 10.11 we can deduce that

$$\frac{1}{\omega C_{c2}(R_C' + R_L)} = 1$$

and the low-*frequency cutoff*, f_L, can be expressed as

$$f_L = \frac{1}{2\pi C_{c2}(R_C' + R_L)} \tag{10.12}$$

10.3.3   PHASE CHANGE AT LOW FREQUENCY

Replacing f_L from Equation 10.12 into Gv_{LF}, Equation 10.10, we can calculate the phase change at the frequency cutoff. At f_L we have

$$Gv_{LF}(f = f_L) = \frac{Gv_{mf}}{1 + (1/j)} = \frac{Gv_{mf}}{1 - j} \tag{10.13}$$

Gv_{mf} can be written in polar form as

$$Gv_{mf} = \left|\frac{h_{fe}R_T}{h_{ie}}\right| \angle 180°$$

and the denominator of Equation 10.13 can be written in polar form as

$$(1 - j) = \sqrt{2}\angle -45°$$

Then, the voltage gain at the low-frequency cutoff can be expressed as

$$Gv_{LF} = \frac{Gv_{mf}}{1 - j} = \frac{\left|h_{fe}R_T/h_{ie}\right|\angle 180°}{\sqrt{2}\angle -45°} = \left|\frac{h_{fe}R_T}{\sqrt{2}h_{ie}}\right|\angle 180° + 45° = \left|\frac{h_{fe}R_T}{\sqrt{2}h_{ie}}\right|\angle 225° \tag{10.14}$$

Hence, the voltage gain at the low-frequency cutoff has a phase of 225°.

10.3.3.1   Effect of the Other Capacitance on the Low-Frequency Response

The capacitor C_{c1} is in series and it will affect the low-frequency response. If we consider the total input resistance of the amplifier R_{in}, then Figure 10.3 represents an equivalent circuit for the input of the amplifier. This equivalent circuit can be used to analyze the gain of the amplifier as the frequency is reduced.

As the frequency is reduced, the impedance of C_{c1} is increased and the voltage gain between v_{in} and v_i is reduced. The combination series of C_{c1} and R_{in} will produce a low-frequency cutoff, f_{o1}, equal to

$$f_{o1} = \frac{1}{2\pi C_{c1}R_{in}} \tag{10.15}$$

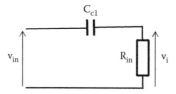

FIGURE 10.3   Circuit representing the first coupling capacitor response to low frequency.

The deduction of this equation and detailed analysis on this RC circuit is included in Chapters 16 and 20.

The effect of the emitter capacitor C_E in Figure 10.2 will also affect the frequency response as the frequency is reduced. In this case, some feedback signal to the input will occur. This is analyzed in Chapter 17.

In general, the higher low-frequency cutoff will have a larger influence on the low-frequency response.

10.4   CE AMPLIFIER AT HIGH FREQUENCY

As the frequency increases the impedance of any capacitor decreases and small capacitances in parallel, negligible at low frequency, can affect the gain at high frequency. In particular, any capacitance in parallel to input or output will be likely to reduce the gain. In a bipolar transistor the capacitance between the collector and base will have a large effect on the gain at high frequency. The collector–base capacitance, C_{cb}, has an influence on the voltages in the input and output of an amplifier and can be represented by the circuit shown in Figure 10.4, where G_v is the amplifier voltage gain, the multiplication of the C_{cb} by G_v at the input is known as the Miller effect. When the input voltage is applied from a source with series impedance then the capacitor G_vC_{cb} reduces the input voltage and the overall voltage gain. But if the voltage source has a very small or zero impedance, then the capacitor G_vC_{cb} only reduces the input impedance and the capacitor at the output, C_{cb}, reduces the voltage gain.

10.4.1   VOLTAGE GAIN AT HIGH FREQUENCY

Considering the case with no source impedance in series at the input and adding all parallel capacitances at the output to C_{cb} we can obtain an expression for the

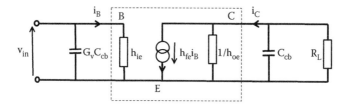

FIGURE 10.4   High-frequency CE amplifier equivalent circuit.

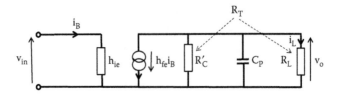

FIGURE 10.5   Simplified high-frequency CE amplifier equivalent circuit.

voltage gain at high frequency. Capacitances in parallel at the output will include stray capacitance of wiring and capacitance of components. We can combine all the capacitance in parallel at the output, including C_{cb}, into one capacitance, C_P, then the equivalent circuit at high frequency simplifies to the circuit shown in Figure 10.5.

R_L and R'_C can be combined in parallel into R_T as described earlier. Adding R_T in parallel to the impedance, the capacitor C_P, we can obtain an equivalent impedance, Z_{TT}, at the output.

$$Z_{TT} = \frac{R_T/j\omega C_P}{(R_T + (1/j\omega C_P))} = \frac{R_T}{1 + j\omega C_P R_T} \tag{10.16}$$

Z_{TT} will reduce all impedances in parallel at the output circuit shown in Figure 10.5. Multiplying the Z_{TT} impedance by the current $h_{fe}i_b$ we can obtain the output voltage as

$$v_o = -h_{fe}i_b \times Z_{TT} = -h_{fe}i_b \left[\frac{R_T}{1 + j\omega C_P R_T} \right] \tag{10.17}$$

The input voltage can be obtained from the input circuit of Figure 10.5 as

$$v_{in} = i_b h_{ie} \tag{10.18}$$

Then, the voltage gain at high frequency, Gv_{HF}, can be obtained as the ratio of the output voltage to the input voltage as

$$Gv_{HF} = \frac{v_o}{v_{in}} = \frac{-h_{fe}i_b[(R_T/(1 + j\omega C_P R_T))]}{i_b h_{ie}} = -\frac{h_{fe}}{h_{ie}} \left(\frac{R_T}{1 + j\omega C_P R_T} \right) \tag{10.19}$$

Gv_{HF} can also be expressed in terms of the voltage gain at mid-frequency $Gv_{mf} = -h_{fe}R_T/h_{ie}$ as

$$Gv_{HF} = \left(\frac{Gv_{mf}}{1 + j\omega C_P R_T} \right) \tag{10.20}$$

At medium frequencies the term $j\omega C_p R_T$ will be smaller than 1 and the high-frequency gain will reach Gv_{mf}.

10.4.2 HIGH-FREQUENCY CUTOFF

Rearranging Equation 10.20 we have

$$\frac{Gv_{HF}}{Gv_{mf}} = \left(\frac{1}{1 + j\omega C_P R_T}\right)$$

and the high-frequency cutoff, f_H, can be obtained when the magnitude of the gain is reduced by a $1/\sqrt{2}$ factor. Then,

$$\left|\frac{Gv_{HF}}{Gv_{mf}}\right| = \frac{1}{\sqrt{2}} = \left|\frac{1}{1 + j\omega C_P R_T}\right| = \frac{1}{\sqrt{1^2 + \omega^2 C_P^2 R_T^2}}$$

The parallel bar denotes absolute value. To satisfy this equation the term $\omega C_P R_T$, at the frequency cutoff, must be equal to 1 and the high-frequency cutoff, f_H, can be expressed as

$$f_H = \frac{1}{2\pi C_P R_T} \tag{10.21}$$

10.4.3 PHASE CHANGE AT HIGH FREQUENCY

Replacing f_H from Equation 10.21 into Gv_{HF} in Equation 10.20 we can calculate the phase change at the frequency cutoff. At f_H we have

$$Gv_{HF}(\text{at } f = f_H) = \frac{Gv_{mf}}{1 + j}$$

Gv_{mf} can be written in polar form as $Gv_{mf} = |(h_{fe}R_T/h_{ie})|\angle 180°$ and $(1 + j)$ can be written in polar form as $(1 + j) = \sqrt{2}\angle 45°$ and the phase at high-frequency cutoff can be obtained as

$$Gv_{HF}(\text{at } f = f_H) = \frac{Gv_{mf}}{1 + j} = \frac{|(h_{fe}R_T/h_{ie})|\angle 180°}{\sqrt{2}\angle + 45°} = \left|\frac{h_{fe}R_T}{\sqrt{2}h_{ie}}\right|\angle(180° - 45°)$$

$$= \left|\frac{h_{fe}R_T}{\sqrt{2}h_{ie}}\right|\angle 135° \tag{10.22}$$

Therefore, the phase at the high-frequency cutoff is equal to 135°.

10.5 TOTAL FREQUENCY RESPONSE

In general, the frequency response of amplifiers containing transistors will have the behavior indicated in Figure 10.6.

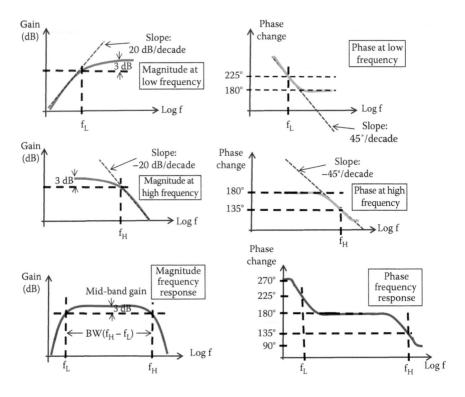

FIGURE 10.6   Amplifier magnitude and phase frequency response.

The fall in voltage gain at low frequency is mainly due to capacitances in series with input and output, while the fall in voltage at high frequencies is mainly due to capacitance in parallel increased by the Miller effect.

Table 10.1 includes a summary of expressions and values of different parameters at different frequencies.

TABLE 10.1
Frequency Response Parameters and Expressions for a CE Amplifier

	Voltage Gain	Cutoff Frequency	Phase at Cut off
Low frequency	$Gv_{LF} = -\dfrac{h_{fe}}{h_{ie}}\left[\dfrac{R_T}{1 + 1/j\omega C_{c2}(R'_C + R_L)}\right]$	$f_L = \dfrac{1}{2\pi C_{c2}(R'_C + R_L)}$	225°
Medium frequency	$Gv_{mf} = -\dfrac{h_{fe}R_T}{h_{ie}}$	—	—
High frequency	$Gv_{HF} = -\dfrac{h_{fe}}{h_{ie}}\left(\dfrac{R_T}{1 + j\omega C_P R_T}\right)$	$f_H = \dfrac{1}{2\pi C_P R_T}$	135°

10.6 KEY POINTS

- Reactive components such as inductors and capacitors are affected by the frequency variation of the applied signal while resistors are unaffected by frequency variation.
- Since the characteristics of reactive components change with frequency, the behavior of circuits using these components will also change.
- The way in which an amplifier or a circuit changes with frequency is termed its frequency response. The frequency response considers the variations in the magnitude of the gain and in the phase response.
- In an amplifier, the gain is usually represented in decibels and the frequency as logarithmic of frequency.
- Frequency cutoffs determine the low-, high-, and mid-frequency response in an amplifier. The low- and high-frequency cutoffs are determined as the points where the power gain is half the mid-frequency power gain. The low- and high-frequency cutoffs can be determined at the frequency where the voltage gain has reduced by 3 dB from the voltage gain at mid-frequency.
- A cutoff frequency can also be determined at the point where the voltage gain has dropped by a factor of $1/\sqrt{2}$ from the voltage gain at mid-frequencies.
- In general, capacitors in series with inputs and outputs will affect the low-frequency response and the capacitors in parallel with inputs and outputs affect the high-frequency response.
- The BW represents the range of frequencies between the low-frequency cutoff, f_L, and the high-frequency cutoff, f_H, giving the range of frequencies over which an amplifier can be used.

11 Common Collector Amplifier/Emitter Follower

The common collector configuration, also known as emitter follower, provides the main applications of impedance matching and buffer amplifier. In this amplifier, the output is almost identical to its input, but its input impedance is high and output impedance is low.

11.1 VOLTAGE GAIN, Gv_{CC}

The collector terminal of a transistor can be made common to input and output. Figure 11.1 shows a transistor used as a common collector amplifier.

Coupling capacitors act as a short circuit at medium and high frequencies. The biasing source V_{cc} presents very low impedance to an ac signal making the collector common to input and output as indicated in Figure 11.2. In this configuration, the emitter resistor R_E is not bypassed by a capacitor and it appears in the medium-frequency equivalent circuit.

The equivalent circuit at medium frequencies, shown in Figure 11.2, can be rearranged as indicated in Figure 11.3 in order to analyze its behavior.

We can add the two resistors, R_E, and $1/h_{oe}$ in parallel to create R_T as

$$R_T = \frac{R_E/h_{oe}}{R_E + 1/h_{oe}} = \frac{R_E}{1 + h_{oe}R_E} \tag{11.1}$$

The current passing through R_T, i_T, can be obtained by applying KCL at node E in Figure 11.3 as

$$i_T = i_B + h_{fe}i_B \tag{11.2}$$

Then, the output voltage, v_o, can be obtained by multiplying i_T and R_T as given below:

$$v_O = R_T i_T = i_B(1 + h_{fe})\frac{R_E}{1 + h_{oe}R_E} \tag{11.3}$$

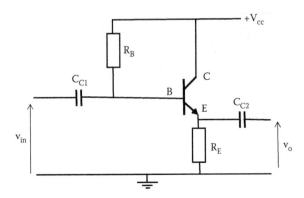

FIGURE 11.1 Circuit diagram for a common collector amplifier.

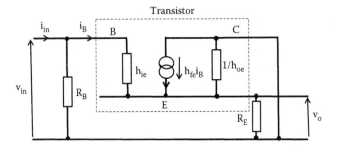

FIGURE 11.2 Ac medium frequencies equivalent circuit for a common collector amplifier.

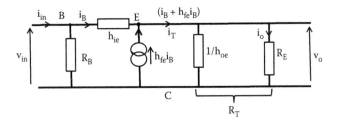

FIGURE 11.3 Rearranged version of Figure 11.2 for an equivalent circuit of a common collector amplifier.

From the equivalent circuit shown in Figure 11.3, the input voltage v_{in} can be obtained using KVL in the close circuit that include v_{in}, the voltage across h_{ie} and v_o, as

$$v_{in} = h_{ie}i_B + v_O \tag{11.4}$$

Replacing v_o from Equation 11.3 into Equation 11.4, we have

$$v_{in} = h_{ie}i_B + v_O = h_{ie}i_B + i_B(1+h_{fe})\frac{R_E}{1+h_{oe}R_E} \qquad (11.5)$$

and with v_o (Equation 11.3) and v_{in} (Equation 11.5) the voltage gain in common collector configuration, Gv_{CC}, can be written as

$$Gv_{CC} = \frac{v_O}{v_{in}} = \frac{i_B(1+h_{fe})R_E/(1+h_{oe}R_E)}{i_Bh_{ie} + i_B(1+h_{fe})R_E/(1+h_{oe}R_E)} \qquad (11.6)$$

Multiplying numerator and denominator by $(1 + h_{oe}R_E)$ and cancelling i_B, Gv_{CC} can be expressed as

$$Gv_{CC} = \frac{R_E(1+h_{fe})}{h_{ie}(1+h_{oe}R_E)+R_E(1+h_{fe})} = \frac{1}{1+h_{ie}(1+h_{oe}R_E)/R_E(1+h_{fe})} \qquad (11.7)$$

Now the term $h_{ie}(1 + h_{oe}R_E)$ is always much smaller than $R_E(1 + h_{fe})$ and the ratio $h_{ie}(1 + h_{oe}R_E)/R_E(1 + h_{fe})$ can be neglected compared to 1 given an approximation for Gv_{CC} as 1. Then,

$$Gv_{CC} \approx 1 \qquad (11.8)$$

Therefore, one of the characteristics of a common collector configuration is that the input and output are the same with no phase shift.

EXAMPLE 11.1

A semiconductor manufacturer provided the following h parameters data for a general purpose transistor: $h_{ie} = 1$ kΩ, $h_{fe} = 110$, and $h_{oe} = 80$ µS. Calculate the voltage gain of a common collector amplifier when the emitter resistor R_E is 1 and 50 kΩ.

SOLUTION

At $R_E = 1$ kΩ, the common collector gain is

$$Gv_{CC} = \frac{1}{1+h_{ie}(1+h_{oe}R_E)/R_E(1+h_{fe})}$$

$$= \frac{1}{1+1\times10^3(1+80\times10^{-6}\times1\times10^3)/1000(1+110)} = \frac{1}{1+1080/111000}$$

and $Gv_{CC} = 0.9904$.
    For $R_E = 50$ kΩ, the common collector gain is

$$Gv_{CC} = \frac{1}{1+1\times10^3(1+80\times10^{-6}\times50\times10^3)/50\times10^3(1+110)}$$

$$= \frac{1}{1+4000/5.55\times10^6}$$

and $Gv_{CC} = 0.9993$.

The common collector amplifier provides a voltage gain of ~1, independent of the value of R_E.

11.2   CURRENT GAIN, Gi_{CC}

The output current is the current passing through the resistor R_E as shown in Figure 11.3. This current can be obtained using the current divider rule as we know the current $i_T = (i_B + h_{fe} i_B)$ enters the two resistors in parallel R_E and $1/h_{oe}$. Therefore,

$$i_O = i_{RE} = \frac{R_E/h_{oe}}{R_E + 1/h_{oe}}(i_B + h_{fe}i_B) \qquad (11.9)$$

Now the ac current through resistor R_B is very small and can be neglected. Then,

$$i_{in} = i_B \qquad (11.10)$$

with the input and output currents we can obtain the gain for the common collector configuration as

$$Gi_{CC} = \frac{i_O}{i_{in}} = \frac{i_{RE}}{i_B} = \frac{R_E(1+h_{fe})/h_{oe}}{R_E + 1/h_{oe}} = \frac{R_E(1+h_{fe})}{R_E h_{oe} + 1} \qquad (11.11)$$

11.3   INPUT IMPEDANCE, Zi_{CC}

The input impedance, Zi, is the ratio of the input voltage to the input current. The input voltage is given by Equation 11.4 and the input current is i_B. Then,

$$v_{in} = h_{ie}i_B + v_O$$

and

$$i_{in} \approx i_B$$

The current through R_B is very small and can be neglected. With the input voltage, input current, and output voltage (Equation 11.3) we can obtain an expression for the input impedance as under common collector configuration as

$$Zi_{CC} = \frac{v_{in}}{i_{in}} = \frac{h_{ie}i_B + v_O}{i_B} = h_{ie} + \frac{v_O}{i_B} = h_{ie} + (1+h_{fe})\frac{R_E}{1 + h_{oe}R_E} \qquad (11.12)$$

Then, common collector configuration provides a large value of input impedance.

In the common collector amplifier shown in Figures 11.1 and 11.3 there is no emitter capacitor; hence, the emitter resistor is not bypassed by a capacitor. As a consequence the input resistance is large.

Now this input impedance Zi_{CC} was calculated without the effect of R_B. To include the effect of R_B then the total input impedance, Zi_T, will be the parallel combination of R_B and Zi_{CC} as

$$Zi_T = \frac{\{h_{ie} + R_E(1 + h_{fe})/(1 + h_{oe}R_E)\}R_B}{R_B + h_{ie} + R_E(1 + h_{fe})/1 + h_{oe}R_E} \qquad (11.13)$$

R_B is usually large, but it may reduce the overall input impedance as Zi_{CC} is much larger.

The input impedance is high and the voltage gain is 1, this amplifier can be used as buffer amplifier or as impedance matching.

11.4   POWER GAIN, Gp_{CC}

The power gain can be expressed as the product of the voltage and current gains or as the ratio of the output power to the input power. The power gain for a common collector amplifier at medium frequencies can be expressed as

$$Gp_{CC} = Gi_{CC}Gv_{CC} = \left[\frac{R_E(1 + h_{fe})}{R_E h_{oe} + 1}\right]\left[\frac{1}{1 + h_{ie}(1 + h_{oe}R_E)/R_E(1 + h_{fe})}\right] \qquad (11.14)$$

But in a common collector configuration the voltage gain is almost 1, therefore the power gain is approximately equal to the current gain

$$Gp_{CC} \approx \frac{R_E(1 + h_{fe})}{R_E h_{oe} + 1} \qquad (11.15)$$

An alternative biasing for a common collector amplifier is shown in Figure 11.4.

For this circuit the analysis to obtain the voltage and current gains, input and output impedances, and power gain is the same. The only difference is that the resistor R_B in the equivalent circuit shown in Figure 11.3 is replaced by the parallel combination of R_1 and R_2.

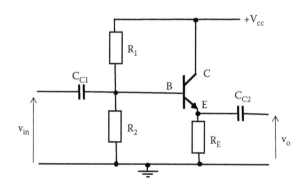

FIGURE 11.4   Potential divider common collector amplifier circuit.

11.5   KEY POINTS

- The voltage gain in a common collector amplifier is almost 1.
- The common collector configuration is also known as emitter follower.
- In a common collector configuration the input voltage and output voltage are in phase, that is, there is no phase shift.
- In a common collector configuration the input impedance is very high.
- Main application of common collector amplifier is as impedance matching or as buffer amplifier.

12 Common Base Amplifier

12.1 COMMON BASE AMPLIFIERS UNDER h PARAMETERS

The common base amplifier is sometimes used for matching impedances; it has low-input impedance and high-output impedance. The circuit diagram of a common base amplifier is shown in Figure 12.1. The base is common to the input and the output.

An ac equivalent circuit at medium frequency for the common base amplifier is shown in Figure 12.2. The resistor R_E is usually much larger than h_{ib} and the parallel combination of R_E and h_{ib} can be approximated to h_{ib} as indicated in Figure 12.2. The equivalent circuit of Figure 12.2 is identical in form to the equivalent circuit for a common emitter amplifier used in Chapter 10, but in this case the appropriate parameters are the h base parameters. An expression for the voltage gain under common base configuration can be deduced in the same way as the voltage gain under common emitter configuration.

From Figure 12.2, we can combine R_C and $1/h_{ob}$ into one resistor, R_C'. The combined resistances of R_C and $1/h_{ob}$ in parallel, R_C', becomes

$$\frac{1}{R_C'} = \frac{1}{R_C} + \frac{1}{1/h_{ob}} = \frac{1}{R_C} + h_{ob} = \frac{1+R_C h_{ob}}{R_C}$$

Therefore,

$$R_C' = \frac{R_C}{1+R_C h_{ob}} \tag{12.1}$$

This circuit can be simplified further by combining R_C' and R_L in parallel into one resistor R_T as

$$\frac{1}{R_T} = \frac{1}{R_L} + \frac{1}{R_C'} = \frac{1}{R_L} + \frac{1+R_C h_{ob}}{R_C} = \frac{R_C + R_C(1+R_C h_{ob})}{R_L R_C}$$

and

$$R_T = \frac{R_L R}{R_C + R_C(1+R_C h_{ob})} \tag{12.2}$$

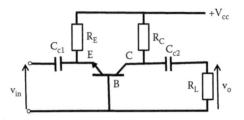

FIGURE 12.1   Circuit diagram common base amplifier.

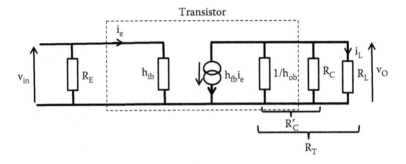

FIGURE 12.2   Common base amplifier medium frequency ac equivalent circuit using h parameters.

12.2   VOLTAGE GAIN, Gv_{CB}

From the input of circuit shown in Figure 12.2 the input voltage, v_{in}, can be written as

$$v_{in} = h_{ib}i_e \tag{12.3}$$

and from the output circuit of Figure 12.2 the output voltage, v_o, can be written as

$$v_O = -h_{fb}i_e R_T \tag{12.4}$$

and the voltage gain of a common base amplifier at medium frequencies, Gv_{CB}, can be expressed as

$$Gv_{CB} = \frac{v_O}{v_i} = \frac{-h_{fb}i_e R_T}{h_{ib}i_e} = -\frac{h_{fb}}{h_{ib}}R_T \tag{12.5}$$

Here the common base parameter h_{fb} typically has a value just under unity and is negative. Typically, $h_{fb} = -0.98$, therefore there is no phase shift.

12.3   CURRENT GAIN, Gi_{CB}

The output current passing through R_L, i_L, can be deduced using Kirchhoff's laws or the current divider rule, between R_L and R_C' as

$$i_L = -\frac{R_C'}{R_C'+R_L}(h_{fb}i_e) \tag{12.6}$$

The negative sign is due to the opposite direction of the currents. The input current is i_e and the current gain under common base configuration at medium frequencies can be expressed as

$$Gi_{CB} = \frac{i_L}{i_e} = -\frac{h_{fb}R_C'}{R_C' + R_L} \tag{12.7}$$

The current gain, Gi_{CB}, can also be expressed in terms of h_{ob} and R_C by replacing the expression for into Equation 12.7.

12.4   POWER GAIN, Gp_{CB}

As indicated earlier the power gain can be expressed as the product of the voltage and current gains or as the ratio of the output power to the input power. The power gain for a common base amplifier at medium frequencies can be expressed as

$$Gp_{CB} = Gi_{CB}Gv_{CB} = \frac{P_o}{P_i} = \frac{v_o^2/R_L}{v_{in}^2/h_{ib}} = Gv_{CB}^2\frac{h_{ib}}{R_L} \tag{12.8}$$

In the same way, it is possible to obtain an equivalent expression for the power gain in terms of the current gain.

EXAMPLE 12.1

For the amplifier shown in Figure 12.3, draw a mid-band frequency equivalent circuit using h parameters and calculate the ac voltage, current, and power gains of the amplifier. The transistor has $h_{ib} = 75\ \Omega$, $h_{fb} = -0.95$, $h_{ob} = 0$, and $h_{rb} = 0$.

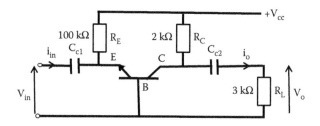

FIGURE 12.3   Circuit diagram of a common base amplifier for Example 12.1

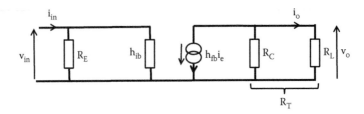

FIGURE 12.4 Equivalent circuit for amplifier of Example 12.1 using h_b parameters.

SOLUTION

At mid-band frequency, the coupling capacitors can be assumed to have zero impedance. As $h_{ob} = 0$ and $h_{rb} = 0$, the ac mid-band equivalent circuit is shown in Figure 12.4.

As R_E is much larger than h_{ib}, it can be neglected, or alternatively, it can be combined in parallel with h_{ib}, which provides a value approximate to that of h_{ib}.

The value of R_T, in this case, is the parallel combination of R_C and R_L. Therefore,

$$R_T = \frac{R_L \times R_C}{R_L + R_C} = \frac{2 \times 10^3 \times 3 \times 10^3}{2 \times 10^3 + 3 \times 10^3} = 1.2\,k\Omega$$

and the voltage gain is

$$Gv_{CB} = \frac{v_O}{v_{in}} = -\frac{h_{fb}}{h_{ib}} R_T = -\frac{(-0.95)}{75} 1.2 \times 10^3 = 15.2$$

The current gain in this case is given by

$$Gi_{CB} = \frac{i_o}{i_{in}} = -\frac{h_{fb} R_C}{R_C + R_L} = \frac{(-0.95) \times 2 \times 10^3}{2 \times 10^3 + 3 \times 10^3} = 0.38$$

The power gain is the product of the voltage gain and the current gain as given by

$$G_P = G_i G_v = 15.2 \times 0.38 = 5.78$$

12.5 KEY POINTS

- In a common base configuration the input voltage and output voltage are in phase, that is, there is no phase shift.
- In a common base configuration the input impedance is low.
- In a common base configuration the output impedance is high.
- One of the applications of common base is as impedance matching.
- Common base amplifiers have low input capacitance at high frequency.

13 Common Emitter Amplifier in Cascade

13.1 INTRODUCTION

In many cases, a single-stage amplifier, such as the one seen in previous chapters, does not provide all the gain required by a system. In some cases, if the overall gain required is provided by a single stage it may produce too much distortion. A solution to these problems can be to increase the gain by adding stages. The output of the first stage becomes the input of the next stage to increase the overall gain; this is to connect stages in cascade.

In this chapter, we attempt to see the effect of cascading on the overall gain and on the frequency response.

13.2 OVERALL GAIN OF AMPLIFIERS IN CASCADE

A connection of n auto bias common emitter amplifier stages in cascade could be represented as indicated in Figure 13.1. The input signal v_{in} from the source through R_S is amplified by transistor T1 of the first stage, the output signal of the first stage v_{o1} becomes the input signal of second stage and is amplified by transistor T2. The process continues in each stage to amplify the previous output until the last n stage, which provides its output, v_o, to the load R_L.

As before, we can obtain an ac equivalent circuit for all stages in cascade; this is shown in Figure 13.2.

As in previous chapters, R'_C is the parallel combination of R_C and $1/h_{oe}$ given by

$$R'_C = R_C/(1 + R_C h_{oe})$$

as deduced in Chapter 9 (Equation 9.19). The biasing resistors, R_1 and R_2, have been omitted as they are usually much larger than h_{ie}.

The voltage gain of each stage will depend on the impedance connected as load. The last stage has R_L for load, while the other stages will have the input impedance of the next stage for load. Also, the first stage input is connected to a source with a series resistance R_S. Then we have three different voltage gains: voltage gain from the last n stage, voltage gain of all $(n - 1)$ stages, and "voltage gain" due to the voltage source.

The general voltage gain for any stage, deduced in Chapter 9 (Equation 9.23), is given as

$$G_v = -\frac{h_{fe}}{h_{ie}} R_T$$

where R_T is the parallel combination of R'_C and the load resistor.

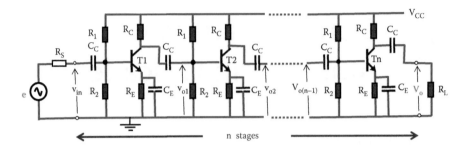

FIGURE 13.1   Circuit diagram of n stages common emitter amplifiers connected in cascade.

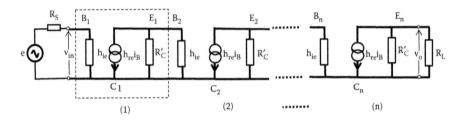

FIGURE 13.2   Ac equivalent circuit using h parameters for n stages common emitter amplifiers connected in cascade.

13.2.1   Voltage Gain of the Last Stage n

The voltage gain of the last stage Gv_n has R_L as load, then the total output resistance R_{Tn} will be

$$R_{Tn} = \frac{R_L R_C}{R_c + R_L(1 + R_c h_{oe})}$$

as calculated in Chapter 9 (Equation 9.20) and the voltage gain for all the n stage Gv_n is given by

$$Gv_n = -\frac{h_{fe}}{h_{ie}} R_{Tn}$$

13.2.2   Voltage Gain for the (n − 1) Stages

The rest of the stages from 1 to $(n-1)$ have h_{ie} from the next stage as load. An expression for their gain can be obtained by replacing R_L for h_{ie}. In the voltage gain expression we can use $R_{T(n-1)}$ as the parallel combination of R'_C and h_{ie}. Then,

$$R_{T(n-1)} = \frac{h_{ie} R_C}{R_c + h_{ie}(1 + R_c h_{oe})} \tag{13.1}$$

and the voltage gain for all the $(n - 1)$ stages $Gv_{(n-1)}$ is given by

$$Gv_{(n-1)} = -\frac{h_{fe}}{h_{ie}} R_{T(n-1)} \qquad (13.2)$$

13.2.3 "Voltage" Gain of the Source

Now, to obtain the overall voltage gain from the source input e to output v_o we have to include the attenuation voltage drop across the source resistance R_S. Figure 13.3 provides an equivalent circuit for the input voltage source connected to the first-stage amplifier.

The input current i_{in} can be obtained as

$$i_{in} = \frac{e}{R_S + h_{ie}} \qquad (13.3)$$

Then, the input voltage v_{in} is

$$v_{in} = \frac{e h_{ie}}{R_S + h_{ie}} \qquad (13.4)$$

and the "voltage gain" (attenuation) of the source can be written as

$$\frac{v_{in}}{e} = \frac{h_{ie}}{R_S + h_{ie}} \qquad (13.5)$$

When stages are cascaded, the overall voltage gain is the product of all the stage's voltage gains. Due to the properties of logarithms the overall voltage gain of stages in cascade is the sum of all the voltage gains in decibels. In our case, we have n stages plus the source attenuation, and then the overall gain from the source e to the voltage across R_L can be expressed as

$$Gv_{(TOTAL)} = \frac{v_O}{e} = \left[\frac{h_{ie}}{R_S + h_{ie}}\right] \times \left[-\frac{h_{fe}}{h_{ie}} R_{T(n-1)}\right]^{n-1} \times \left[-\frac{h_{fe}}{h_{ie}} R_{Tn}\right] \qquad (13.6)$$

The number of stages will depend on the overall gain required.

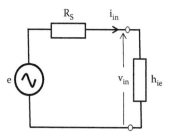

FIGURE 13.3   Input circuit of first stage of amplifiers n stages connected in cascade.

EXAMPLE 13.1

A two-stage amplifier is shown in Figure 13.4. Draw a mid-band equivalent circuit and calculate the overall voltage gain. Q1 and Q2 have $h_{ie} = 2$ kΩ, $h_{fe} = 55$, and $h_{oe} = 25$ μS.

SOLUTION

We can assume that the capacitors have zero impedance at mid-frequency range. An equivalent circuit can be represented as in Figure 13.5.

The total output resistance for the second stage, R_{T_2}, is the parallel combination of $1/h_{oe}$, R_{C_2}, and R_L, given by

$$R_{T_2} = \frac{R_L R_{C_2}}{R_{C_2} + R_L(1 + R_{C_2}h_{oe})} = \frac{50 \times 10^3 \times 10 \times 10^3}{10 \times 10^3 + 50 \times 10^3 (1 + 10 \times 10^3 \times 25 \times 10^{-6})}$$

$$= 6896.55\,\Omega$$

and the voltage gain of the second stage is given by

$$Gv_2 = -\frac{h_{fe}}{h_{ie}}R_{T_2} = -\frac{55}{2 \times 10^3} \times 6896.55 = -189.66$$

For the first stage R_{T_1}, neglecting R_{21} and R_{22}, is the parallel combination of $1/h_{oe}$, R_{C_1}, and h_{ie}, given by

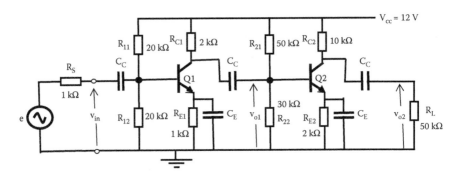

FIGURE 13.4 Two-stage amplifier circuit diagram for Example 13.1.

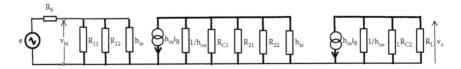

FIGURE 13.5 Mid-frequency equivalent circuit of two common emitter amplifiers connected in cascade for Example 13.1.

$$R_{T1} = \frac{h_{ie}R_{C_1}}{R_{C_1} + h_{ie}(1 + R_{C_1}h_{oe})} = \frac{2\times10^3 \times 2\times10^3}{2\times10^3 + 2\times10^3(1 + 2\times10^3 \times 25\times10^{-6})} = 975.61\,\Omega$$

A more accurate value for the total output resistance of first stage will include R_{21} and R_{22} in parallel with R_{T1}, but as R_{T1} is much smaller than R_{21} or R_{22} their influence will be very small.

And the voltage gain of the first stage is given by

$$Gv_1 = -\frac{h_{fe}}{h_{ie}}R_{T1} = -\frac{55}{2\times10^3}\times975.61 = -26.83$$

The attenuation G_S of resistance R_S in series with the source e is given by

$$G_S = \frac{h_{ie}}{R_S + h_{ie}} = \frac{2\times10^3}{1\times10^3 + 2\times10^3} = 0.67$$

The overall voltage gain, G_T, will be the product of the three gains as follows:

$$G_T = \frac{v_O}{e} = G_S \times G_{V_1} \times G_{V_2} = 0.67\times(-26.83)\times(-189.66) = 3409.35$$

13.3   FREQUENCY RESPONSE

As indicated in the previous section we have three types of different stages. Different type of stages connected in cascade will have different frequency cutoff. The frequency response will be affected by the dominant frequency cutoff. For the low-frequency response the larger low-frequency cutoff will be the dominant. For the high-frequency response the lower high-frequency cutoff will be the dominant cutoff.

13.3.1   Low-Frequency Cutoff

In general, a capacitance in series will reduce the gain at low frequencies. This will affect the low-frequency cutoff. In cascading amplifiers we will couple stages, load, and source with a series capacitor.

13.3.1.1   Frequency Cutoff of Last Stage

The coupling capacitor in series with the load R_L will cause a reduction of gain in the last stage. Figure 13.6 shows the equivalent circuit at low frequency for the output last stage.

This is the same circuit analyzed in Chapter 10 where the low-frequency cutoff was found to be given by Equation 10.12. Then, for the last stage n the low-frequency cutoff f_{Ln} is

$$f_{Ln} = \frac{1}{2\pi C_c(R'_C + R_L)} \tag{13.7}$$

This is a cutoff frequency generated by the last coupling capacitor and R_L.

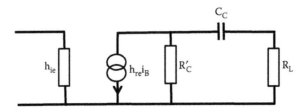

FIGURE 13.6 Equivalent circuit at low frequency for the output of the last n stage.

13.3.1.2 Frequency Cutoff for 1 to (n − 1) Stages

For the other (n − 1) stages the output circuit from each stage can be represented by the equivalent circuit shown in Figure 13.7, where h_{ie} from the next stage replaces the resistor R_L, then the cutoff frequency of these stages $f_{L(n-1)}$ is given by

$$f_{L(n-1)} = \frac{1}{2\pi C_c (R'_C + h_{ie})} \tag{13.8}$$

13.3.1.3 Frequency Cutoff Input Circuit First Stage

The first coupling capacitor at the input of the first amplifier and the source resistance R_S in series with h_{ie} will also produce a cutoff frequency $f_{L(source)}$. This cutoff frequency can be obtained in the same way as before using the equivalent circuit shown in Figure 13.8.

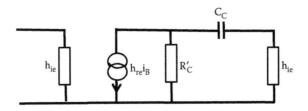

FIGURE 13.7 Equivalent circuit at low-frequency output of any (n − 1) stage connected in cascade.

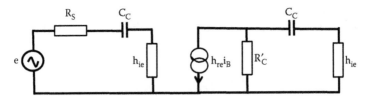

FIGURE 13.8 Equivalent circuit at low frequency for the source stage of n amplifiers connected in cascade.

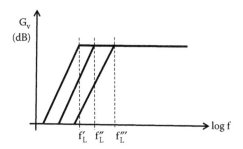

FIGURE 13.9   Three different low-frequency cutoffs for n stages in cascade.

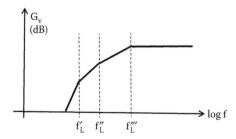

FIGURE 13.10   Overall frequency response at low frequency of n stages connected in cascade.

Now we have two resistors in series with the first coupling capacitor, R_S and h_{ie}. Then, the cutoff frequency due to the source plus input coupling capacitor is

$$f_{L(source)} = \frac{1}{2\pi C_c (R_S + h_{ie})} \tag{13.9}$$

In general, these three cutoff frequencies will be different f_L', f_L'', and f_L''', assuming that $f_L' < f_L'' < f_L'''$ as indicated in Figure 13.9.

Adding the three different frequency responses at low frequency we can represent the overall low-frequency response of the n stages. Figure 13.10 shows the overall low-frequency response of the n stages in cascade.

The value of each frequency cutoff, f_L', f_L'', and f_L''' will depend on the different values of components affecting the cutoff frequency. The dominant frequency will be the larger f_L, then the dominant cutoff frequency will depend on the values of h_{ie}, R_L, or the value of the series source resistance R_S.

EXAMPLE 13.2

For Example 13.1, assume that all coupling capacitors have a value of 800 nF. Calculate the low-frequency cutoff of the different stages.

Solution

For the second stage, the parallel combination of R_{C2} and $1/h_{oe}$ will produce R'_{C2} as

$$R'_{C2} = \frac{R_{C2}}{1+R_{C2}h_{oe}} = \frac{10\times10^3}{1+10\times10^3\times25\times10^{-6}} = 8\,k\Omega$$

and the frequency cutoff for the second stage, f_{L2}, is given by

$$f_{L2} = \frac{1}{2\pi C_c(R'_{C2}+R_L)} = \frac{1}{2\pi\times800\times10^{-9}(8\times10^3+50\times10^3)} = 3.43\,Hz$$

For the first stage, the parallel combination of R_{C1} and $1/h_{oe}$ will produce R'_{C1} as

$$R'_{C1} = \frac{R_{C1}}{1+R_{C1}h_{oe}} = \frac{2\times10^3}{1+2\times10^3\times25\times10^{-6}} = 1904.76\,\Omega$$

and the frequency cutoff, f_{L1}, for the first stage is given by

$$f_{L1} = \frac{1}{2\pi C_c(R'_{C1}+h_{ie})} = \frac{1}{2\pi\times800\times10^{-9}(1904.76+2\times10^3)} = 50.95\,Hz$$

and the low-frequency cutoff due to the series coupling capacitor between the voltage source and the first stage is given by

$$f_{L(source)} = \frac{1}{2\pi C_c(R_S+h_{ie})} = \frac{1}{2\pi\times800\times10^{-9}(1\times10^3+2\times10^3)} = 66.31\,Hz$$

The larger dominant low-frequency cutoff is due to the input coupling capacitor in this case.

13.3.2 High-Frequency Cutoff

In general, coupling capacitors in parallel will cause drop in gain affecting the high-frequency cutoff. As in the case of low-frequency cutoff in common emitter amplifiers in cascade there are three different high-frequency cutoffs involved.

13.3.2.1 Frequency Cutoff of Last Stage

Figure 13.11 shows the equivalent circuit at high frequency for the output of the last stage. The coupling capacitor in parallel C_{Sn} represents the sum of all capacitances in parallel at the output such as the capacitance between base and collector, the stray capacitance of wiring, and capacitance of components connected at the output. All output capacitance included in C_{Sn} cause a reduction of gain in the last stage.

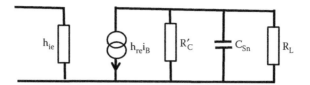

FIGURE 13.11   Equivalent circuit at high frequency for the output of the last n stage.

This is the same circuit analyzed in Chapter 10 (Equation 10.20), then the high-frequency cutoff for the last stage is

$$f_{Hn} = \frac{1}{2\pi C_{Sn} R_{Tn}} \tag{13.10}$$

This cutoff frequency is created by the last parallel capacitor Cs_n, R'_C, and R_L. R_{Tn} is the parallel combination of R'_C and R_L.

13.3.2.2   High-Frequency Cutoff for 1 to (n − 1) Stages

For the other (n − 1) stages the output circuit from each stage can be represented as Figure 13.12 where h_{ie} from the next stage replaces the resistor R_L, then the cutoff frequency of these stages $f_{H(n-1)}$ is given by

$$f_{H(n-1)} = \frac{1}{2\pi Cs_{(n-1)} R_{T(n-1)}} \tag{13.11}$$

This cutoff frequency is generated by the parallel capacitor of $Cs_{(n-1)}$, R'_C, and h_{ie}. $R_{T(n-1)}$ is the parallel combination of R'_C and h_{ie}.

13.3.2.3   Frequency Cutoff Input Circuit First Stage

The first parallel capacitor at the input and the source resistance R_S in series with h_{ie} will also produce a cutoff frequency $f_{H(source)}$. This cutoff frequency can be obtained in the same way as before using the equivalent circuit shown in Figure 13.13.

The cutoff frequency due to the source plus input parallel capacitor C_{SI} is

$$f_{H(source)} = \frac{1}{2\pi C_{SI} R_{T(source)}} \tag{13.12}$$

where $R_{T(source)}$ is the combination of R_S and h_{ie} in parallel.

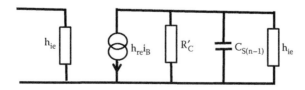

FIGURE 13.12   Equivalent circuit at high frequency for any (n − 1) stage of n amplifiers connected in cascade.

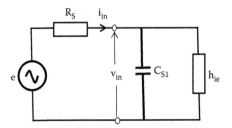

FIGURE 13.13 Equivalent circuit at high frequency for the source stage.

The value of each frequency cutoff, f_H', f_H'', and f_H''' will depend on the different values of components affecting the cutoff frequency. The dominant frequency will be the smaller f_H, then the dominant cutoff frequency will depend on the values of h_{ie}, R_L, or the value of the series source resistance R_S.

In general, these three cutoff frequencies will be different as indicated in Figure 13.14.

When added, assuming that $f_H' < f_H'' < f_H'''$, they produce an overall response as indicated in Figure 13.15.

The high-frequency cutoff will depend on the values of h_{ie}, R_L, or the value of the series source resistance R_s.

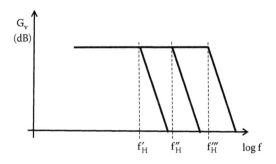

FIGURE 13.14 Three different high-frequency cutoffs for stages in cascade.

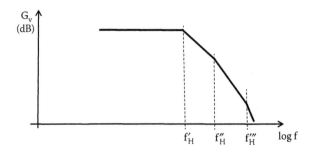

FIGURE 13.15 Overall frequency response at high frequency for stages in cascade.

EXAMPLE 13.3

For Example 13.1 assume that the addition of all capacitances in parallel due to wiring and Miller effect can be represented by a capacitance C_P and that this capacitance produces a high-frequency cutoff in each stage. If C_P has a value of 10 nF, determine the high-frequency cutoff of the different stages.

SOLUTION

For the second stage, the parallel combination of R'_{C2} and R_L will produce R_{T2} as

$$R_{T2} = \frac{R'_{C2} \times R_L}{R'_{C2} + R_L} = \frac{8 \times 10^3 \times 50 \times 10^3}{8 \times 10^3 + 50 \times 10^3} = 6896.55\,\Omega$$

and the high-frequency cutoff, f_{H2}, for the second stage is given by

$$f_{H2} = \frac{1}{2\pi C_P R_{T2}} = \frac{1}{2\pi \times 10 \times 10^{-9} \times 6896.55} = 2307.7 \text{ Hz}$$

For the first stage, the parallel combination of R'_{C1} and h_{ie} will produce R_{T1} as

$$R_{T1} = \frac{R'_{C1} \times h_{ie}}{R'_{C1} + h_{ie}} = \frac{1904.76 \times 2 \times 10^3}{1904.76 + 2 \times 10^3} = 975.61\,\Omega$$

and the high-frequency cutoff for the first stage, f_{H1}, is given by

$$f_{H1} = \frac{1}{2\pi C_P R_{T1}} = \frac{1}{2\pi \times 10 \times 10^{-9} \times 975.61} = 16{,}313.38 \text{ Hz}$$

and for the source stage the parallel combination of R_S and h_{ie} will produce $R_{T(source)}$ as

$$R_{T(source)} = \frac{R_S \times h_{ie}}{R_S + h_{ie}} = \frac{1 \times 10^3 \times 2 \times 10^3}{1 \times 10^3 + 2 \times 10^3} = 666.66\,\Omega$$

and the high-frequency cutoff due to the voltage source, $f_{H(source)}$ is given by

$$f_{H(source)} = \frac{1}{2\pi C_P R_{T(source)}} = \frac{1}{2\pi \times 10 \times 10^{-9} \times 666.66} = 23{,}873.48 \text{Hz}$$

Therefore, in this case the dominant high-frequency cutoff is the cut off due to the parallel capacitance, C_P, in the last stage.

In this example, we have simplified the calculations. In practice, the parallel capacitance will be different at different stages and Miller effects will have different influences at different stages.

13.4   KEY POINTS

- Amplifiers can be cascaded to achieve a desired gain.
- Cascading amplifiers will affect the cutoff frequency.
- In a cascade of two or more amplifier stages; at low frequencies, there are different low-frequency cutoffs; the larger low-frequency cutoff is the dominant one.
- In a cascade of two or more amplifier stages at high frequencies, there are different high-frequency cutoffs; the smaller high-frequency cutoff is the dominant one.

14 Field Effect Transistor Biasing

14.1 INTRODUCTION

As with the BJT, in order to use an FET as part of an electronic system it is necessary to fix a proper operation point Q. The aim is to provide the appropriate dc gate to source voltage in order to setup a desired value of drain current. A difference with the bipolar transistors is that the FET has high input impedance that can be considered to be an open circuit in many cases. The examples included in this chapter are based on an n-channel transistor. For the case of a p-channel transistor the biasing circuits are the same but the polarities of applied voltages are reversed. In this chapter, we will analyze certain types of biasing for the different types of field effect transistors.

14.2 MOSFET BIASING

As we have seen in Chapter 6, there are two types of MOS; they can be enhancement type and depletion type. The biasing for each of these two types of MOS transistor is slightly different.

14.2.1 DEPLETION MOSFET

The depletion MOS will have a physical channel created at fabrication, that is, they have a thin layer of semiconductor material of the same type as the drain and source that connects them. As the channel is there, with no voltage applied to the gate, there can be conduction between drain and source. This transistor will work with negative or positive voltage polarity at the gate with respect to the voltage at the source. Figure 14.1 shows a biasing circuit with one resistor connected to the drain. R_G is not part of the biasing, but isolates an ac signal from ground.

R_G is very large and its function is to isolate an ac signal from ground as there is no dc current at the gate. Let us assume a dc voltage at the gate equal to zero with respect to the source $V_{GS} = 0$. Then I_D is equal to the drain saturation current I_{DSS}.

Applying KVL to the closed circuit consisting of the voltage across R_D, the voltage between source and drain, $V_{DS,}$ and V_{DD} we can write their relationship as:

$$V_{DD} = V_{DS} + I_D R_D \tag{14.1}$$

and we can find the drain-to-source voltage, V_{DS}, as

$$V_{DS} = V_{DD} - I_D R_D \tag{14.2}$$

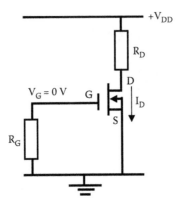

FIGURE 14.1   Depletion n-channel MOSFET biasing circuit.

This equation contains I_D and V_{DS} to select the operating point and determine the value of R_D required, satisfying Q.

EXAMPLE 14.1

The FET shown in Figure 14.2 has a $V_{DS} = 3$ V. Determine the value of R_D required in order to achieve a drain current $I_D = 16$ mA.

SOLUTION

Applying KVL we can write

$$V_{DD} = V_{DS} + I_D R_D = 12 = 3 + 16 \times 10^{-3} \times R_D$$

Then,

$$R_D = 562.5 \ \Omega$$

14.2.2   ENHANCEMENT MOSFET

In the case of an enhancement MOS the channel is not physically present. It needs to be created by applying a voltage to the gate. In order to bias an enhancement MOS

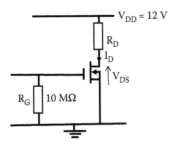

FIGURE 14.2   Depletion MOSFET biasing circuit diagram for Example 14.1.

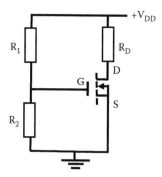

FIGURE 14.3 Enhancement n-channel MOSFET voltage divider bias circuit.

we need a voltage at the gate V_{GS} greater than the threshold voltage, V_T. Applying zero gate voltage, the channel does not exist and no current will circulate between drain and source. Biasing this type of transistor will require a gate voltage more positive than the source and larger than V_T. This voltage can be applied using a voltage divider R_1 and R_2 dividing V_{DD} as indicated in Figure 14.3.

The voltage between gate and source can be found from the voltage divider as

$$V_{GS} = \left(\frac{R_2}{R_1 + R_2}\right) V_{DD} \tag{14.3}$$

As in the case of the depletion FET applying KVL to the closed circuit consisting of the voltage across R_D, the voltage between source and drain, V_{DS}, and V_{DD} we can find V_{DS} (Equation 14.2) as

$$V_{DS} = V_{DD} - I_D R_D$$

and the drain current can be obtained from the I–V characteristic for an FET (Equation 6.1) as

$$I_{DS} = \frac{W}{L} \mu_n C_{ox} \left[(V_{GS} - V_T)V_{DS} - \frac{V_{DS}^2}{2} \right] \tag{14.4}$$

Here W, L, μ_n, and C_{ox} are constant at a particular temperature and they can be represented by one constant K. Then in designing FET amplifier I_{DS} usually can be approximated to

$$I_{DS} \approx K(V_{GS} - V_T)^2 \tag{14.5}$$

EXAMPLE 14.2

For the enhancement MOSFET biasing circuit shown in Figure 14.4, calculate the voltage between gate and source, V_{GS}, and the voltage between drain and source, V_{DS}, when a drain current of 5 mA is achieved.

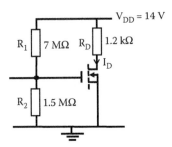

FIGURE 14.4   Enhancement MOSFET biasing circuit for Example 14.2.

Solution

By KVL we have

$$V_{DD} = V_{DS} + I_D R_D = 14 = V_{DS} + 1.2 \times 10^{+3} \times 5 \times 10^{-3}$$

This gives

$$V_{DS} = 8 \text{ V}$$

The voltage V_{GS} is provided by the potential divider formed by R_1 and R_2, then

$$V_{GS} = \left(\frac{R_2}{R_1 + R_2} \right) V_{DD} = \frac{1.5 \times 10^6}{1.5 \times 10^6 + 7 \times 10^6} \times 14 = 2.47 \text{ V}$$

14.2.2.1 Drain-Feedback Bias

Alternative biasing for enhancement FET biasing is to use a feedback resistor at the drain as shown in Figure 14.5.

For the drain-feedback bias circuit, R_G is large and the gate current is very small. The gate current can be neglected and the voltage drop across R_G can be assumed zero, then

$$V_{GS} = V_{DS} \tag{14.6}$$

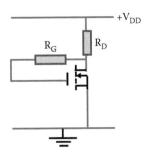

FIGURE 14.5   Enhancement n-channel MOSFET drain-feedback bias circuit.

Also, Equations 14.1 and 14.2 will apply for this drain-feedback biasing, then

$$V_{DS} = V_{DD} - I_D R_D$$

14.3   JFET BIASING

There are two common types of biasing JFET amplifiers: self-bias and voltage divider bias.

14.3.1   SELF-BIAS

To control the size of the depletion region, the self-bias JFET will require the gate–source reverse biased. For an n-channel transistor this requires a negative voltage at the gate V_{GS}, the inclusion of a resistor at the source will help to achieve this as indicated in Figure 14.6.

R_G does not affect the biasing as there is essentially no voltage drop across it. The gate voltage therefore remains at 0 V. R_G is large to isolate an ac signal from ground to ac signals.

The voltage at the source terminal V_S with respect to ground is equal to the voltage across R_S as

$$V_S = I_S R_S \qquad (14.7)$$

Here, V_S is positive with respect to ground and $I_S = I_D$ then V_S can be written as

$$V_S = I_D R_S \qquad (14.8)$$

As the voltage at the gate is equal to zero V_{GS} has the same magnitude as V_s, but with reverse polarity. Then,

$$V_{GS} = -I_D R_S \qquad (14.9)$$

Applying KVL to the closed circuit consisting of the voltage across R_S, the voltage between source and drain V_{DS}, the voltage across R_D and V_{DD} we can write

$$V_{DD} = I_D R_D + V_{DS} + I_S R_S \qquad (14.10)$$

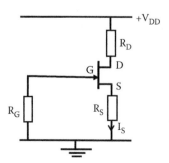

FIGURE 14.6   JFET self-bias circuit

We can assume $I_D = I_S$ then the V_{DS} can be expressed in terms of I_D as

$$V_{DS} = V_{DD} - I_D(R_D + R_S) \tag{14.11}$$

Now the voltage at the drain terminal with respect to ground, V_D, can be expressed as

$$V_D = V_{DS} + I_S R_S \tag{14.12}$$

or replacing V_D into Equation 14.10 and rearranging we can write

$$V_D = V_{DD} - I_D R_D \tag{14.13}$$

EXAMPLE 14.3

For the self-biased circuit shown in Figure 14.7, calculate V_{DS} and V_{GS} when the potential difference across R_D is equal to 3.5 V.

SOLUTION

The voltage across R_D is 3.5 V, then by Ohm's law the drain current, I_D, is

$$I_D = \frac{V_{RD}}{R_D} = \frac{3.5\,V}{800\,\Omega} = 4.375\text{ mA}$$

and by KVL

$$V_{DD} = V_{RD} + V_{DS} + I_S R_S = 16 = 3.5 + V_{DS} + 300 \times 4.375 \times 10^{-3}$$

Then,

$$V_{DS} = 11.19\text{ V}$$

and by KVL

$$V_{RG} = V_{GS} + V_{RS}$$

But the dc voltage at the gate with respect to ground is zero, then

$$V_{GS} = -V_{RS} = -I_S R_S = -4.375 \times 10^{-3} \times 300 = -1.3125\text{ V}$$

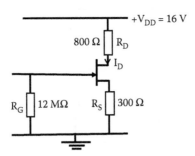

FIGURE 14.7   Self-bias biasing circuit diagram for Example 14.3.

14.3.2 Voltage Divider Bias

A voltage divider formed with R_1 and R_2 can be used to apply a voltage to the gate as indicated in Figure 14.8. With the resistor at the source R_S and drain R_D this circuit bias a JFET in the same way as a BJT auto bias.

The voltage at the gate and source are positive, then the voltage at the source needs to be larger than the voltage at the gate to produce a gate–source reverse bias.

The voltage at the source terminal with respect to ground V_S is equal to the voltage drop across R_S as

$$V_S = I_D R_S \tag{14.14}$$

The gate voltage with respect to ground V_G is provided by voltage divider through R_1 and R_2 as

$$V_G = \left(\frac{R_2}{R_1 + R_2} \right) V_{DD} \tag{14.15}$$

This is the voltage across the resistor R_2.

Now applying KVL to the closed circuit consisting of the voltage across R_2 (V_G), the voltage between gate and source (V_{GS}), and the voltage across R_S (V_S), we can write

$$\left(\frac{R_2}{R_1 + R_2} \right) V_{DD} = V_{GS} + I_D R_S \tag{14.16}$$

or

$$V_{GS} = \left(\frac{R_2}{R_1 + R_2} \right) V_{DD} - I_D R_S \tag{14.17}$$

Rearranging we can deduce an expression for the drain current I_D as

$$I_D = \frac{V_{DD} R_2 / (R_1 + R_2) - V_{GS}}{R_S} \tag{14.18}$$

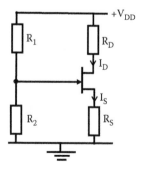

FIGURE 14.8 JFET voltage divider bias circuit.

With equation for V_{GS} and I_D we can fix the operation point Q and determine values of resistors to achieve this operating point.

EXAMPLE 14.4

For the biasing voltage divider circuit shown in Figure 14.9, if the operating point is given by $I_D = 1.2$ mA and $V_{DS} = 4$ V, calculate the voltage across R_D, R_S, voltage V_{DD}, and the voltage at the gate with respect to ground, V_G.

SOLUTION

By KVL

$$V_{DD} = R_D \times I_D + V_{DS} + I_S R_S = 4 \times 10^3 \times 1.2 \times 10^{-3} + 4 + 1.5 \times 10^3 \times 1.2 \times 10^{-3}$$

and
$$V_{DD} = 10.6 \text{ V}$$

$$V_{RS} = R_S \times I_S = 1.5 \times 10^3 \times 1.2 \times 10^{-3} = 1.8 \text{ V}$$

$$V_{RD} = R_D \times I_D = 4 \times 10^3 \times 1.2 \times 10^{-3} = 4.8 \text{ V}$$

and the voltage at the gate V_G is provided by the potential divider formed by R_1, R_2, and V_{DD}. Then,

$$V_G = \left(\frac{R_2}{R_1 + R_2} \right) V_{DD} = \frac{180 \times 10^3}{180 \times 10^3 + 500 \times 10^3} \times 10.6 = 2.81 \text{ V}$$

14.4 KEY POINTS

- FET is biased to fix the operating point Q in a similar way as to BJT.
- The operating point Q in an FET is fixed by the voltage between drain and source V_{DS} and the drain current I_D.

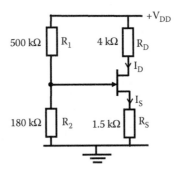

FIGURE 14.9 Voltage divider biasing circuit diagram for Example 14.4.

- In a depletion MOS transistor the dc voltage at the gate can be assumed equal to zero and R_G is very large to isolate an ac signal from ground.
- To bias an enhancement MOS transistor a voltage divider can be used to make the gate voltage more positive than the source and larger than V_T.
- There are two common types of biasing for JFET transistors: self-bias and voltage divider bias.

15 Field Effect Transistor as Amplifiers

15.1 INTRODUCTION

FETs are used to amplify ac signals. The biasing of an FET using dc voltages determines the operating point. Superimposed to the dc current an ac input signal can be applied in order to be amplified. It means that the operating point, Q, will change along with the ac voltage variations. FETs are used in the same way as BJT to amplify voltages or currents. As in BJT, capacitors are used to separate the dc biasing of a transistor and the amplification of an ac signal. In this chapter, we will see the FET as part of an amplifier and deduce the characteristics of these amplifiers.

15.2 COMMON SOURCE AMPLIFIER

Figure 15.1 shows a common source FET amplifier. The coupling capacitors C_C block the dc biasing at input and output and the C_S bypass R_S at ac to make the source common to input and output.

15.2.1 AC VOLTAGE GAIN AT MEDIUM FREQUENCIES

Figure 15.2 shows an equivalent circuit for this amplifier using the mutual conductance parameters model for the transistor under medium ac frequencies. It is assumed that at the working frequencies, medium frequencies, the impedance for the capacitors are very small and can be neglected. Also note that the resistor R_G is represented as an open circuit. R_G is a very large resistor and in this type of transistor the input impedance is very large as we have seen in previous chapters.

We can deduce the output voltage, v_O, by adding the two resistors r_d and R_D in parallel to form R_T' as

$$R_T' = \frac{R_D r_d}{R_D + r_d} \tag{15.1}$$

and then, this equivalent resistor multiplied by the current from the source, $g_m v_{GS}$ will provide v_O as

$$v_O = -\frac{R_D r_d}{R_D + r_d}(g_m v_{GS}) \tag{15.2}$$

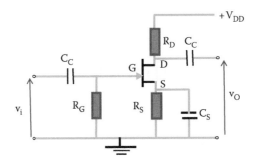

FIGURE 15.1   Circuit diagram for a MOSFET common source amplifier circuit.

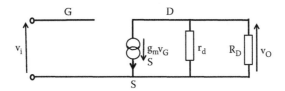

FIGURE 15.2   FET common source amplifier ac equivalent circuit at medium frequencies.

The negative sign is due to the opposite direction of the currents. The input voltage is the voltage applied between gate and source. Therefore,

$$v_i = v_{GS}$$

and the voltage gain at medium frequencies, Gv_{mf}, can be expressed as the ratio of output voltage to input voltage as

$$Gv_{mf} = \frac{v_O}{v_i} = \frac{-(R_D r_d / R_D + r_d)(g_m v_{GS})}{v_{GS}} = -\frac{R_D r_d}{R_D + r_d} g_m \qquad (15.3)$$

In order to obtain a high gain, R_D needs to be larger than r_d.

EXAMPLE 15.1

The common source amplifier shown in Figure 15.3 has an FET with $r_d = 100$ kΩ and $g_m = 20$ mS. Draw an equivalent circuit at medium frequencies and calculate the voltage gain at mid-frequencies and the power dissipated in R_L when the input voltage is 25 mV.

SOLUTION

At medium frequencies an equivalent circuit, using mutual conductance model for the transistor is shown in Figure 15.4.

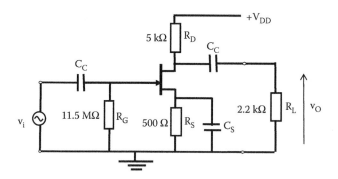

FIGURE 15.3   Common source amplifier for Example 15.1.

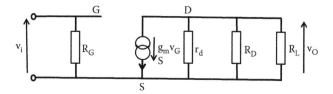

FIGURE 15.4   Medium frequency equivalent circuit for the common source amplifier of Example 15.1.

The resistors R_D and R_L can be combined into one resistor, R'_D as

$$R'_D = \frac{R_D R_L}{R_D + R_L} = \frac{5 \times 10^3 \times 2.2 \times 10^3}{5 \times 10^3 + 2.2 \times 10^3} = 1527.78\ \Omega$$

and the voltage gain at medium frequencies is

$$Gv_{mf} = \frac{v_O}{v_i} = -\frac{R'_D r_d}{R'_D + r_d} g_m - \frac{1527.78 \times 100 \times 10^3}{1527.78 + 100 \times 10^3} \times 20 \times 10^{-3} = 30.10$$

The output voltage is

$$v_O = v_i \times Gv_{mf} = 30.10 \times 25 \times 10^{-3} = 752.5\ \text{mV}$$

and the power dissipated in R_L is

$$P_{R_L} = \frac{v_0^2}{R_L} = \frac{(752.5 \times 10^{-3})^2}{2.2 \times 10^3} = 0.2574\ \text{mW}$$

EXAMPLE 15.2

The voltage divider amplifier shown in Figure 15.5 has an FET with $r_d = 100\ \text{k}\Omega$ and $g_m = 20\ \text{mS}$. Draw an equivalent circuit at medium frequencies for this amplifier and calculate the current passing through R_L when the input voltage is 30 mV.

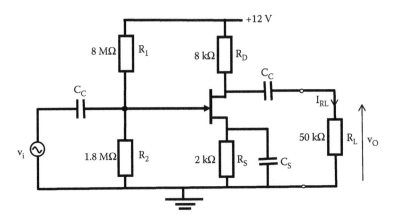

FIGURE 15.5   Amplifier circuit diagram of Example 15.2.

SOLUTION

At medium frequencies an equivalent circuit, using mutual conductance model for the transistor is shown in Figure 15.6.

The resistors R_D and R_L can be combined into one resistor, R'_D as

$$R'_D = \frac{R_D R_L}{R_D + R_L} = \frac{8 \times 10^3 \times 50 \times 10^3}{8 \times 10^3 + 50 \times 10^3} = 6.897 \text{ k}\Omega$$

and the voltage gain at medium frequencies is

$$Gv_{mf} = \frac{v_O}{v_i} = -\frac{R_D r_d}{R_D + r_d} g_m = \frac{6.897 \times 10^3 \times 100 \times 10^3}{6.897 \times 10^3 + 100 \times 10^3} \times 20 \times 10^{-3} = 129.04$$

The output voltage is

$$v_O = v_i \times Gv_{mf} = 30 \times 10^{-3} \times 129.04 = 3.87 \text{ V}$$

and the current through R_L is

$$I_{RL} = \frac{v_O}{R_L} = \frac{3.87}{50 \times 10^3} = 7.74 \times 10^{-5} \text{ A}$$

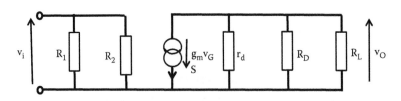

FIGURE 15.6   Medium frequency equivalent circuit for the common source amplifier of Example 15.2.

15.2.2 AC VOLTAGE GAIN AT HIGH FREQUENCIES

At high frequencies R_G can still be neglected, but the capacitances between the gate and source C_{GS}, between gate and drain C_{GD}, and between drain and source C_{DS} will affect the voltage gain. Figure 15.7 shows an equivalent circuit for the common source amplifier of Figure 15.1 at high frequencies including these interelectrodes capacitances.

We can deduce the output voltage, v_O, by adding R'_T (parallel combination of R_D and r_d) to the impedance of C_{DS} in parallel to equivalent impedance Z'_T as

$$Z'_T = \frac{R'_T 1/j\omega C_{DS}}{R'_T + 1/j\omega C_{DS}} = \frac{R'_T}{1 + j\omega C_{DS} R'_T} \tag{15.4}$$

where Z'_T is the parallel combination of the impedances C_{DS}, r_d, and R_D.

The equivalent impedance Z'_T multiplied by the current i_2 will provide the output voltage v_O as

$$v_O = -i_2 Z'_T \tag{15.5}$$

The negative sign is due to the opposite direction of the currents.

The input voltage is the voltage applied between gate and source, so

$$v_i = v_{GS} \tag{15.6}$$

In order to obtain an expression for the voltage gain we need to find an expression to relate i_2 to input and output voltages. Now if we apply KCL at the node formed by currents i_1, i_2, and $g_m v_{GS}$ we find the following relationship:

$$g_m v_{GS} = i_1 + i_2 \tag{15.7}$$

The current i_1 can be found by dividing the voltage across the capacitor C_{GD} by the impedance of C_{GD} capacitor. The voltage across C_{GD} is $(v_{GS} - v_O)$ and the current i_1 can be found as

$$i_1 = \frac{v_{GS} - v_O}{1/j\omega C_{GD}} = (v_{GS} - v_O)j\omega C_{GD} \tag{15.8}$$

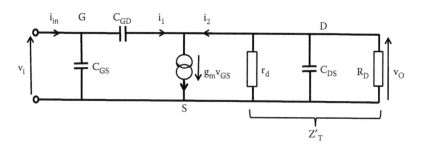

FIGURE 15.7 FET common source amplifier ac equivalent circuit at high frequencies.

Now replacing i_2 from Equation 15.5 and i_1 from Equation 15.8 into Equation 15.7, we get

$$g_m v_{GS} = (v_{GS} - v_O)j\omega C_{GD} - \frac{v_O}{Z'_T} = v_{GS}j\omega C_{GD} - v_O j\omega C_{GD} - \frac{v_O}{Z'_T}$$

or

$$v_{GS}(j\omega C_{GD} - g_m) = v_O\left(j\omega C_{GD} + \frac{1}{Z'_T}\right) \qquad (15.9)$$

Equation 15.9 contains the input and output voltages, rearranging this equation we find an expression for the voltage gain at high frequencies as

$$Gv_{HF} = \frac{v_O}{v_i} = \frac{v_O}{v_{GS}} = -\frac{g_m Z'_T(1 - j\omega C_{GD}/g_m)}{1 + j\omega C_{GD}Z'_T} \qquad (15.10)$$

If $1/g_m$ is much smaller than Z'_T, then the voltage gain at high frequency can be approximated to

$$Gv_{HF} = \frac{v_O}{v_i} \approx -\frac{g_m Z'_T}{1 + j\omega C_{GD}Z'_T} = -\frac{g_m R'_T}{1 + j\omega R'_T(C_{GD} + C_{DS})} \qquad (15.11)$$

As the frequency increases, the voltage gain at high frequencies reduces according to Equation 15.11.

15.2.3 INPUT IMPEDANCE AT HIGH FREQUENCY

The high frequencies equivalent circuit for the common source shown in Figure 15.7 allows us to obtain an expression for the input impedance at high frequency. Applying KCL to the node formed by i_{in}, i_1, and current through the capacitor C_{GS}, an expression for the input current can be obtained as

$$i_{in} = i_1 + v_{GS}(j\omega C_{GS}) \qquad (15.12)$$

Now replacing i_1 from Equation 15.8 into Equation 15.12, we can write

$$i_{in} = (v_{GS} - v_O)j\omega C_{GD} + v_{GS}(j\omega C_{GS}) \qquad (15.13)$$

From Figure 15.7, we can obtain the output voltage v_O as

$$v_O = -v_{GS} g_m Z'_T \qquad (15.14)$$

Replacing the expression of Z'_T from Equation 15.4 the output voltage v_O is

$$v_O = -\frac{v_{GS} g_m R'_T}{1 + j\omega C_{GD} R'_T} \qquad (15.15)$$

Replacing v_O from Equation 15.15 into Equation 15.13 and rearranging we can write

$$i_{in} = v_{GS}\left[j\omega C_{GS} + j\omega C_{GD} + \frac{j\omega C_{GD} g_m R'_T}{1 + j\omega C_{GD} R'_T} \right] \qquad (15.16)$$

The term $j\omega C_{GD} R'_T$ is much less than 1 and Equation 15.16 can be approximated to

$$i_{in} \approx v_{GS}[j\omega C_{GS} + j\omega C_{GD} + j\omega C_{GD} g_m R'_T] = v_{GS} j\omega\{C_{GS} + C_{GD}(1 + g_m R'_T)\} \quad (15.17)$$

Now we can write that the input impedance Z_{in} as the input voltage, v_{in}, divided by the input current, i_{in}, to be

$$Z_{in} = \frac{v_i}{i_{in}} = \frac{v_{GS}}{v_{GS} j\omega\{C_{GS} + C_{GD}(1 + g_m R'_T)\}} = \frac{1}{j\omega\{C_{GS} + C_{GD}(1 + g_m R'_T)\}} \qquad (15.18)$$

From this expression for the input impedance we can deduce that the total input capacitance, C_{in}, can be expressed as

$$C_{in} = \{C_{GS} + C_{GD}(1 + g_m R'_T)\} \qquad (15.19)$$

The term multiplying the C_{GD} indicates the Miller effect. The input impedance can be written in terms of the input capacitance as

$$Z_{in} = \frac{v_i}{i_{in}} = \frac{1}{j\omega C_{in}} \qquad (15.20)$$

The miller capacitance is large thereby reducing the gain at high frequency.

15.3   COMMON DRAIN (SOURCE FOLLOWER)

Figure 15.8 shows a common drain FET amplifier, the coupling capacitors C_C block the dc biasing at input and output.

There is no capacitor in parallel with R_S making the drain common to input and output. As described earlier, R_G is very large for small ac signals resulting in an almost open circuit.

15.3.1   AC VOLTAGE GAIN AT MEDIUM FREQUENCY

Figure 15.9 shows an equivalent circuit for this amplifier using the mutual conductance parameters model for the transistor under medium frequencies ac. R_G is

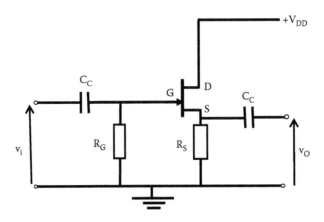

FIGURE 15.8 MOSFET common drain amplifier circuit.

represented as an open circuit; R_G is a very large resistor and in this type of transistor the input impedance is very large as we have seen in previous chapters.

From Figure 15.9 we can see that the output voltage v_O can be expressed as the current $g_m v_{GS}$ multiplied by the parallel combination of r_d and R_S $[R_s r_d/(r_d + R_S)]$ giving

$$v_O = g_m v_{GS} \frac{R_S r_d}{r_d + R_S} \tag{15.21}$$

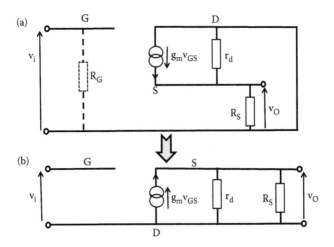

FIGURE 15.9 FET: (a) Common drain amplifier ac equivalent circuit at medium frequencies including the gate resistance R_G. (b) Simplified equivalent circuit for a common drain amplifier ac equivalent circuit at medium frequencies.

and the voltage between gate and source, v_{GS}, relates input and output voltages as

$$v_{GS} = v_i - v_O \tag{15.22}$$

Replacing v_{GS} from Equation 15.22 into v_O in Equation 15.21, we get

$$v_O = g_m(v_i - v_O)\frac{R_S r_d}{r_d + R_S} = g_m v_i \frac{R_S r_d}{r_d + R_S} - v_O g_m \frac{R_S r_d}{r_d + R_S} \tag{15.23}$$

Rearranging, we have

$$v_O\left(1 + g_m \frac{R_S r_d}{r_d + R_S}\right) = g_m v_i \frac{R_S r_d}{r_d + R_S} \tag{15.24}$$

and the voltage for a common drain amplifier at medium frequencies, Gv_{CDmf}, can be expressed as

$$Gv_{CDmf} = \frac{v_O}{v_i} = \frac{g_m R_S r_d/(r_d + R_S)}{1 + g_m R_S r_d/(r_d + R_S)} \tag{15.25}$$

Now $g_m R_S r_d/(r_d + R_S) \gg 1$ and the gain for a common drain amplifier is almost unity, then

$$Gv_{CDmf} \approx 1 \tag{15.26}$$

EXAMPLE 15.3

The common drain amplifier shown in Figure 15.10 has FET with $r_d = 100$ kΩ and $g_m = 200$ mS. Calculate the voltage gain at mid-frequencies.

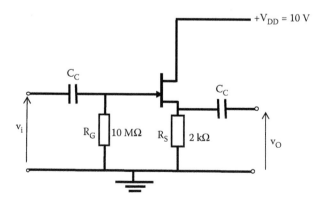

FIGURE 15.10 Circuit diagram for the common drain amplifier of Example 15.3.

SOLUTION

At medium frequencies, the voltage gain using mutual conductance model for the transistor is

$$Gv_{CDmf} = \frac{v_O}{v_i} = \frac{g_m R_S r_d/(r_d + R_S)}{1 + g_m R_S r_d/(r_d + R_S)}$$

$$= \frac{20 \times 10^{-3} \times 2 \times 10^3 \times 100 \times 10^3/(100 \times 10^3 + 2 \times 10^3)}{1 + 20 \times 10^{-3} \times 2 \times 10^3 \times 100 \times 10^3/(100 \times 10^3 + 2 \times 10^3)}$$

and the voltage gain is $G_v = 0.975$.

Therefore, the voltage gain of a common drain amplifier can be approximated to 1.

15.3.2 AC VOLTAGE GAIN AT HIGH FREQUENCY

At high frequencies R_G can still be neglected, but the capacitances between the gate and source C_{GS}, between gate and drain C_{GD}, and between drain and source C_{DS} will affect the gain. Figure 15.11 shows an equivalent circuit at high frequencies including these interelectrodes capacitances.

Note that the voltage between gate and source v_{GS} is not the input voltage. The output voltage v_O can be written as

$$v_O = (g_m v_{GS} + i_1)Z_T = Z_T g_m v_{GS} + Z_T i_1 \tag{15.27}$$

where Z_T is the parallel combination of the impedances: r_d, $1/j\omega C_{DS}$, and R_S. The voltage across gate and source, v_{GS}, can be written as

$$v_{GS} = v_i - v_O = \frac{i_1}{j\omega C_{GS}} \tag{15.28}$$

or

$$i_1 = j\omega C_{GS}(v_i - v_O) \tag{15.29}$$

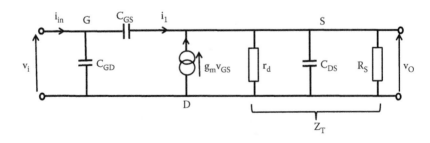

FIGURE 15.11   FET common drain amplifier ac equivalent circuit at high frequencies.

Replacing v_{GS} and i_1 into Equation 15.27, we have

$$v_O = Z_T g_m (v_i - v_O) + Z_T j\omega C_{GS}(v_i - v_O) \tag{15.30}$$

or

$$v_O = Z_T g_m v_i - Z_T g_m v_O + Z_T j\omega C_{GS} v_i - Z_T j\omega C_{GS} v_O$$

or

$$v_O[1 + Z_T(g_m + j\omega C_{GS})] = Z_T v_i(g_m + j\omega C_{GS})$$

and the voltage for a common drain amplifier at high frequencies, Gv_{CDhf}, can be expressed as

$$Gv_{CDhf} = \frac{v_O}{v_i} = \frac{Z_T(g_m + j\omega C_{GS})}{1 + Z_T(g_m + j\omega C_{GS})} = \frac{g_m + j\omega C_{GS}}{1/Z_T + g_m + j\omega C_{GS}} \tag{15.31}$$

Now, $1/Z_T = 1/R_S + 1/r_d + j\omega C_{DS}$ and the common drain voltage gain at high frequency can be written as

$$Gv_{CDhf} = \frac{v_O}{v_i} = \frac{(g_m + j\omega C_{GS})R_S}{1 + [g_m + 1/r_d + j\omega(C_{GS} + C_{DS})]R_S} \tag{15.32}$$

At low frequency in the limit $\omega = 0$ and the voltage gain can be approximated to

$$Gv_{CDlf}(\omega = 0) = \frac{g_m R_S}{1 + [g_m + 1/r_d]R_S} \approx \frac{g_m}{g_m + 1/r_d} \tag{15.33}$$

Therefore, the voltage gain at a low frequency is positive and less than 1.

15.3.3 INPUT IMPEDANCE AT HIGH FREQUENCY

In the equivalent circuit shown in Figure 15.11, we can apply KCL at the node created by currents through capacitors C_{GS}, C_{GD}, and the input current to obtain an expression for the input current i_{in} as

$$i_{in} = i_1 + v_i j\omega C_{GD} \tag{15.34}$$

where i_1 is given in Equation 15.29 and $v_o = G_v v_i$. Then, the input current can be expressed as

$$i_{in} = j\omega C_{GS}(v_i - v_O) + v_i j\omega C_{GD} = j\omega C_{GS}(v_i - G_v v_i) + v_i j\omega C_{GD}$$

or

$$i_{in} = v_i\{j\omega[C_{GD} + C_{GS}(1 - G_v)]\} \tag{15.35}$$

and the input impedance at high frequency, Z_{infh}, is then

$$Z_{inhf} = \frac{v_i}{i_{in}} = \frac{1}{j\omega[C_{GD} + C_{GS}(1 - G_v)]} \tag{15.36}$$

Therefore, the input capacitance is

$$C_{in} = C_{GD} + C_{GS}(1 - G_v) \tag{15.37}$$

Due to typical values of common drain at high frequency the input capacitance can be approximated to $C_{in} \approx C_{GD}$.

The input capacitance at high frequency is not increased by the Miller effect and the common drain amplifier presents high input impedance and low input capacitance.

15.3.4 OUTPUT IMPEDANCE AT HIGH FREQUENCY

To obtain the output impedance we can set the input voltage to zero and apply a voltage v_L at the output. The short circuit at the input will then bypass the capacitor C_{GD}. Figure 15.12 shows the equivalent circuit used to obtain the output impedance.

As the input is short circuited, the output voltage v_O will be equal to the voltage between gate and source:

$$v_{GS} = -v_L = v_O \tag{15.38}$$

The current source in the equivalent circuit will be $g_m v_O$ with the direction indicated in Figure 15.12. Applying KCL, the current i_o can be expressed as

$$i_o = g_m v_O + v_O[1/r_d + j\omega C_{DS} + j\omega C_{GS}] = v_O[g_m + 1/r_d + j\omega(C_{DS} + C_{GS})] \tag{15.39}$$

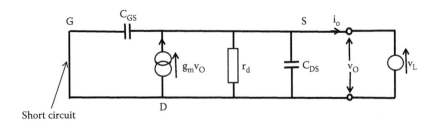

FIGURE 15.12 Equivalent circuit for a common drain amplifier with an input short circuited.

The output impedance at high frequency, Zo_{hf}, is then given by

$$Zo_{hf} = \frac{v_O}{i_o} = \frac{1}{[g_m + 1/r_d + j\omega(C_{DS} + C_{GS})]} \qquad (15.40)$$

At low frequencies, the dominant term is g_m. Then,

$$Zo_{hf} \approx \frac{1}{g_m} \qquad (15.41)$$

The common drain amplifier can be used as buffer amplifier.

15.4   KEY POINTS

- The FET transistor connected in common source mode has relative high voltage gain.
- FET amplifier has high input impedance.
- The common source amplifier has a poor high-frequency response due to high Miller effect capacitance.
- In the common drain amplifier, the gain approximate unity has low input capacitance and high input impedance. It can be used as a buffer amplifier.

16 Transfer Function and Bode Diagrams

16.1 INTRODUCTION

The relationship between the input and the output of a network provides information about the behavior of electronic systems. The transfer function considers the ratio of output to input in an electrical or electronic system. Electronic systems usually contain reactive components, and the characteristics of these components change with frequency. Bode diagrams provide graphical response of the magnitude and phase in an electronic system as the frequency is varied. The use of idealized straight-line Bode plots provides a quick frequency response of a system. Bode plots are quite commonly used in investigating frequency response of amplifiers and filters. This chapter deals with transfer functions and graphical representation of their frequency response as a Bode diagram.

16.2 TRANSFER FUNCTIONS

A transfer function is the ratio between an output quantity and an input quantity. In a passive or active system,

$$\text{Transfer function} = \frac{\text{Output quantity}}{\text{Input quantity}} \quad (16.1)$$

The quantity can be impedances, admittances, currents, voltages, powers, etc. Our concern here is with voltages transfer functions.

A representation of an electronics system as a block is shown in Figure 16.1. The electronic system can be an active or passive network.

The voltage transfer function (TF) in this case will be the ratio of the output voltage to the input voltage as

$$\text{Transfer function} = \text{TF} = \frac{v_O}{v_i} \quad (16.2)$$

The ratio input value to output value is termed inverse TF.

16.2.1 EXAMPLES OF TRANSFER FUNCTIONS PASSIVE COMPONENTS

In various electronic systems, the transfer function presents certain forms of expressions that are repeated. Some common examples are:

A constant value TF = K
First-order low-pass filter response TF = $1/(1 + j(f/f_p))$

237

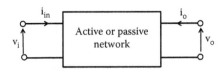

FIGURE 16.1   Block diagram transfer function of an electronic system.

First-order high-pass filter response TF $= 1/(1 + j(f_z/f))$
Straight-line TF $= j2\pi f$

This can be illustrated with examples.

16.2.2   First-Order Low-Pass Transfer Function

Consider the circuit shown in Figure 16.2 where the system is a series RC circuit and the output is taken as the voltage across the capacitor.
The input voltage is given by

$$v_i = i\left(R + \frac{1}{j\omega C}\right)$$

and the output voltage is

$$v_O = i\left(\frac{1}{j\omega C}\right)$$

Then, the transfer function is

$$TF = \frac{v_O}{v_i} = \frac{i/j\omega C}{i/(R + 1/j\omega C)} = \frac{1}{1 + j\omega RC} = \frac{1}{1 + j2\pi fRC} \qquad (16.3)$$

The term $1/2\pi CR$ has a dimension of frequency in Hertz. If we let f_p be the frequency cut-off given by

$$f_p = \frac{1}{2\pi CR} \qquad (16.4)$$

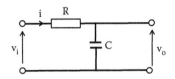

FIGURE 16.2   Circuit diagram of an RC passive network.

then in Equation 16.3, the TF can be written as

$$TF = \frac{1}{1+ j(f/(1/2\pi CR))} = \frac{1}{1+ j(f/f_p)} \tag{16.5}$$

This is the usual form for analyzing frequency response of a single low-pass frequency response.

In polar form this TF can be written as

$$TF = \frac{1}{\sqrt{1+(f/f_p)^2}} \angle -\tan^{-1}\left(\frac{f}{f_p}\right) \tag{16.6}$$

and the magnitude and phase change as the frequency is varied. This behavior of the transfer function with frequency can be analyzed graphically using what is known as *Bode plots*.

16.2.3 FIRST-ORDER HIGH-PASS TRANSFER FUNCTION

Consider now the circuit shown in Figure 16.3, a series CR circuit and the output is taken as the voltage across the resistor.

The input voltage is given by

$$v_i = i\left(R + \frac{1}{j\omega C}\right)$$

and the output voltage is

$$v_O = iR$$

Then, the transfer function is

$$TF = \frac{v_O}{v_i} = \frac{iR}{i(R + 1/j\omega C)} = \frac{1}{1 + (1/j\omega CR)} \tag{16.7}$$

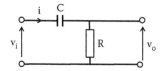

FIGURE 16.3 Circuit diagram for a CR passive network.

If we let f_o be the frequency cut-off given by

$$f_o = \frac{1}{2\pi CR} \tag{16.8}$$

then the TF can be written as

$$TF = \frac{1}{1+(1/j\omega CR)} = \frac{1}{1-j(f_o/f)} \tag{16.9}$$

or in polar form

$$TF = \frac{1}{\sqrt{1+(f_o/f)^2}} \angle \tan^{-1}(f_o/f) \tag{16.10}$$

and the magnitude and phase change can be analyzed as the frequency is varied. By combining passive components it is possible to obtain transfer functions with a variety of behaviors.

EXAMPLE 16.1

For the circuit shown in Figure 16.4, obtain a transfer function between input, v_i, and output, v_o.

By using KVL, the input voltage can be expressed in terms of the voltages across the inductor and resistance as

$$v_i = j\omega L i + R i$$

and the output voltage is the voltage across the resistance as given below:

$$v_O = R i$$

Then, the transfer function for this circuit is

$$TF = \frac{v_O}{v_i} = \frac{R i}{j\omega L i + R i} = \frac{R}{j\omega L + R} = \frac{1}{1+j\omega L/R} \tag{16.11}$$

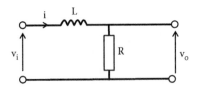

FIGURE 16.4 LR circuit diagram for Example 16.1.

This transfer function also provides a single-pole response.
If

$$f_c = \frac{1}{2\pi L/R} \tag{16.12}$$

then Equation 16.11 can be written as

$$TF = \frac{1}{1 + j(f/f_c)} \tag{16.13}$$

This transfer function also provides a low-pass frequency response.

EXAMPLE 16.2

For the circuit shown in Figure 16.5, obtain a transfer function between input, v_i, and output, v_o.

By using KVL, the input voltage can be expressed in terms of the voltages across the resistance and inductor as

$$v_i = Ri + j\omega Li$$

and the output voltage is the voltage across the inductor as given below:

$$v_O = j\omega Li$$

Then the transfer function, TF, for this circuit is

$$TF = \frac{v_O}{v_i} = \frac{j\omega Li}{Ri + j\omega Li} = \frac{j\omega L}{R + j\omega L} = \frac{1}{1 + R/j\omega L} = \frac{1}{1 + (1/j\omega L/R)} \tag{16.14}$$

This transfer function also provides a high-pass frequency response.

EXAMPLE 16.3

For the circuit shown in Figure 16.6, obtain a transfer function between input, v_i, and output, v_o.

By using KVL, the input voltage can be expressed in terms of the voltages across the capacitor and the two resistances as

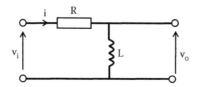

FIGURE 16.5   RL circuit diagram for Example 16.2.

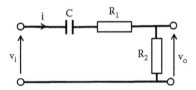

FIGURE 16.6   Two resistors and a capacitor circuit diagram for Example 16.3.

$$v_i = R_1 i + R_2 i + \frac{i}{j\omega C}$$

and the output voltage is the voltage across the resistor R_2 as given below:

$$v_O = R_2 i$$

Then the transfer function, TF, for this circuit is

$$TF = \frac{v_O}{v_i} = \frac{R_2 i}{R_1 i + R_2 i + (i/j\omega C)} = \frac{R_2}{(R_1 + R_2) + (1/j\omega C)} = \frac{R_2/(R_1 + R_2)}{1 + (1/j\omega C(R_1 + R_2))}$$

$$(16.15)$$

This transfer function also provides a high-pass frequency response plus a constant response.

EXAMPLE 16.4

For the circuit shown in Figure 16.7, obtain a transfer function between input, v_i, and output, v_o.

By using KVL, the input voltage can be expressed in terms of the voltages across the resistor, inductor, and capacitor as

$$v_i = R_1 i + j\omega L i + \frac{i}{j\omega C}$$

and the output voltage is the voltage across the capacitor as follows:

$$v_O = \frac{i}{j\omega C}$$

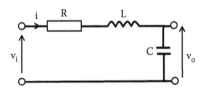

FIGURE 16.7   RLC circuit diagram for Example 16.4.

Then the transfer function, TF, for this circuit is

$$TF = \frac{v_O}{v_i} = \frac{i/j\omega C}{R_i i + j\omega L i + (i/j\omega C)} = \frac{1/j\omega C}{R_1 + j\omega L + (1/j\omega C)} = \frac{1/j\omega C}{R_1 + j(\omega L - 1/\omega C)} \quad (16.16)$$

This transfer function provides a straight-line due to the function of the numerator and low-pass-type frequency response due to the denominator. The overall frequency response is the combination of these two responses.

16.3   BODE PLOTS

The frequency response of network is indicated by plotting two curves: the magnitude of the transfer voltage gain as a function of frequency and the phase-lead angle as a function of frequency. These characteristics are called Bode plots. The magnitude is plotted as 20 log $|G_v|$ dB versus logarithmic of frequency and the phase as phase-shift angle in degrees versus logarithmic of frequency.

For a single low-pass frequency response, the expression for the transfer function was found as indicated in Equation 16.5 or Equation 16.6 as

$$TF = \frac{1}{1 + j(f/f_p)}$$

or in polar form

$$TF = \frac{1}{\sqrt{1 + (f/f_p)^2}} \angle - \tan^{-1}(f/f_p)$$

16.3.1   MAGNITUDE BODE PLOT

We can express the magnitude of the transfer function in decibel as

$$|Gv(f)| = 20 \log\left(\frac{1}{\sqrt{1 + (f/f_p)^2}}\right) = 20 \log(1) - 20 \log\left(\sqrt{1 + (f/f_p)^2}\right)$$

As 20 log (1) = 0 we can write an expression for magnitude of the transfer function in decibels as a function of the frequency as given below:

$$|Gv(f)| = -20 \log\left(\sqrt{1 + (f/f_p)^2}\right) \quad (16.17)$$

For the phase Bode diagram we use the phase angle in degrees. From the expression of the transfer function the phase angle will be

$$\text{Phase-shift angle } \theta = -\tan^{-1}(f/f_p) \quad (16.18)$$

Using Equations 16.17 and 16.18 we can vary the frequency to obtain a Bode plot of the magnitude and phase. These are shown in Figure 16.8.

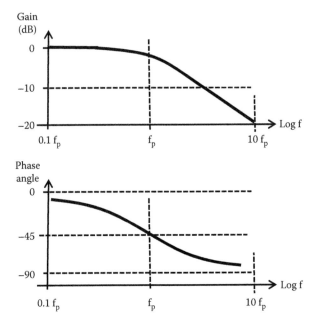

FIGURE 16.8 Bode plot of a single low-pass frequency response.

The input signal is amplified from zero frequency up to the cut-off frequency, f_p. As the frequency increases above cut off the gain decreases linearly at a rate of -20 dB/decade.

There are other transfer functions with simple Bode plot construction.

16.3.2 Transfer Function = TF = K (Constant)

As the constant does not change with variation of frequency, the magnitude will be a constant value. If the transfer constant is a constant value K (constant), then

$$TF = K \qquad (16.19)$$

The Bode plot of its magnitude is a horizontal line and the phase will be $0°$ or $-180°$ at all frequencies. Therefore, the magnitude Bode plot is given by

$$\text{Magnitude of TF} = |TF| = 20\log(K) \qquad (16.20)$$

and

$$\text{Phase of TF} = \angle(TF) = 0° = \text{if K positive} \qquad (16.21)$$

or

$$= \angle(TF) = -180° \text{ if K negative}$$

16.3.3 TRANSFER FUNCTION = TF = $j2\pi f$

In this case, the transfer function changes linearly with variation of frequency

$$TF = j2\pi f \qquad (16.22)$$

The magnitude Bode plot is given by

$$\text{Magnitude of TF} = |TF| = 20\log(2\pi f) \qquad (16.23)$$

Then, this function will provide, as magnitude, a straight-line with 20 dB/decade for slope. In this case, the real part of the TF is zero, then the phase angle for the TF is

$$\text{Phase of TF} = \angle(TF) = \tan^{-1}(2\pi f/0) = 90° \qquad (16.24)$$

16.3.4 TRANSFER FUNCTION = TF = $(1 + jf/f_p)$

This function is the reciprocal of the pole function. Due to the property of logarithms where $\log(1/Z) = -\log(Z)$ the response is the same as the pole, but with negative values. The phase response is the negative of the pole phase response and it proceeds from 0 at low frequencies to +90° at high frequencies. More complex transfer functions can be constructed combining simple transfer functions using the properties of logarithms.

16.4 IDEALIZED BODE PLOTS

The Bode curves can be approximated by straight linear regions. These interconnected straight-line characteristics are referred to as idealized Bode plots (or linear Bode plot, or asymptotic curves Bode plot).

16.4.1 MAGNITUDE ASYMPTOTIC

The expression for the magnitude of a single low-pass TF is shown in Equation 16.17.

$$|Gv(f)| = -20\log\left(\sqrt{1+\left(\frac{f}{f_p}\right)^2}\right)$$

When the frequency is small compared to f_p then $(f/f_p)^2 \ll 1$ and the gain becomes $|Gv(f)| = -20\log(1) = 0$ a constant straight-line. Now at high frequencies with respect to f_p, $(f/f_p)^2 \gg 1$ and the gain becomes $|Gv(f)| = -20\log(f/f_p)$ a straight-line whose slope is −20 dB/decade. That is, the gain magnitude decreases by 20 dB every decade, for example, at $f = 10f_p$ the gain becomes

$$|Gv(f = 10f_p)| = -20\log\left(\frac{10f_p}{f_p}\right) = -20\log(10) = -20\,dB$$

These two asymptotic lines are drawn in Figure 16.9a together with the actual Bode plot. Note that the two asymptotic lines intersect at $f = f_p$. This point is called the corner frequency or the 3 dB frequency as the Bode plot and the idealized Bode plot differ by 3 dB at this point; therefore, it is the cut-off frequency.

16.4.2 BODE PLOT ASYMPTOTIC PHASE

For the single low-pass TF we found that the phase is given by Equation 16.18 as given below:

$$\text{Phase-shift angle } \theta = -\tan^{-1}(f/f_p)$$

If frequencies are much less than the cut-off frequency $f \ll f_p$ then the phase $\theta = 0°$.

If frequencies are much larger than the cut-off frequency $f \gg f_p$ then the phase $\theta = -90°$.

If frequencies are equal to the cut-off frequency $f = f_p$ then the phase $\theta = -45°$.

The idealized phase Bode plot is constructed as follows: one horizontal line for $0 < f/f_p < 0.1$ $\theta = 0°$, one horizontal line for $f/f_p > 0$ $\theta = -90°$. These two lines are joined by a straight-line of slope $-45°$/decade passing through $\theta = -45°$ at f_p as shown in Figure 16.9b.

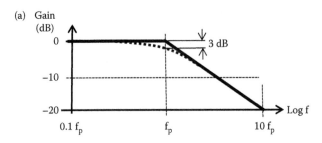

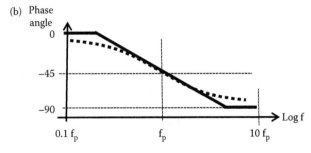

FIGURE 16.9 Single pole idealized Bode plot. (a) Asymptotic magnitude of a low-pass response and (b) asymptotic phase of a single low-pass response.

16.5 CONSTRUCTION OF BODE PLOTS

Bode plots of complex frequency response functions are constructed by summing the magnitude and phase angle contributions of each term in the transfer function. This is due to the property of logarithms. The idealized Bode plots are often sufficient to analyze the frequency response of a transfer function. If more accurate plots are desired, errors introduced by the asymptotic approximations may be determined at selected frequencies.

An example for a transfer function terms with different cut-off frequencies is represented by the following equation

$$Gv(f) = \frac{k(1 + jf_{o1}/f)(1 + jf_{o2}/f)}{(1 + jf/f_{p1})(1 + jf/f_{p2})} \tag{16.25}$$

The magnitude is given by

$$20\log|Gv(f)| = 20\log(k) + 20\log\left[\left|1 + \frac{f_{o1}}{f}\right|\right] + 20\log\left[\left|1 + \frac{f_{o2}}{f}\right|\right]$$

$$+ 20\log\left[\left|\frac{1}{(1 + f/f_{p1})}\right|\right] + 20\log\left[\left|\frac{1}{(1 + f/f_{p2})}\right|\right] \tag{16.26}$$

and the phase

$$< Gv(f) = < [k] + < \left[1 + \frac{f_{o1}}{f}\right] + < \left[1 + \frac{f_{o2}}{f}\right] + < \left[\frac{1}{(1 + f/f_{p1})}\right] + < \left[\frac{1}{(1 + f/f_{p2})}\right] \tag{16.27}$$

16.6 BODE PLOT EXAMPLE

We want to draw an idealized (asymptotic) Bode plot for the transfer function of the circuit shown in Figure 16.10.

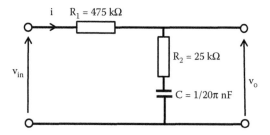

FIGURE 16.10   Circuit diagram for idealized Bode plot example.

The current i can be expressed as

$$i = \frac{v_{in}}{R_1 + R_2 + 1/j\omega C}$$

Then, the output voltage can be written as the current i times the impedance at the output as

$$v_O = (R_2 + 1/j\omega C)\frac{v_{in}}{R_1 + R_2 + 1/j\omega C} = \frac{v_{in}(1 + j\omega CR_2)}{1 + j\omega C(R_1 + R_2)}$$

Then the transfer function voltage gain for this network is

$$\frac{v_O}{v_{in}} = \frac{(1 + j\omega CR_2)}{1 + j\omega C(R_1 + R_2)} \tag{16.28}$$

Now, if

$$f_1 = \frac{1}{2\pi C(R_1 + R_2)}$$

and

$$f_2 = \frac{1}{2\pi CR_2}$$

replacing f_1 and f_2 the voltage gain becomes

$$\frac{v_O}{v_{in}} = \frac{(1 + j(f/f_2))}{(1 + j(f/f_1))} \tag{16.29}$$

Using the values of resistors and capacitance of this example the values of f_1 and f_2 can be calculated as

$$f_1 = \frac{1}{2\pi(10^{-9}/20\pi)(475 \times 10^3 + 25 \times 10^3)} = 20\,\text{kHz}$$

$$f_2 = \frac{1}{2\pi(10^{-9}/20\pi) \times 25 \times 10^3} = 400\,\text{kHz}$$

and the bode plot for the magnitude is

$$20\log|Gv(f)| = 20\log\left|1 + \frac{f}{f_2}\right| + 20\log\left[\frac{1}{(1 + f/f_1)}\right]$$

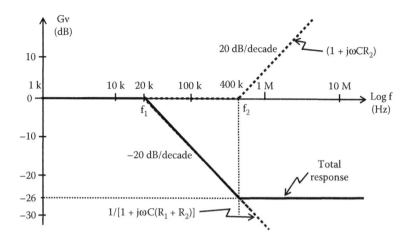

FIGURE 16.11 Asymptotic magnitudes Bode diagram for the transfer function of the network shown in Figure 16.10. Bode plot example.

Figure 16.11 shows the idealized Bode plot of this circuit voltage transfer function. For the $1/[1 + j\omega C(R_1 + R_2)]$ term the magnitude response is 0 dB until f_1 then as the frequency increases above f_1 the amplitude decreases linearly at a 20 dB/decade rate. For the $(1 + j\omega CR_2)$ term response, the voltage gain magnitude is 0 dB until f_2 then as the frequency increase above f_2 the amplitude increases linearly at a 20 dB/decade rate. The combined total response can be obtained by adding the responses of $1/[1 + j\omega C(R_1 + R_2)]$ and $(1 + j\omega CR_2)$. Figure 16.11 shows the response of $1/[1 + j\omega C(R_1 + R_2)]$ and $(1 + j\omega CR_2)$ plus the total response of the network. We can see from the overall frequency response that from 0 Hz until f_2 there is contribution from $1/[1 + j\omega C(R_1 + R_2)]$ only; above f_2 there are contribution from $(1 + j\omega CR_2)$ and $1/[1 + j\omega C(R_1 + R_2)]$, the response is a constant value from f_2 as the responses from $(1 + j\omega CR_2)$ and $1/[1 + j\omega C(R_1 + R_2)]$ are at the same rate but in opposite direction.

The phase response for the network shown in Figure 16.10 is given by

$$< Gv(f) = < \left(1 + \frac{f}{f_2}\right) + < \frac{1}{(1 + f/f_1)}$$

Figure 16.12 shows the idealized Bode plot phase response for $(1 + j\omega CR_2)$ and $1/[1 + j\omega C(R_1 + R_2)]$ as well as the total phase response for this example.

In general, Bode diagrams present an easy and quick assessment of the frequency response of a network. For this example, the phase changes as the frequency increases; for the function on $1/[1 + j\omega C(R_1 + R_2)]$ the phase angle is 0 from 0 Hz (dc) to about a decade less than f_1 frequency. As the frequency increases, the phase decreases at a rate of 45°/decade reaching −45° at a frequency of 20 kHz; when the phase reaches −90° the phase does not decrease and remains at −90°. The response of the functions on $(1 + j\omega CR_2)$ provides a phase equal to 0 until about one decade less than the cut-off frequency, f_2, then the phase gradually increases at a rate of 45°/decade reaching 45°

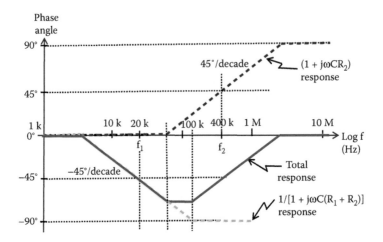

FIGURE 16.12   Asymptotic phase Bode diagram for the Bode plot example.

at the f_2 frequency of 400 kHz. When the phase reaches 90° it does not increase any further and the phase remains at 90°. Combining the two responses we obtain the total response as indicated in Figure 16.12.

16.7   KEY POINTS

- A transfer function is the ratio between an output quantity and an input quantity.
- The frequency response of network is indicated by plotting two curves: the magnitude of the transfer gain and the phase-lead angle as a function of frequency. These characteristics are called Bode plots. The magnitude is plotted as $20 \log |G_v|$ dB versus logarithmic of frequency and the phase as phase-shift angle degree versus logarithmic of frequency.
- Bode diagrams provide a graphical response of the magnitude and phase in an electronic system as the frequency is varied.
- The use of idealized straight-line Bode plots provides a quick frequency response of a system.
- Bode plots are quite commonly used in investigating frequency responses of amplifiers and filters.
- Bode plots of complex frequency response functions are constructed by summing the magnitude and phase-angle contributions of standard Bode functions such as poles and zeros (or pair of complex poles or zeros).

17 Feedback in Amplifiers

17.1 INTRODUCTION

A system with feedback, FB, is a circuit where part of the output signal is added to the input signal. For example, if an amplifier has FB the input becomes dependent on the signal at the output. Part of the output signal can be added as in the following: (i) to reduce the input; then such FB is called negative feedback (NFB) (other names for NFB are inverse FB, degenerative FB, or anti-phase FB); or (ii) to increase the input signal; this is called positive FB (other names for positive FB are regenerative FB and in-phase FB).

To illustrate how FB affects a system, let us consider the block diagram of a system with FB as indicated in Figure 17.1, where G is the gain of the original system without FB and B is the gain of the FB network. A signal sample from the output X_2 is taken to be added or subtracted from the input X_1. The FB signal BX_2 is added (positive FB) or subtracted (NFB) from the input signal X_1 and then amplified by G. Then, output with FB is

$$X_2 = G(X_1 \pm BX_2) = GX_1 \pm GBX_2$$

or

$$GX_1 = X_2(1 \mp GB)$$

Then, the overall gain with FB is

$$\frac{X_2}{X_1} = \frac{G}{(1 \mp GB)} \tag{17.1}$$

The sign of $\mp$ indicates whether there is NFB (+) or positive FB (−).
In this chapter, we will concentrate on the effect of NFB in amplifiers.

17.2 NEGATIVE FEEDBACK

For the NFB case, the output signal is taken from the input signal and the overall gain then becomes

$$\frac{X_2}{X_1} = \frac{G}{(1 + GB)} \tag{17.2}$$

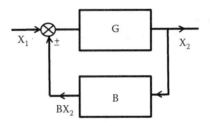

FIGURE 17.1 Block diagram of a system with FB.

If the gain G is large, which is usually the case, then GB ≫ 1 and the overall gain can be approximated to

$$\frac{X_2}{X_1} \approx \frac{1}{B} \tag{17.3}$$

and the overall gain is now independent of G, the gain of the system without feedback. This is a system whose overall gain is no longer defined by the actual gain of the internal amplifier, but only by the FB path. If the FB path is constructed of passive components, and if these components are chosen to be very stable types, then the amplifier overall gain can also be very stable. Therefore NFB produces a gain stability, which is a useful property of FB.

The sample signal to be used as feedback can be a current or a voltage. Figure 17.2 shows a general block diagram of an amplifier with gain G and NFB gain B.

The output may be a voltage or a current. At the input the FB signal can combine voltages (series FB) or can combine currents (parallel FB).

Voltages or currents can be used as feedback in series or parallel. We then have four possible different FB systems:

- Series voltage NFB
- Series current NFB
- Parallel voltage NFB
- Parallel current NFB

In the following sections it is assumed that the amplifier is a non-inverter amplifier. In this amplifier, there is no phase change between output and input.

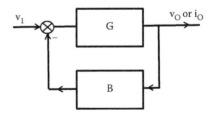

FIGURE 17.2 General block diagram of a system with NFB.

17.2.1 Series Voltage NFB

Figure 17.3 shows a block diagram of series voltage NFB. A sample of the output voltage is fed back as a voltage in series with the input voltage.

An example of the FB network can be a simple voltage divider as shown in Figure 17.4.

The "gain" or transfer function for this FB voltage divider circuit is

$$B = \frac{R_1}{R_1 + R_2} \tag{17.4}$$

Representing the amplifier by a general three-parameter model an equivalent circuit can be drawn to show the network including the feedback as shown in Figure 17.5,

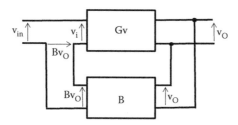

FIGURE 17.3 Block diagram of a system with series voltage NFB.

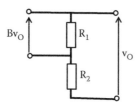

FIGURE 17.4 Diagram of the feedback circuit for a series voltage NFB.

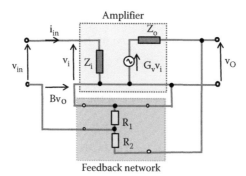

FIGURE 17.5 Circuit diagram of an amplifier with series voltage NFB.

where G_v is the open voltage gain of the amplifier alone, Z_i and Z_o are the input and output impedances, respectively, of the amplifier without NFB.

17.2.1.1 Series Voltage NFB Voltage Gain

For series voltage NFB we have the input to the amplifier, v_i, is equal to

$$v_i = v_{in} - Bv_O \tag{17.5}$$

But $v_i = v_O/G_v$ replacing v_i into Equation 17.5, we have

$$v_O = G_v(v_{in} - Bv_O) = G_v v_{in} - G_v Bv_O$$

Rearranging, we have

$$v_O(1 + G_v B) = G_v v_{in}$$

and the overall voltage gain for series voltage NFB can be written as

$$Gv_{FB} = \frac{v_O}{v_{in}} = \frac{G_v}{1 + G_v B} \tag{17.6}$$

The voltage gain of the overall system had been reduced by a factor $(1 + G_v B)$ when series voltage NFB is applied.

17.2.1.2 Series Voltage NFB Effects on Input Impedance

The input impedance can be calculated using Equation 17.5. Rearranging this equation, we can write

$$v_{in} = v_i + Bv_O = v_i \left(1 + B\frac{v_O}{v_i}\right)$$

The overall input impedance Zi_{FB} when series voltage NFB is applied, can be obtained as

$$Zi_{FB} = \frac{v_{in}}{i_{in}} = \frac{v_i[1 + (B(v_O/v_i))]}{i_{in}} = \frac{v_i}{i_{in}}\left[1 + \left(\frac{v_O}{v_i}\right)B\right]$$

But $Gv = v_O/v_i$ and the input impedance of the amplifier without FB is $Z_i = v_i/i_{in}$ then Zi_{FB} can be expressed as

$$Zi_{FB} = Z_i(1 + G_v B) \tag{17.7}$$

That is, the input impedance with series voltage NFB is increased by a factor of $(1 + G_v B)$.

17.2.1.3 Series Voltage NFB Effects on Output Impedance

To find the output impedance with feedback we short circuit the input ($v_{in} = 0$), apply a voltage v_o at the output, and calculate the output current, I_o. Figure 17.6 shows the circuit of the amplifier with NFB with the input replaced by a short circuit.

Applying KVL at the output of the circuit shown in Figure 17.6, we have

$$v_O - G_v v_i = I_o Z_o \tag{17.8}$$

and applying KVL at the input ($v_{in} = 0$), we can write

$$v_i = -Bv_O \tag{17.9}$$

Replacing v_i from Equation 17.9 into Equation 17.8, we have

$$v_O(1 + G_v B) = I_o Z_o$$

and the output impedance with series voltage NFB Zo_{FB} can be written as

$$Zo_{FB} = \frac{v_O}{I_o} = \frac{Z_o}{(1 + G_v B)} \tag{17.10}$$

The output impedance with series voltage NFB is reduced to a $(1 + G_v B)$ factor.

17.2.1.4 Series Voltage NFB Effects on Frequency Response

At high frequency, the voltage gain was found to be (Equation 10.20, Chapter 10)

$$Gv_{HF} = \left(\frac{Gv_{mf}}{1 + j\omega C_P R_T} \right)$$

or

$$Gv_{HF} = \left(\frac{Gv_{mf}}{1 + jf/f_H} \right) \tag{17.11}$$

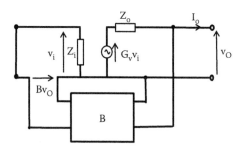

FIGURE 17.6 Circuit diagram of the amplifier with NFB with a short circuit at the input to calculate output impedance.

where

$$f_H = \frac{1}{2\pi C_p R_T}$$

with series voltage NFB the voltage gain at high frequency, $Gv_{HF.FB}$, will be

$$Gv_{HF.FB} = \frac{Gv_{HF}}{1 + BGv_{HF}} \qquad (17.12)$$

and replacing Gv_{HF} from Equation 17.11, we have

$$Gv_{HF.FB} = \frac{Gv_{mf}/(1 + jf/f_H)}{1 + BGv_{mf}/(1 + jf/f_H)} = \frac{Gv_{mf}}{(1 + jf/f_H) + BGv_{mf}}$$

Rearranging this equation, we have the voltage gain at high frequency with series voltage NFB, $Gv_{HF.FB}$, as

$$Gv_{HF.FB} = \frac{Gv_{mf}/(1 + jf/f_H)}{(1 + BGv_{mf})[1 + jf/f_H(1 + BGv_{mf})]} \qquad (17.13)$$

That can be written as

$$Gv_{HF.FB} = \left(\frac{Gv_{FB}}{1 + jf/f_{H.FB}} \right) \qquad (17.14)$$

where Gv_{FB} is the voltage gain with series voltage NFB at medium frequency and

$$f_{H.FB} = f_H(1 + BGv_{mf}) \qquad (17.15)$$

is the new frequency cut-off at high frequency with series voltage NFB.

That is, the high frequency cut-off is increased by a factor of $(1 + BGv_{mf})$ when series voltage NFB is applied.

Similarly, the low-frequency cut-off can be obtained as

$$f_{L.FB} = \frac{f_L}{(1 + BGv_{mf})} \qquad (17.16)$$

The low-frequency cut-off is reduced by $(1 + G_v B)$.

The bandwidth $BW = f_H - f_L \approx f_H$ is increased to

$$BW_{NFB} = f_{H.FB} - f_{L.FB} \approx f_{H.FB} \qquad (17.17)$$

Therefore, the bandwidth is increased by $(1 + G_vB)$.

Note that the product gain times bandwidth is constant.

17.2.1.5   Series Voltage NFB Effects on Internal Distortion

Output stages with large voltages swing may no longer have linear characteristics. This may introduce harmonic distortions D, then the output voltage will be

$$v_O = G_v v_i + D \tag{17.18}$$

where D is the distortion signal.

The distortion D will depend on the magnitude of v_O. The input voltage with NFB including the distortion D will be

$$v_i = v_{in} - B(G_v v_i + D)$$

or

$$v_i(1 + BG_v) = v_{in} - BD$$

Then,

$$v_i = \frac{v_{in} - BD}{(1 + BG_v)} \tag{17.19}$$

Replacing v_i from Equation 17.19 into Equation 17.18, we have

$$v_O = \frac{G_v(v_{in} - BD)}{1 + BG_v} + D = \frac{G_v v_{in}}{1 + BG_v} - \frac{G_v BD}{1 + BG_v} + \frac{(1 + BG_v)}{(1 + BG_v)}D \tag{17.20}$$

and the output voltage with series voltage NFB is

$$v_O = \frac{G_v v_{in}}{1 + BG_v} + \frac{D}{1 + BG_v} \tag{17.21}$$

The distortion D is reduced by $(1 + BG_v)$, but v_O is also reduced by $(1 + BG_v)$. If the input signal can be amplified by $(1 + BG_v)$ without introducing more distortion then the original distortion D is reduced.

17.2.1.6   Series Voltage NFB Current Gain

Without NFB we found in Chapter 9 that the current gain (Equation 9.35) and voltage gain (Equation 9.34), written in terms of Z_0 and Z_1, are

$$G_i = \frac{A_v Z_i}{Z_o + R_L} \quad \text{and} \quad G_v = \frac{A_v R_L}{Z_o + R_L}$$

where R_L is a load connected at the output

Now with NFB we have

$$i_O = \frac{v_O}{R_L} = \frac{v_{in}G_v/(1+BG_v)}{R_L} \qquad (17.22)$$

and

$$v_{in} = i_{in}Zi_{FB} = i_{in}Z_i(1+BG_v)$$

Therefore,

$$i_{in} = \frac{v_{in}}{Z_i(1+BG_v)} \qquad (17.23)$$

With output current, i_O, from Equation 17.22 and the input current, i_{in}, from Equation 17.23, we can write the current gain with NFB as

$$Gi_{FB} = \frac{i_O}{i_{in}} = \frac{(v_{in}G_v/(1+BG_v))/R_L}{v_{in}/Z_i(1+BG_v)} = \frac{Z_iG_v}{R_L} = \frac{Z_i}{R_L}\left[\frac{A_vR_L}{Z_o+R_L}\right]$$

and the current gain with NFB can be written as

$$Gi_{FB} = \frac{A_vZ_i}{Z_o+R_L} = G_i \qquad (17.24)$$

The current gain with series voltage NFB is unchanged.

Thus, applying series voltage NFB offers a few advantages in improving the efficiency of an amplifier. It reduces the voltage gain but the gain is more stable and can be controlled by external FB components. It increases the input impedance. It reduces the low-frequency cut-off, hence increases the bandwidth of the amplifier making a better overall amplifier.

17.2.2 Series Current NFB

A sample of the output current can be picked up by the FB network and feedback as indicated in Figure 17.7 where a block diagram of series current NFB is shown.

An example of realization of series current NFB circuit is a resistor R_f connected as indicated in Figure 17.8. The amplifier is represented by its three-parameter model equivalent circuit; Figure 17.8 includes the FB circuit and load R_L.

Now the FB sample taken by the FB network is

$$i_LR_f = Bv_O \qquad (17.25)$$

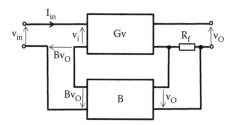

FIGURE 17.7   Block diagram of a system with series current NFB.

and the transfer function for this FB voltage divider circuit is

$$B = \frac{R_f}{R_L} \qquad (17.26)$$

17.2.2.1   Series Current NFB Voltage Gain

Examining the equivalent circuit shown in Figure 17.8 and applying KVL we can write an expression for the input voltage v_i at the amplifier as

$$v_i = v_{in} + Bv_O \qquad (17.27)$$

and the voltage gain of the amplifier is

$$G_v = \frac{v_O}{v_i} \qquad (17.28)$$

Replacing v_i from Equation 17.28 into Equation 17.27, we have

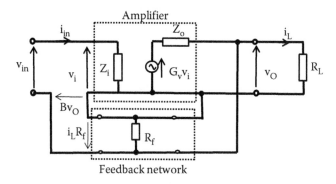

FIGURE 17.8   Circuit diagram of an amplifier with series current NFB.

$$\frac{v_O}{G_v} = v_{in} + Bv_O$$

Rearranging this equation, the voltage gain for the amplifier with series current NFB can be written as

$$Gv_{FB} = \frac{v_O}{v_{in}} = \frac{G_v}{1 - BG_v} \qquad (17.29)$$

In this type of series current NFB the voltage gain is reduced by $(1 - BG_v)$.

17.2.2.2 Series Current NFB Current Gain

As it is a series current FB, the current at the output is the same with feedback. The input current is also the same; therefore, the current gain remains unchanged:

$$Gi_{FB} = Gi \qquad (17.30)$$

17.2.2.3 Series Current NFB Effects on Input Impedance

The input impedance is given by the ratio of the input voltage v_{in} to the input current i_{in} as given below:

$$Zi_{FB} = \frac{v_{in}}{i_{in}} = \frac{v_i - Bv_O}{v_i / Z_i} = \left(\frac{v_i - Bv_O}{v_i}\right) Z_i = \left(1 - B\frac{v_O}{v_i}\right) Z_i$$

Then, the input impedance with series current NFB is

$$Zi_{FB} = (1 - BG_v)Z_i \qquad (17.31)$$

The input impedance with series current NFB is increased by $(1 - BG_v)$.

17.2.2.4 Series Current NFB Effects on Output Impedance

Similarly, the output impedance can be obtained as

$$Zo_{FB} = (1 - BG_v)Z_o \qquad (17.32)$$

The output impedance with series current NFB is increased by $(1 - BG_v)$.

Note that the application of series current NFB increases both the input and output impedances.

17.2.3 PARALLEL VOLTAGE NFB

This time the output voltage is sampled through the FB network as shown in Figure 17.9 for parallel voltage NFB.

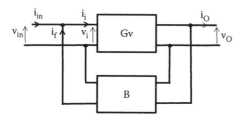

FIGURE 17.9 Block diagram of a system with parallel voltage NFB.

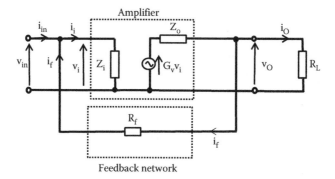

FIGURE 17.10 Circuit diagram of an amplifier with parallel voltage NFB.

An example of realization of parallel voltage NFB circuit is the resistor R_f connected as indicated in Figure 17.10. Representing the amplifier by its equivalent circuit we can have the circuit shown in Figure 17.10 including the FB circuit and load where the output current can circulate.

The transfer function for this FB voltage divider circuit is

$$B = \frac{R_L}{R_f} \tag{17.33}$$

17.2.3.1 Parallel Voltage NFB Voltage Gain

As it is a parallel voltage FB the voltage at the output is the same with feedback. The input voltage is also the same, then the voltage gain remains unchanged:

$$Gv_{FB} = Gv \tag{17.34}$$

17.2.3.2 Parallel Voltage NFB Current Gain

The FB current is $i_f = Bi_O$ and the current gain without feedback is $G_i = i_O/i_i$. Now the input current can be obtained applying KCL as

$$i_{in} = i_i - i_f = i_i - Bi_O \tag{17.35}$$

Then, the current gain for parallel voltage NFB is

$$Gi_{FB} = \frac{i_O}{i_{in}} = \frac{i_O}{i_i - Bi_O} = \frac{i_O/i_i}{1 - Bi_O/i_i}$$

and the current gain with parallel voltage NFB becomes

$$Gi_{FB} = \frac{i_O}{i_{in}} = \frac{G_i}{1 - BG_i} \qquad (17.36)$$

The current gain with parallel voltage NFB is reduced by $(1 - BG_i)$.

17.2.3.3 Parallel Voltage NFB Effects on Input Impedance

The FB voltage, v_{in}, is the same as the voltage without feedback v_i

$$v_{in} = v_i$$

and the impedance without feedback is given by

$$Z_i = \frac{v_i}{i_i}$$

Then, the input impedance with parallel voltage NFB is

$$Zi_{FB} = \frac{v_{in}}{i_{in}} = \frac{v_{in}}{i_i - Bi_O} = \frac{v_{in}/i_i}{1 - Bi_O/i_i}$$

and the input impedance with parallel voltage NFB becomes

$$Zi_{FB} = \frac{v_{in}}{i_{in}} = \frac{Z_i}{1 - BG_i} \qquad (17.37)$$

The input impedance with parallel voltage NFB is reduced by $(1 - BG_i)$.

17.2.3.4 Parallel Voltage NFB Effects on Output Impedance

Similarly, it can be demonstrated that the output impedance with parallel voltage NFB becomes

$$Zo_{FB} = \frac{Z_o}{(1 - BG_v)} \qquad (17.38)$$

The output impedance with parallel voltage NFB is decreased by $(1 - BG_v)$.

Note that the application of parallel (or shunt) voltage NFB decreases both the input and output impedances.

17.2.4 PARALLEL CURRENT NFB

Figure 17.11 shows a block diagram of parallel current NFB. A current sample is taken from the output and fed in parallel to the input as current.

An example of realization of parallel current NFB circuit is the resistor R_f connected as shown in Figure 17.12. The amplifier is represented by its three-parameter model equivalent circuit; Figure 17.12 includes the FB circuit and load R_L where the output current can circulate.

17.2.4.1 Parallel Current NFB Current Gain

The FB signal is a current i_f given as

$$i_f = Bi_O \qquad (17.39)$$

Examining Figure 17.12 applying KCL we can find the input current i_i as

$$i_i = i_{in} + i_f \qquad (17.40)$$

From the current gain expression we can find the output current as $i_O = G_i i_i$ replacing i_i from Equation 17.40 into the equation for i_O, we have

$$i_O = G_i(i_{in} + i_f)$$

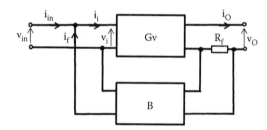

FIGURE 17.11   Block diagram of a system with parallel current NFB.

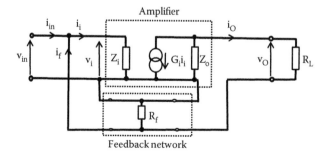

FIGURE 17.12   Circuit diagram of an amplifier with parallel voltage NFB.

and replacing i_f from Equation 17.39 into the expression for i_O, we have

$$i_O = G_i(i_{in} + Bi_O)$$

or

$$i_O(1 - BG_i) = G_i i_{in}$$

Rearranging we can obtain the current gain for this parallel current NFB as

$$G_{i.FB} = \frac{i_O}{i_{in}} = \frac{G_i}{(1 - BG_i)} \qquad (17.41)$$

when we apply parallel current NFB the current gain is reduced by a factor $(1 - BG_i)$ but stabilized.

17.2.4.2 Parallel Current NFB Effects on Input Impedance

Examining Figure 17.12, the input voltage v_{in} can be expressed as

$$v_{in} = Z_i(i_{in} + i_f) = Z_i(i_{in} + Bi_O) = Z_i i_{in} + Z_i Bi_O$$

where Z_i is the input impedance without feedback.

But $i_O = G_i i_i$ and $v_{in} = Z_i i_i$ then the input voltage v_{in} can be written as

$$v_{in} = Z_i i_i = Z_i i_{in} + Z_i BG_i i_i = Z_i i_{in} + v_{in} BG_i$$

Rearranging, we obtain an expression in terms of input voltage and current as

$$v_{in}(1 - BG_i) = Z_i i_{in}$$

and the input impedance with parallel current NFB can be obtained as

$$Z_{i.FB} = \frac{v_{in}}{i_{in}} = \frac{Z_i}{(1 - BG_i)} \qquad (17.42)$$

Parallel current NFB reduces the input impedance by factor $(1 - BG_i)$.

17.2.4.3 Parallel Current NFB Effects on Output Impedance

To calculate the output impedance let us short circuit the input and apply a voltage v_O at the output. This is shown in the equivalent circuit in Figure 17.13.

The output current is

$$i_O = G_i i_i + \frac{v_O}{Z_o} \qquad (17.43)$$

where Z_o is the output impedance without feedback.

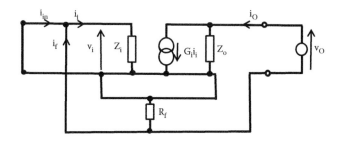

FIGURE 17.13 Circuit diagram to calculate the output impedance with parallel current NFB.

As we have applied a short circuit at the input, the input current i_{in} is zero and at the input we only have the FB current i_f. Hence,

$$i_i = i_f = Bi_O \qquad (17.44)$$

Replacing i_i in Equation 17.43, it becomes

$$i_O = G_i Bi_O + \frac{v_O}{Z_o}$$

or

$$i_O(1 - G_i B) = \frac{v_O}{Z_o}$$

This equation contains the output voltage and current required to obtain an expression for the output impedance with parallel current NFB as

$$Z_{O.FB} = \frac{v_O}{i_O} = Z_o(1 - G_i B) \qquad (17.45)$$

Parallel current NFB increases the output impedance by factor $(1 - BG_i)$.

17.2.4.4 Parallel Current NFB Voltage Gain

From Figure 17.12 the output voltage v_O can be expressed as the output current times the load resistor R_L as

$$v_O = i_O R_L \qquad (17.46)$$

The output current i_O can also be expressed in terms of the current gain with NFB as

$$i_O = G_{i.FB} i_{in} = [G_i / (1 - BG_i)] i_{in}$$

Replacing i_o into Equation 17.46, we have

$$v_O = R_L \left[\frac{G_i}{(1 - BG_i)} \right] i_{in} = R_L G_i \left[\frac{i_{in}}{(1 - BG_i)} \right] \tag{17.47}$$

But from Equation 17.37 for the input impedance, the term $[i_{in}/(1 - BG_i)]$ can be replaced by v_{in}/Z_i as

$$\frac{v_{in}}{Z_i} = \frac{i_{in}}{(1 - BG_i)}$$

Replacing $[i_{in}/(1 - BG_i)]$ into Equation 17.47, we can write an expression for the output voltage gain in terms of the input impedance and the current as

$$v_O = \frac{R_L G_i v_{in}}{Z_i}$$

Then, the voltage gain for parallel current NFB is

$$G_{v.FB} = \frac{v_O}{v_{in}} = \frac{R_L G_i}{Z_i} = G_v \tag{17.48}$$

The voltage gain is unchanged by parallel current NFB.

17.2.4.4.1 NFB on Amplifiers Summary

Table 17.1 summarizes the effects of NFB on voltage and current gain plus effects on input and output impedances.

TABLE 17.1
Summary of NFB to Amplifiers

Type of NFB	Voltage Gain Gv_{NFB}	Current Gain Gi_{NFB}	Output Impedance Zo_{NFB}	Input Impedance Zin_{NFB}
Series voltage	$\dfrac{G_v}{1 + GB}$	Unchanged $G_{INFB} = G_I$	$\dfrac{Z_0}{1 + GB}$	$Z_I(1 + GB)$
Series current	$\dfrac{G_v}{1 - GB}$	Unchanged $G_{INFB} = G_I$	$Z_0(1 - GB)$	$Z_I(1 - GB)$
Parallel voltage	Unchanged $G_{VNFB} = G_V$	$\dfrac{G_I}{1 - GB}$	$\dfrac{Z_0}{1 - GB}$	$\dfrac{Z_I}{1 - GB}$
Parallel current	Unchanged $G_{VNFB} = G_V$	$\dfrac{G_I}{1 - GB}$	$Z_0(1 - GB)$	$\dfrac{Z_I}{1 - GB}$

(a)   (b)

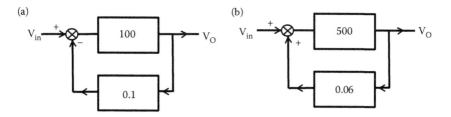

FIGURE 17.14 Two electronic systems showing feedback for Example 17.1 (a) showing negative FB and (b) showing positive FB.

EXAMPLE 17.1

For the negative FB and positive FB networks shown in Figure 17.4, calculate the new voltage gains with feedback.

SOLUTION

For the FB system (a) the voltage gain without feedback is 100, the gain including feedback is given by

$$Gv_{FB(a)} = \frac{V_O}{V_{in}} = \frac{G_v}{1 + G_v B} = \frac{100}{1 + 100 \times 0.1} = 9.09$$

and the new gain with feedback for system (b) shows

$$Gv_{FB(b)} = \frac{V_O}{V_{in}} = \frac{G_v}{1 - G_v B} = \frac{500}{1 - 500 \times 0.06} = -17.24$$

EXAMPLE 17.2

An amplifier has a gain of 8000 NFB that is applied to achieve an overall gain with feedback of 600. Determine the FB ratio required to obtain the required voltage gain.

SOLUTION

The FB system has a voltage gain without feedback of 8000; the voltage gain of 600 including feedback is given by

$$600 = Gv_{FB} = \frac{G_v}{1 + G_v B} = \frac{8000}{1 + 8000 \times B}$$

Then the FB ratio, B, required to achieve an overall voltage gain of 600 is

$$B = \frac{8000 - 600}{8000 \times 600} = 1.54 \times 10^{-3}$$

EXAMPLE 17.3

A two-stage amplifier has a voltage gain of $[50/(1 + jf/f_C)]^2$. This amplifier has a value of $f_C = 2$ kHz and a NFB of $B = 3 \times 10^{-3}$. Calculate the magnitude and phase angle of the voltage gain with NFB at 6 kHz.

SOLUTION

The total voltage gain without feedback is

$$Gv = \left[\frac{50}{1+j6\times10^3/2\times10^3}\right]^2 = \left[\frac{50}{3.162\angle71.57°}\right]^2 = 250.04\angle-143.14°$$

and the voltage gain after NFB is

$$G_{VFB} = \frac{G_v}{1+G_vB} = \frac{250.04\angle-143.14°}{1+(250.04\angle-143.14°)\times3\times10^{-3}} = \frac{250.04\angle-143.14°}{1+(0.75\angle-143.14°)}$$

$$G_{VFB} = \frac{250.04\angle-143.14°}{1+(-0.60-j0.45)} = \frac{250.04\angle-143.14°}{0.40-j0.45}$$

$$= \frac{250.04\angle-143.14°}{0.6\angle-48.36°} = 416.73\angle-94.78°$$

Then, the magnitude of the voltage gain with feedback is 416.73 and the phase angle of the voltage gain with feedback is –94.78°.

EXAMPLE 17.4

An amplifier without feedback has a voltage gain of 40 dB. Calculate the new gain if 0.03 of the output is fed back into the input in opposition to the input signal.

SOLUTION

The voltage gain without feedback is

$$G_V = 40\,dB = 20\log\left(\frac{V_O}{V_i}\right)$$

Then,

$$G_V = \frac{V_O}{V_i} = 10^2 = 100$$

After feedback is applied the gain becomes

$$G_{VFB} = \frac{G_v}{1+G_vB} = \frac{100}{1+100\times0.03} = 25 = 27.959\,dB$$

and the voltage gain with feedback is reduced from 40 to 27.959 dB.

17.3   KEY POINTS

- A system with feedback has part of the output signal added to the input signal.
- There are two types of feedback: NFB and positive FB. NFB is likely to *reduce* the input signal; positive FB tends to *increase* the input.
- Applying series voltage NFB: the voltage gain is reduced but stabilized, the input impedance is increased, the output impedance is reduced, the bandwidth is increased, the internal noise is reduced, and the current gain remains unchanged.
- Applying series current NFB: the voltage gain is reduced but stabilized, the current gain is unchanged, the input impedance is increased, and the output impedance is increased.
- Applying parallel voltage NFB: the voltage gain is unchanged, the current gain is reduced, the input impedance is reduced, and the output impedance is reduced.
- Applying parallel current NFB: the voltage gain is unchanged, the current gain is reduced, the input impedance is reduced, and the output impedance is increased.

18 Differential Amplifiers

18.1 INTRODUCTION

In previous chapters we discussed the biasing, behaviors, and design of a common emitter amplifier. The main structure that provided amplification was based on a transistor and resistors at the emitter and collector, as indicated in Figure 18.1.

The input signal is applied to the base and the output is taken at the collector. Now it is possible to share the emitter resistor R_E with another basic transistor to produce two amplifiers, as shown in Figure 18.2.

We can assume that the two transistors and the collector resistances are identical. This can be achieved as this circuit is usually fabricated in IC form; hence the components are fabricated at the same time with the same conditions. In this combination we have two inputs and two outputs, but we can also obtain another output between the outputs A and B of each amplifier. Then this is a *differential amplifier*. This is also known as long-tailed pair or long-pair amplifier and today is widely used in designing linear ICs. In terms of biasing in IC form, it usually uses two dc sources, one with negative polarity and one with positive polarity.

18.2 SINGLE INPUT VOLTAGE

Let us consider the case when we apply a voltage at input v_{i1} and zero voltage to input v_{i2} (connected to ground). This will create a single input voltage v_i amplifier. Figure 18.3 shows the applied voltage v_i in this single input voltage.

Using the simplified T-model two parameters equivalent circuit for the transistor, as shown in Figure 18.4, we can obtain an ac equivalent circuit small signal for this differential amplifier with a single input voltage.

Now the resistor R_E is very large and for the ac small signal input almost no current will circulate through it, then

$$(\beta+1)i_b' \approx -(\beta+1)i_b$$

and

$$i_b' = -i_b \tag{18.1}$$

Then the output voltages v_A and v_B are

$$v_A = -\beta i_b R_C \tag{18.2}$$

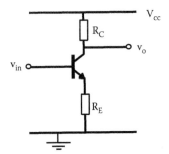

FIGURE 18.1   Circuit diagram of a basic single common emitter amplifier.

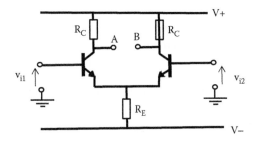

FIGURE 18.2   Circuit diagram of a basic differential amplifier.

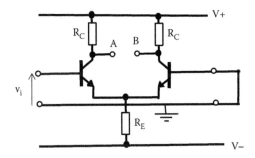

FIGURE 18.3   Differential amplifier with single input voltage.

$$v_B = +\beta i_b R_C \qquad (18.3)$$

As R_E is very large, for ac it is as an open circuit. Neglecting signal current through R_E, we can apply KVL at the loop created by the input voltage v_i and the two r_e resistors to obtain the input voltage as

$$V_i = (\beta + 1)i_b 2r_e \qquad (18.4)$$

Then, the single-ended output voltage gain at A, G_A is

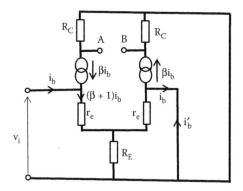

FIGURE 18.4 Ac small-signal equivalent circuit of differential amplifier with a single input voltage.

$$G_A = \frac{V_A}{V_i} = -\frac{\beta i_b R_C}{(\beta+1)i_b 2r_e} = -\frac{\beta R_C}{(\beta+1)2r_e} \tag{18.5}$$

As usually $\beta \gg 1$ then the voltage gain at A can be approximated to

$$G_A \approx -\frac{R_C}{2r_e} \tag{18.6}$$

Then the single-ended output voltage gain at B, G_B is

$$G_B = \frac{V_B}{V_i} = \frac{\beta i_b R_C}{(\beta+1)i_b 2r_e} = +\frac{\beta R_C}{(\beta+1)2r_e} \tag{18.7}$$

$\beta \gg 1$ then the voltage gain at B can be approximated to

$$G_B \approx \frac{R_C}{2r_e} \tag{18.8}$$

or using h parameters these two single output voltage gains can be written as

$$G_A = -\frac{R_C h_{fe}}{2h_{ie}} \tag{18.9}$$

and

$$G_B = \frac{R_C h_{fe}}{2h_{ie}} \tag{18.10}$$

In this amplifier, it is possible to have two outputs one inverting and the other non-inverting output voltage.

18.3 DIFFERENTIAL-MODE VOLTAGE GAIN

This amplifier also gives you the option of selecting an output voltage between terminals A and B; this is the *differential voltage gain*.

The differential output voltage gain at A–B can be obtained as

$$G_{AB} = \frac{v_A - v_B}{v_i} = \frac{v_{AB}}{v_i} = \frac{-\beta i_b R_C - \beta i_b R_C}{(\beta+1)i_b 2r_e} = \frac{-2\beta i_b R_C}{(\beta+1)i_b 2r_e} \qquad (18.11)$$

As indicated earlier, β is usually much larger than 1 and we can approximate the *differential voltage gain* as

$$G_{AB} \approx -\frac{R_C}{r_e} \qquad (18.12)$$

or expressing this differential voltage gains in terms of h parameters, we can write

$$G_{AB} = -\frac{R_C h_{fe}}{h_{ie}} \qquad (18.13)$$

As there are no coupling capacitors at the input or output we can have a large gain at low frequency including dc.

18.4 COMMON MODE VOLTAGE GAIN: EFFECT ON NOISE

Suppose we have a signal containing noise, the signal will then have an extra noise voltage v_n. This noise will be picked up at both inputs and hence the same voltage v_n will appear at both inputs. Figure 18.5 shows a differential amplifier with a signal at one input plus a noise source at the two inputs.

Now let us apply the same noise voltage v_n at the two inputs, as indicated in the equivalent circuit for this differential amplifier shown in Figure 18.6.

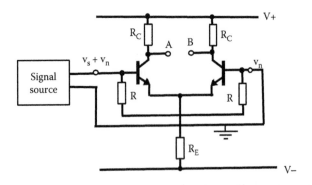

FIGURE 18.5 Differential amplifier with a noise source at both inputs.

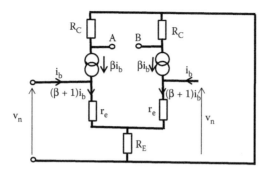

FIGURE 18.6 Ac small-signal equivalent circuit of differential amplifier common mode with a noise voltage, V_n, at both inputs.

Applying KVL at the input closed circuit created by noise voltage, V_n, voltage across r_e and voltage across R_E, we can write

$$v_n = (\beta+1)i_b r_e + 2(\beta+1)i_b R_E = (\beta+1)i_b[r_e + 2R_E] \qquad (18.14)$$

In this case, the current passing through both R_C is in the same direction, so the output voltage at A and B will have the same polarity. Then, the output voltage at A and B will be

$$v_A = -\beta i_b R_C \qquad (18.15)$$

$$v_B = -\beta i_b R_C \qquad (18.16)$$

and the output voltage between A and B is

$$V_{AB} = v_A - v_B = -\beta i_b R_C - (-\beta i_b R) = 0$$

and the common mode voltage gain in this case is

$$G_{AB} = \frac{v_A - v_B}{v_n} = \frac{0}{v_n} = 0 \qquad (18.17)$$

The common mode amplifies the difference of the inputs. In this case, both noise signal inputs are the same so the voltage gain is zero. The common mode noise is eliminated in the output A–B.

The voltage gain at one output is

$$G_A = G_B = \frac{v_A}{v_n} = \frac{v_B}{v_n} = \frac{-\beta i_b R_C}{(\beta+1)i_b[r_e + 2R_E]} = \frac{-\beta R_C}{(\beta+1)[r_e + 2R_E]} \qquad (18.18)$$

As $\beta \gg 1$ and $R_E \gg r_e$ then we can approximate the gain as

$$G_A = G_B = \approx \frac{-R_C}{2R_E} \qquad (18.19)$$

Then the noise signal will be rejected. The single input will be amplified if taken at a single output.

18.5   COMMON MODE REJECTION RATIO

The component of voltage, which is identical at the two inputs, is usually an unwanted signal; therefore, the common mode gain is made as small as possible. The parameter used to express this quality of an amplifier is the common mode rejection ratio (CMRR). CMRR is defined as

$$CMRR = \frac{\text{Signal gain (differential output)}}{\text{Common mode noise gain}} \qquad (18.20)$$

For the circuit shown in Figure 18.5, for the differential outputs the CMRR is

$$CMRR = \frac{\text{Signal gain (differential output)}}{\text{Common mode noise gain}} = \frac{R_C/2r_e}{0} \Rightarrow \infty \qquad (18.21)$$

The CMRR is infinitive.
For the single-ended output the CMRR is

$$CMRR = \frac{\text{Signal gain (differential output)}}{\text{Single-ended output voltage gain}} = \frac{R_C/2r_e}{R_C/2R_E} = \frac{R_E}{r_e} \qquad (18.22)$$

To maximize the CMRR we need a large R_E, without affecting the bias of the transistor. A possible solution is the circuit shown in Figure 18.7.

The emitter resistor is replaced by a circuit with a transistor, T_3 biased by three resistors and the connection of the collector of T_3 replaces the emitter resistor R_E. This circuit behaves as a current source, where the current required to bias the transistor T_1 and T_2 is set as the collector current of transistor T_3. In this circuit, any fluctuation of voltage at the collector of T_3 will not affect the biasing current. For the fluctuation of signal voltage we have what is effectively an infinite resistance R_E. This will increase the CMRR without affecting the bias for T_2 and T_1.

Then we have an amplifier with high CMRR, high gain, and amplification from dc to high frequencies. All these characteristics form the basis for an ideal amplifier, which can be implemented as a unit in integrated form and create an operational amplifier (Op-Amp) to be studied in Chapter 19.

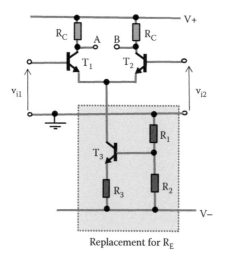

Replacement for R_E

FIGURE 18.7 Differential amplifier with emitter resistance replaced by a current source.

EXAMPLE 18.1

For the differential amplifier shown in Figure 18.8, calculate the output voltage at A, V_{OA}, output voltage at B, V_{OB}, and the voltage between A and B, V_{AB}, when an input voltage of 10 mV is applied and the emitter current is 4 mA. If a small noise signal is contained in the input signal, and it is applied to both inputs, calculate the CMRR for a single ended output of this amplifier. State any assumption made.

SOLUTION

We can assume that both transistors are identical, $\beta \gg 1$ and the amplifier is used at room temperature of 300° K. The emitter resistance, r_e can be determined as

$$r_e = \frac{25 \times 10^{-3} \text{ V}}{I_E} = \frac{25 \times 10^{-3} \text{ V}}{4 \times 10^{-3} \text{ A}} = 6.25 \ \Omega$$

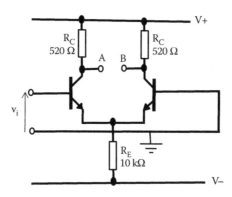

FIGURE 18.8 Circuit diagram for a long pair amplifier use in Example 18.1.

and the voltage gain for the single output at A is

$$GV_A = -\frac{R_C}{2r_e} = -\frac{520}{2 \times 6.25} = -41.6$$

Then, the output voltage at output A is

$$Vo_A = -GV_A \times V_{in} = -41.6 \times 10 \times 10^{-3} = -416 \, mV$$

and the voltage gain for the single output at B is

$$GV_B = \frac{R_C}{2r_e} = \frac{520}{2 \times 6.25} = 41.6$$

Then the output voltage at output B is

$$Vo_B = +416 \, mV$$

and the voltage gain between A and B, GV_{AB} is

$$GV_{AB} = \frac{-R_C}{r_e} = -\frac{520}{6.25} = -83.2$$

Then, the output voltage between A and B, Vo_{AB} is

$$Vo_{AB} = -GV_{AB} \times V_{in} = -83.2 \times 10 \times 10^{-3} = -832 \, mV$$

If noise is introduced into both inputs, the CMRR for the single-ended output is

$$CMRR = \frac{Signal \; gain(differential \; output)}{Single \; ended \; output \; voltage \; gain}$$

$$= \frac{R_L/2r_e}{R_L/2R_E} = \frac{520/2 \times 6.25}{520/2 \times 10 \times 10^3} = \frac{41.6}{0.026} = 1600$$

A high rejection in response to noise.

18.6 KEY POINTS

- Differential amplifiers are very versatile amplifiers, they can provide high gain.
- A differential amplifier can provide gain as single-ended output by taking the output from any of the two outputs or as common mode by taking the output as the difference of the two single-ended outputs.

- A parameter used to express noise reduction in an amplifier is the CMRR. CMRR is an indication of how well an amplifier can reduce or eliminate noise.
- Differential amplifiers have a very high CMRR.
- Differential amplifiers can amplify a signal from a large range of frequencies from dc to high frequency.
- Differential amplifiers form the basis for an Op-Amp.

19 Operational Amplifiers

19.1 INTRODUCTION

A further improvement of the differential amplifier shown in Figure 18.7, Chapter 18, will be to replace the resistor connected to collector R_C by a piece of circuitry including two transistors as seen in Figure 19.1.

Transistors T_4 and T_5 connected as indicated in Figure 19.1 act as the collector load for the differential amplifier. They form what is known as a *current mirror* and in this circuit behave as a dynamic load. This dynamic load increases the voltage gain of the differential amplifier. It also provides the same biasing current for T_1 and T_2. One advantage of the differential amplifier is that they can be directly coupled without capacitors.

A further improvement in creating an amplifier that is nearer to ideal is to reduce the output impedance. This can be achieved by connecting another stage of a common collector amplifier as indicated in Figure 19.2. The circuit shown in Figure 19.2 is a directed coupled version of a common collector amplifier. In this common collector, as in the case of the long pair amplifier, the emitter resistance has been replaced by a current source that replaces R_E.

By adding stages we gradually achieve the characteristics similar to an ideal amplifier. The differential amplifier will provide high voltage gain, very high rejection to noise, amplification from dc to high frequency, and options for inputs and outputs. Adding stages in cascade of common emitter amplifiers will provide a very large voltage gain of several thousand. A final stage can be a common collector amplifier to decrease the output impedance.

Due to the advances of semiconductor technologies thousands or millions of transistors can be fabricated at the same time in integrated form. IC allows complex electronics systems to be fabricated as a unit that can be used as a component.

The operational amplifier (*Op-Amp*) is the unit that meets all the requirements of a near ideal amplifier.

Operational amplifiers (Op-Amps) are among the most widely used components for the construction of analog electronic circuits. They are used by specialist electronic engineers in complex systems and also to implement simple electronic circuits.

Op-Amps were first designed in the 1970s. An Op-Amp consists of an electronic system based on three main functions: (1) an input stage that provides high-input impedance, differential inputs, and large CMRR; (2) a high gain stage to include large amplification from dc to relatively high frequencies; and (3) an output stage to provide low-output impedance and some output power. Figure 19.3 shows a block diagram of a simple basic Op-Amp.

The modern Op-Amp system also includes circuitry to compensate for offset voltage, compensation of frequency response, and compensation of temperature change. Biasing of transistor in an Op-Amp is achieved using circuitries that behave

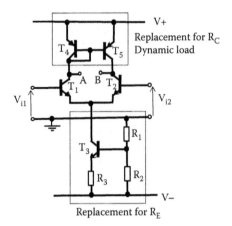

FIGURE 19.1 Differential amplifier with the collector load resistor R_C replaced by a current mirror.

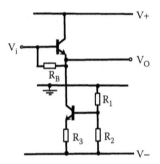

FIGURE 19.2 Circuit diagram for a dc-coupled common collector amplifier circuit.

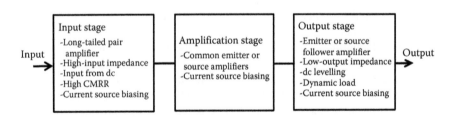

FIGURE 19.3 Block diagram of a simple basic Op-Amp component.

as current sources and dynamic loads. A modern Op-Amp can include more than a hundred transistors.

Modern Op-Amps can include more transistors to improve its characteristics. They can involve a mix of technologies to make use of the advantages of FET and BJT to get even closer to an ideal amplifier.

19.2 OP-AMPS CHARACTERISTICS

Originally, the Op-Amp was developed to perform mathematical operations in analog computing, hence the name. There are a variety of Op-Amps manufactured with different specifications. One of the most common is the general purpose Op-Amp, the 741. The Op-Amp is a highly adaptable device. With suitable external connections and a few extra components it is possible to construct ac and dc amplifiers. It can be used to create complex waveform generators, integrators and differentiators, and a vast number of other applications. Op-Amps are among the most widely used building blocks in electronics. They are ICs often DIL (dual in line) or SMT (surface mount) components as shown in Figure 19.4.

The circuit symbol for an Op-Amp is shown in Figure 19.5, where V– is the inverting input terminal. For a positive input signal connected to the inverting terminal the Op-Amp produces a negative output. The V+ is the noninverting input terminal. The input connected to the noninverting input produces an output signal that has the same polarity (or phase) to its input.

Most Op-Amps are biased using two dc power supplies. Connections to these power supplies are usually not included in circuit diagrams.

The pin connections of a general 741 Op-Amp are indicated in Figure 19.6.

Typical biasing supply voltage arrangement uses voltages of +15 and –15 V. The 741 can use voltages in the range ±5 to ±18 V. Some Op-Amps allow biasing voltages up to ±30 V or low-biasing voltages of ±1.5 V. Single voltage supply operation typically uses biasing voltages between 4 and 30 V.

FIGURE 19.4 IC Op-Amp packages: (a) DIL and (b) SMT.

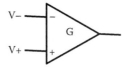

FIGURE 19.5 Op-Amp circuit symbol.

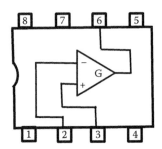

FIGURE 19.6 741 Op-Amp pin connections. Pin 1: offset null, Pin 2: inverting input, Pin 3: noninverting input, Pin 4: bias voltage (negative), Pin 5: offset null, Pin 6: output, Pin 7: bias voltage (positive), and Pin 8: ground.

The *ideal operational amplifier* characteristics:

1. An infinite input impedance, so that it does not affect the source that is providing the input signal (typical value for a general purpose 741 the input impedance is 2 MΩ or higher).
2. Zero output impedance, so that variations in load impedance do not affect the operation of the Op-Amp. The output from the Op-Amp will thus change by a negligible amount when current flows (typical value for the 741 the output impedance is 75 Ω or less).
3. The Op-Amp must have an extremely high inherent voltage gain in order to maximize the stability and performance of the amplifier ideally equal to infinitive (typical value for voltage gain of an Op-Amp 741 is in the range of 10^5–10^6. High gain devices might have a gain of 10^8).
4. With no input signals applied, the output voltage should be zero (in practice there is a very small output voltage with zero input voltage, Op-Amp has an *offset null circuit* to eliminate this unwanted output).
5. An infinite bandwidth (typical at open-loop gain 1, the bandwidth is 1 MHz from dc).
6. Infinite CMRR (CMRR are in the range 80–120 dB. Typical 741 has a CMRR of about 90 dB).
7. Infinite slew rate (typical value 0.5 V/μs).

This is not a full list of parameters of an Op-Amp; manufacturers usually provide a full set of parameters.

19.2.1 Offset Null Circuit

In a practical circuit with no input signal applied there is usually a very small voltage difference at the inputs due to internal circuit noise. This is known as the "input offset voltage." As a result, the output will drift away from earth potential by an amount equal to this offset voltage multiplied by the closed-loop gain of the Op-Amp.

19.2.2 COMPENSATION CIRCUIT

Capacitances or resistors can be used to compensate the original frequency response. Usually the Op-Amp has integrated a compensation circuit.

19.2.3 SLEW RATE

The capacitance in the amplifier has to be charged and discharged every cycle. With a limited current available there is a maximum rate for this change. The maximum possible rate of change of the output voltage is the slew rate provided in volts per microseconds.

19.3 OP-AMP GAIN

The Op-Amp is a special differential amplifier. The output V_O is proportional to the *difference* between its two inputs

$$V_O = G \ [V(+) - V(-)] \tag{19.1}$$

where G is the voltage gain of the Op-Amp alone (strictly called the open-loop gain, G).

19.4 INVERTING AMPLIFIER

When feedback is applied, the characteristic of the Op-Amp is determined largely by the feedback network. NFB may be applied very easily to an Op-Amp by feeding a proportion of the output back to the input.

Figure 19.7 shows the Op-Amp connected as an inverting amplifier, the input signal is connected through the R_{in} resistor to the inverting (−) terminal of the Op-Amp. G is the voltage gain of the Op-Amp alone (open-loop gain), R_f the feedback resistor connected between the input and the output, and R_{in} is the input resistance connected to the input inverting terminal of the Op-Amp.

19.4.1 INVERTING AMPLIFIER VOLTAGE GAIN

The gain of the Op-Amp G can be expressed in terms of the input and output voltages of the Op-Amp alone from Figure 19.7 as

$$G = \frac{V_O}{V_{ve}} \tag{19.2}$$

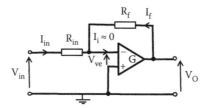

FIGURE 19.7 Circuit diagram for Op-Amp inverting amplifier.

where V_{ve} is the voltage at the input of the Op-Amp between inverting and noninverting terminals. The input current I_{in} can be obtained as the voltage across R_{in} divided by the resistor R_{in}. Then,

$$I_{in} = \frac{V_{in} - V_{ve}}{R_{in}} \tag{19.3}$$

Replacing V_{ve} from Equation 19.2, I_{in} can be written as

$$I_{in} = \frac{V_{in} - V_O/G}{R_{in}}$$

Similarly, the feedback current I_f can be obtained as

$$I_f = \frac{V_O - V_{ve}}{R_f} = \frac{V_O - V_O/G}{R_f} = -\frac{V_O/G - V_O}{R_f} \tag{19.4}$$

Now the current going in the inverting terminal of the Op-Amp, I_i, is almost zero and can be neglected compared to I_f and I_{in}, then we can write

$$I_{in} = -I_f \tag{19.5}$$

Therefore,

$$\frac{V_{in} - V_O/G}{R_{in}} = \frac{V_O/G - V_O}{R_f}$$

Rearranging

$$V_O\left(\frac{1}{G} - 1\right) = V_{in}\left(\frac{R_f}{R_{in}}\right) - \left(\frac{R_f}{R_{in}}\right)\frac{V_O}{G}$$

or

$$V_O\left[\frac{1}{G} - 1 + \left(\frac{R_f}{R_{in}}\right)\frac{1}{G}\right] = V_{in}\left(\frac{R_f}{R_{in}}\right)$$

and the overall gain of the inverting amplifier, Gv_-, is

$$Gv_- = \frac{V_O}{V_{in}} = -\left(\frac{R_f}{R_{in}}\right)\left[\frac{1}{(1 - (1/G)[1 + R_f/R_{in}])}\right] \tag{19.6}$$

The gain G of the Op-Amp G is very large, then 1/G is almost zero and we can approximate the overall voltage gain of the inverting amplifier as

$$Gv_- = \frac{V_O}{V_{in}} = -\frac{R_f}{R_{in}} \tag{19.7}$$

The gain of this amplifier depends only on the external components R_f and R_{in}. Selecting appropriate values for R_f and R_{in}, it is possible to obtain any required gain, this make the design of an amplifier very easy.

19.4.2 VIRTUAL EARTH

The voltage of point x in Figure 19.8 is not connected to ground, but behaves as a virtual ground. Let us calculate the voltage at point x with respect to earth (ground).

Applying KVL around the loop that include voltage across R_f, V_O, and V_{ve} we can write

$$V_{ve} = V_O - R_f I_f \tag{19.8}$$

but

$$V_O = G V_{ve}$$

Replacing V_O into Equation 19.8 we have

$$V_{ve} = G V_{ve} - R_f I_f$$

Rearranging leads to

$$V_{ve}(1 - G) = -R_f I_f$$

and the voltage at point x with respect to ground is

$$V_{ve} = -\frac{R_f I_f}{(1 - G)} \tag{19.9}$$

Now $G \approx \infty$ then $V_{ve} \approx 0$. Therefore, the point x is at 0 V (earthed), but no current circulate to earth. This is known as virtual earth. This characteristic of the Op-Amp can be used to obtain the voltage gain for the inverting amplifier very easy.

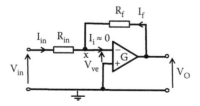

FIGURE 19.8 Op-Amp inverting amplifier voltage at x, virtual ground.

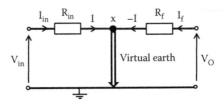

FIGURE 19.9   Equivalent circuit of inverting amplifier using the virtual ground concept.

We can use the concept of *virtual earth* to obtain an approximate equivalent circuit for the inverting amplifier as shown in Figure 19.9.

From Figure 19.9 we can get the voltages V_O and V_{in} as

$$V_O = R_f I_f = -R_f I$$

and

$$V_{in} = R_{in} I_{in} = R_{in} I$$

Then the voltage gain is

$$Gv_- = \frac{V_O}{V_{in}} = -\frac{R_f I}{R_{in} I} = -\frac{R_f}{R_{in}}$$

as seen earlier.

EXAMPLE 19.1

For the amplifier shown in Figure 19.10, determine the value of R_f that will produce a voltage gain of −180.

SOLUTION

In this inverting amplifier the voltage gain is given by

$$Gv_- = \frac{V_O}{V_{in}} = -\frac{R_f}{R_{in}} = -180$$

then

$$R_f = R_{in} \times Gv_- = 3.5 \times 10^3 \times 180 = 630 \text{ k}\Omega$$

and a $R_f = 630$ kΩ will produce a voltage gain of 180.

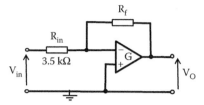

FIGURE 19.10 Circuit diagram of an inverting amplifier for Example 19.1.

EXAMPLE 19.2

Design an inverting amplifier based on an Op-Amp to obtain an output voltage of 255 mV with an input voltage of –3 mV.

SOLUTION

The solution can be achieved using the circuit diagram shown in Figure 19.10, then the voltage gain is

$$Gv_- = \frac{V_O}{V_{in}} = \frac{255\,mV}{-3\,mV} = -85 = -\frac{R_f}{R_{in}}$$

Since it is the ratio of R_f and R_{in} that determine the voltage gain, we can select a suitable value for one of the resistor and calculate the other one. Let us make $R_{in} = 2k\,\Omega$. Then

$$R_f = R_{in} \times Gv_- = 2 \times 10^3 \times 85 = 170\,k\Omega$$

Therefore, using $R_f = 170$ kΩ and $R_{in} = 2$ kΩ an output voltage of 255 mV can be obtained with an input voltage of –3 mV.

19.4.3 INPUT IMPEDANCE

The circuit for an inverting amplifier shown in Figure 19.8 can be used to determine the input impedance.

Since $V_{ve} = 0$ then $V_{in} = R_{in}I_{in}$ and the input impedance $Z_{i/p}$ can be obtained as

$$Z_{i/p} = \frac{V_{in}}{I_{in}} = \frac{R_{in}I_{in}}{I_{in}} = R_{in} \tag{19.10}$$

The input impedance of an inverting amplifier with NFB is equal to the resistance connected to the inverting terminal of the Op-Amp.

19.4.4 OUTPUT IMPEDANCE

The output impedance of an inverting amplifier is

$$Z_{o/p} = \frac{V_O}{I_o} = \frac{Z_o}{G}\left(1 + \frac{R_f}{R_{in}}\right) \tag{19.11}$$

where Z_o is the output impedance of the Op-Amp alone and G is the voltage gain of the Op-Amp alone.

As G is very large, the output impedance of an inverting amplifier with NFB is much less than the output impedance of the Op-Amp alone.

EXAMPLE 19.3

For the amplifier shown in Figure 19.11, determine the input impedance, output impedance, and voltage gain. The manufacturer data sheet provides the following information: $Z_i = 2\ M\Omega$, $Z_o = 70\ \Omega$, and $G = 3 \times 10^5$ for the Op-Amp.

SOLUTION

For the inverting amplifier the input impedance is equal to R_{in}.
Therefore,

$$Z_{i/p} = R_{in} = 5\ k\Omega$$

and for the inverting amplifier the output impedance is equal to

$$Z_{o/p} = \frac{Z_o}{G}\left(1 + \frac{R_f}{R_{in}}\right) = \frac{70}{3 \times 10^5}\left(1 + \frac{180 \times 10^3}{5 \times 10^3}\right) = 8.633\ m\Omega$$

and the voltage gain of this amplifier is

$$Gv_- = \frac{V_O}{V_{in}} = -\frac{R_f}{R_{in}} = -\frac{180 \times 10^3}{5 \times 10^3} = -36$$

19.4.5 BIAS EQUALIZATION

The impedances to ground must be equal in order to equalize bias currents at input terminals. A resistor R_B can be added to the noninverting terminal as shown in

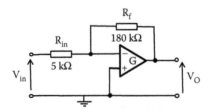

FIGURE 19.11   Circuit diagram of inverting amplifier for Example 19.3.

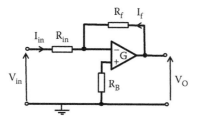

FIGURE 19.12   Circuit diagram for an inverting amplifier with bias equalization resistance.

Figure 19.12. This R_B resistor will make the impedances to ground from both inputs of the amplifier equal.

Assuming that the input voltage is direct coupled then the value of the resistance required for equalization, R_B, is equal to the parallel of R_f and R_{in}. Then

$$R_B = \frac{R_f R_{in}}{R_f + R_{in}} \tag{19.12}$$

Equalizing the bias impedances reduces an output error. If the voltage gain with feedback is small, the error can be very small and can be neglected.

In practice, when designing an amplifier using Op-Amp we make sure to use external components that limit the voltage gain of the amplifier to a value *much less* than the open-loop gain of the Op-Amp. External resistors are small compared with the input resistance and large compared with the output resistance of the Op-Amp. Generally resistors in the ranges 1–100 kΩ are used.

EXAMPLE 19.4

For Example 19.3 modify the circuit to include bias equalization. If the amplifier shown in Example 19.3 has an input bias current of 10 nA at 35°C, determine the output offset voltage due to this input bias current.

Solution

The input bias current will produce a voltage drop across the parallel combination of R_{in} and R_f, which is the impedance that appears at the Op-Amp inverting input, then the offset voltage produced by this input bias current is

$$V_{offset} = 10 \times 10^{-9} \times \frac{R_f R_{in}}{R_f + R_{in}} = 10 \times 10^{-9} \times \frac{180 \times 10^3 \times 5 \times 10^3}{180 \times 10^3 + 5 \times 10^3} = 48.65 \ \mu V$$

and the circuit of Example 19.3 can be equalized by adding a bias equalizing resistor R_B as

$$R_B = \frac{R_f R_{in}}{R_f + R_{in}} = \frac{180 \times 10^3 \times 5 \times 10^3}{180 \times 10^3 + 5 \times 10^3} = 4864.86 \ \Omega$$

as indicated in Figure 19.13.

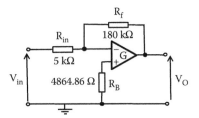

FIGURE 19.13   Circuit for the amplifier of Example 19.3, including a bias equalizing resistance, R_B.

19.5   NONINVERTING AMPLIFIER

In the noninverting amplifier the input voltage is applied to the noninverting terminal (+) of the Op-Amp as shown in Figure 19.14.

A portion of the output is feedback to the inverting input through the feedback resistor R_f. The voltage connected to the inverter terminal is provided by a voltage divider formed by R_1 and R_f.

19.5.1   Noninverting Amplifier Voltage Gain

The voltage across resistor R_1 can be obtained by applying KVL at the input loop as

$$V_{in} - V_{ve} = R_1 I \tag{19.13}$$

The current I can be obtained as

$$I = \frac{V_O}{R_1 + R_f} \tag{19.14}$$

Replacing I from Equation 19.14 into Equation 19.13 and using the expression for V_{ve} from Equation 19.2 we can write

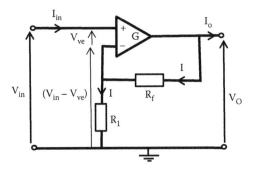

FIGURE 19.14   Circuit diagram for a noninverter amplifier.

$$V_{in} - \frac{V_O}{G} = R_1 I = \frac{R_1}{R_1 + R_f} V_O$$

Rearranging, we have

$$V_{in} = V_O \left[\frac{R_1}{R_1 + R_f} + \frac{1}{G} \right]$$

Then the overall gain of the noninverter amplifier Gv_+ is

$$Gv_+ = \frac{V_O}{V_{in}} = \frac{1}{1/G + R_1/(R_1 + R_f)} \tag{19.15}$$

Now as the gain G for the Op-Amp alone is very large, then

$$\frac{1}{G} \ll \frac{R_1}{(R_1 + R_f)}$$

and the overall voltage gain for a noninverter amplifier is

$$Gv_+ = \frac{V_O}{V_{in}} = \frac{R_1 + R_f}{R_1} = 1 + \frac{R_f}{R_1} \tag{19.16}$$

The voltage gain for a noninverter amplifier depends only on the external components R_f and R_1.

Using Kirchhoff's laws to obtain currents and voltages the input and output impedances can be obtained for the noninverter amplifier.

EXAMPLE 19.5

For the noninverting amplifier shown in Figure 19.15, determine the value of the feedback resistor, R_f, required to obtain a voltage gain of 180 and calculate the output voltage when the input voltage is 2 mV.

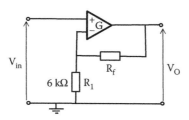

FIGURE 19.15   Circuit diagram for a noninverting amplifier of Example 19.5.

SOLUTION

For the noninverting amplifier the voltage gain is given by

$$Gv_+ = \frac{V_O}{V_{in}} = 1 + \frac{R_f}{R_1} = 180 = 1 + \frac{R_f}{6 \times 10^3}$$

then

$$R_f = 1.074 \ M\Omega$$

and the output voltage is

$$V_O = Gv_+ \times V_{in} = 180 \times 2 \times 10^{-3} = 360 \ mV$$

EXAMPLE 19.6

Design a noninverting amplifier containing an Op-Amp to achieve a voltage gain of 60 dB and determine the input voltage to achieve an output voltage of 1.2 V.

SOLUTION

The solution can be based on a circuit for a noninverting amplifier shown in Figure 19.14. Then the voltage gain is

$$Gv_+ = 20\log\left(\frac{V_O}{V_i}\right) = 60 \ dB$$

then

$$\frac{V_O}{V_i} = 1000$$

$$Gv_+ = 1 + \frac{R_f}{R_1} = 1000$$

The voltage gain depends on the ratio of R_f and R_1, we can choose a value for one of these resistors and calculate the other. Let us make $R_1 = 10 \ k\Omega$, then

$$Gv_+ = 1 + \frac{R_f}{10 \times 10^3} = 1000$$

which gives $R_f = 9.99 \ M\Omega$.

The input voltage can be determined from

$$V_{in} = \frac{V_O}{Gv_+} = \frac{1.2}{1000} = 1.2 \ mV$$

The designed noninverting amplifier is shown in Figure 19.16.

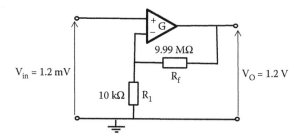

FIGURE 19.16   Circuit diagram for the noninverting amplifier designed in Example 19.6.

19.5.2   INPUT IMPEDANCE

The ratio of the input voltage, V_{in}, to the input current, I_{in}, will provide an expression for the input impedance. The input impedance, $Z_{i/p+}$, for a noninverting amplifier is

$$Z_{i/p+} = \frac{V_{in}}{I_{in}} = \left(1 + \frac{GR_1}{R_1 + R_f}\right) Z_i \qquad (19.17)$$

where Z_i is the input impedance of the Op-Amp and G is voltage gain for the Op-Amp. As G is very large, the input impedance of this amplifier is much greater than the input impedance of the Op-Amp alone.

Typical input resistance of a 741 Op-Amp is 2 MΩ for devices based on bipolar transistors. Op-Amps based on field-effect transistors (FETs) generally have a much higher input resistance in the order of 10^{12} Ω.

19.5.3   OUTPUT IMPEDANCE

An expression for the output impedance of a noninverter amplifier can be developed by applying Kirchhoff's laws to the equivalent circuit of the amplifier. The output impedance for the noninverter amplifier is

$$Z_{o/p+} = \frac{V_O}{I_O} = \frac{Z_o}{1 + GR_1/(R_1 + R_f)} \qquad (19.18)$$

where Z_o is the output impedance of the Op-Amp and G is voltage gain for the Op-Amp. As the Op-Amp gain G is very large, the output impedance of noninverter amplifier is much less than the output impedance of the Op-Amp alone.

Typical output resistance of a 741 Op-Amp is around 75 Ω. The maximum current that an Op-Amp can deliver is usually an important parameter in the design of an amplifier. The 741 supply around 20 mA; high-power devices may supply more than 1 A.

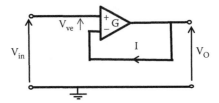

FIGURE 19.17   Circuit diagram for a voltage follower amplifier.

19.5.4   VOLTAGE FOLLOWER

The voltage follower is a special case of the noninverting amplifier where R_f is equal to zero (short circuit) and R_l is infinitive (open circuit). The circuit for a voltage follower amplifier is shown in Figure 19.17.

In this case

$$V_{in} - V_{ve} = V_O$$

or

$$V_{in} - \frac{V_O}{G} = V_O$$

Rearranging

$$V_{in} = V_O\left(1 + \frac{1}{G}\right)$$

Therefore, the gain of a voltage follower amplifier is

$$Gv_+ = \frac{V_O}{V_{in}} = \frac{1}{(1 + 1/G)} \tag{19.19}$$

As $G \gg 1$

$$Gv_+ = 1 \tag{19.20}$$

Their main use is to isolate circuits to avoid reaction between driving and driven stages.

19.6   THE DIFFERENTIAL AMPLIFIER

It is also possible to create a differential amplifier using an Op-Amp. Figure 19.18 shows a circuit of a differential amplifier.

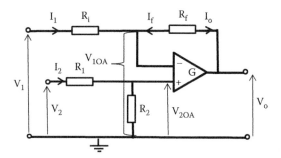

FIGURE 19.18 Op-Amp differential amplifier.

In this amplifier we have inputs signal to the inverter terminal V_1 and noninverter terminal V_2 of the Op-Amp.

19.6.1 DIFFERENTIAL AMPLIFIER VOLTAGE GAIN

The inputs to the Op-Amp V_{1OA} and V_{2OA} for common mode voltages are equal

$$V_{1OA} = V_{2OA}$$

At the input V_1 we have

$$I_1 = \frac{V_1 - V_{1OA}}{R_1} = I_f = \frac{V_{1OA} - V_O}{R_f} \tag{19.21}$$

Rearranging this equation for V_O we have

$$V_O = V_{1OA}\left(1 + \frac{R_f}{R_i}\right) - V_1\left(\frac{R_f}{R_i}\right) \tag{19.22}$$

At the input V_2 we can calculate the input to the noninverting terminal of the Op-Amp as

$$V_{2OA} = \frac{R_2}{R_1 + R_2} V_2 \tag{19.23}$$

As V_{1OA} and V_{2OA} for common mode voltage are equal we can replace V_{2OA} from Equation 19.23 for V_{1OA} in Equation 19.22. Then

$$V_O = \frac{R_2}{R_1 + R_2} V_2\left(1 + \frac{R_f}{R_i}\right) - V_1\left(\frac{R_f}{R_i}\right)$$

Rearranging, we have

$$V_O = V_2 \left[\frac{R_i R_2}{R_i (R_1 + R_2)} + \frac{R_f R_2}{R_i (R_1 + R_2)} \right] - V_1 \left(\frac{R_f}{R_i} \right)$$

or

$$V_O = V_2 \left[\frac{R_2}{R_i} \right] \left[\frac{R_i + R_f}{(R_1 + R_2)} \right] - V_1 \left(\frac{R_f}{R_i} \right) \tag{19.24}$$

This equation provides the gain for inputs V_1 and V_2. If we select the resistor $R_1 = R_i$ and $R_2 = R_f$ then expression for the output voltage given by Equation 19.24 becomes

$$V_O = \frac{R_f}{R_i} (V_2 - V_1) \tag{19.25}$$

and the differential amplifier amplifies the difference of input voltages V_1 and V_2.

EXAMPLE 19.7

For the differential amplifier shown in Figure 19.19, calculate the output voltage when $V_1 = 80$ mV and $V_2 = 40$ mV. If we change the value of the resistors $R_f = R_2 = R_i = R_1 = R$, calculate the new output voltage.

SOLUTION

The output voltage is given by

$$V_O = V_2 \left[\frac{R_2}{R_i} \right] \left[\frac{R_i + R_f}{(R_1 + R_2)} \right] - V_1 \left(\frac{R_f}{R_i} \right)$$

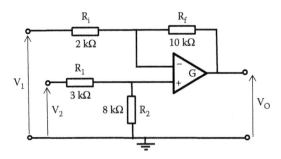

FIGURE 19.19 Differential amplifier circuit for Example 19.7.

$$V_O = 40 \times 10^{-3} \left[\frac{8 \times 10^{-3}}{2 \times 10^{-3}} \right] \left[\frac{2 \times 10^3 + 10 \times 10^3}{(3 \times 10^3 + 8 \times 10^3)} \right] - 80 \times 10^{-3} \left(\frac{10 \times 10^3}{2 \times 10^3} \right)$$

$$V_O = 40 \times 10^{-3}[4][1.0909] - 80 \times 10^{-3}(5) = -225.456 \text{ mV}$$

If $R_f = R_2 = R_i = R_1 = R$. The output voltage is

$$V_O = V_2 \left[\frac{R}{R} \right] \left[\frac{R+R}{(R+R)} \right] - V_1 \left(\frac{R}{R} \right) = V_2 - V_1 = 40 \text{ mV} - 80 \text{ mV} = -40 \text{ mV}$$

The amplifier provides the difference between the two inputs with a change of polarity.

19.6.1.1 Op-Amp as Comparator

A variation of the differential amplifier is to connect two signals directly to the input terminals of an Op-Amp as shown in Figure 19.20.

The output voltage, V_O, for this amplifier will depend on values of V_1 and V_2 as it amplifies the difference between V_1 and V_2 as

$$V_O = G(V_2 - V_1)$$

where G is the voltage gain of the Op-Amp.

If $V_2 > V_1$ the output voltage is positive with a value close to $+V_{s+}$ (the voltage of the positive power supply). This is because G in an Op-Amp is very large and any small difference in voltage at the input will be largely amplified taking the Op-Amp to saturation.

If $V_2 < V_1$ the output is negative with a value close to $-V_{s-}$ (the voltage of the negative power supply).

If $V_2 = V_1$ the difference is zero and $V_O = 0$.

Then this amplifier can be used as comparator of the two input signals.

$$\text{If} \quad V_2 > V_1 \quad \text{then } V_O \approx +V_{s+}$$
$$\text{If} \quad V_2 < V_1 \quad \text{then } V_O \approx -V_{s-}$$
$$\text{If} \quad V_2 = V_1 \quad \text{then } V_O = 0$$

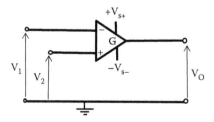

FIGURE 19.20 Circuit diagram showing an Op-Amp as comparator.

19.6.1.2 The Summing Amplifier

A variation of the inverter amplifier is shown in Figure 19.21 where three input voltages are applied to the noninverting input of the Op-Amp.

This circuit can be analyzed using the property of virtual earth at X (Section 19.4.2), then an equivalent circuit for this amplifier is shown in Figure 19.22.

Applying KCL at X we have

$$-I_f = I_1 + I_2 + I_3 \tag{19.26}$$

Written in terms of their voltages, we have

$$-\frac{V_O}{R_f} = \frac{V_1}{R_1} + \frac{V_2}{R_2} + \frac{V_3}{R_3} \tag{19.27}$$

and the output voltage V_O becomes

$$V_O = -R_f \left(\frac{V_1}{R_1} + \frac{V_2}{R_2} + \frac{V_3}{R_3} \right) \tag{19.28}$$

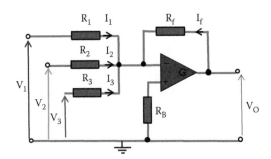

FIGURE 19.21 Circuit diagram showing an Op-Amp as a summing amplifier.

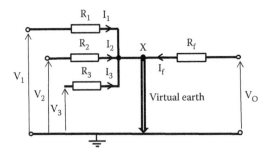

FIGURE 19.22 Equivalent circuit for a summing amplifier using the virtual earth concept.

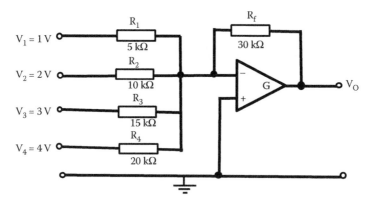

FIGURE 19.23 Summing amplifier with four inputs used in Example 19.8.

If $R_1 = R_2 = R_3 = R_f$, Equation 19.28 becomes

$$V_O = -(V_1 + V_2 + V_3)$$

The output voltage will add the input voltages and change the total phase. The summing amplifier can add two or more input voltages.

EXAMPLE 19.8

For the summing amplifier shown in Figure 19.23, calculate the output voltage, V_O.

SOLUTION

For this summing amplifier the output voltage is

$$V_O = -R_f \left[\frac{V_1}{R_1} + \frac{V_2}{R_2} + \frac{V_3}{R_3} + \frac{V_4}{R_5} \right]$$

Then

$$V_O = -30 \times 10^3 \left[\frac{1}{5 \times 10^3} + \frac{2}{10 \times 10^3} + \frac{3}{15 \times 10^3} + \frac{4}{20 \times 10^3} \right] = -24 \text{ V}$$

19.7 OP-AMP FREQUENCY RESPONSE

The internal circuitry of an Op-Amp will contain various stages with differential, common emitter, and common collector amplifiers with different frequency cutoffs. Internal frequency compensation circuits make an Op-Amp to have a general frequency response of a dominant single pole response. A typical frequency response of an Op-Amp is indicated in Figure 19.24.

One of the main differences with the frequency response from a BJT or FET amplifier is that the Op-Amp can amplify a signal from zero frequency (dc). The

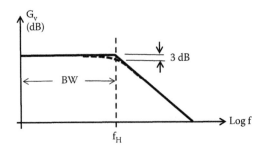

FIGURE 19.24 Typical frequency response of an Op-Amp.

voltage gain remains constant until its high-frequency cutoff (f_H). Increasing the frequency above f_H the voltage gain decreases at 20 dB/decade rate. Therefore, the BW is equal to f_H.

There are some special Op-Amps that have a faster rate of fall in frequencies, but most Op-Amp will have this frequency response.

Manufacturers usually specify the high-frequency cutoff, f_H; the frequency at a voltage gain equal to 1 (0 dB) or the product Gv times frequency when the voltage gain is equal to 1.

The product *voltage gain* × *BW* is approximately constant. As NFB is applied the gain reduces and the BW increases. Then by changing the voltage gain the high-frequency cutoff can be changed and the BW can be changed.

EXAMPLE 19.9

Determine the bandwidth of the amplifier shown in Figure 19.25. The Op-Amp used in this amplifier has a voltage gain of 120 dB and a unity-gain-BW of 8 MHz.

SOLUTION

The amplifier has a voltage gain, Gv, equal to

$$Gv_+ = 1 + \frac{R_f}{R_1} = 1 + \frac{95 \times 10^3}{5 \times 10^3} = 20$$

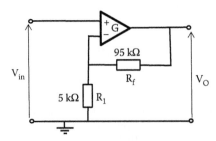

FIGURE 19.25 Noninverting amplifier circuit for Example 19.9.

Now the unity gain BW is equal to BW $\times$ G$_v$
Then,

$$BW = \frac{\text{Unity gain BW}}{G_v} = \frac{8 \times 10^6}{20} = 400 \text{ kHz}$$

and the bandwidth for this amplifier is 400 kHz.

19.8   KEY POINTS

- Op-Amps are extensively used components in analog electronic circuits.
- Op-Amps can be constructed using bipolar transistors, FETs, or a combination of the two.
- Most bipolar Op-Amps use long-tailed pair differential amplifiers. While the circuitry used in real Op-Amps is often fairly complex, many of the stages used are relatively easy to analyze.
- Most Op-Amps operate using two power supplies. Connections to these power supplies are usually not included in circuit diagrams.
- By combining the best features of bipolar and field effect technologies it is possible to produce a very efficient Op-Amp.
- An ideal Op-Amp would have infinite voltage gain, infinite input resistance, and zero output resistance. Real Op-Amps will have characteristics close to ideal Op-Amps.
- With NFB the voltage gain is reduced in exchange for improvements in other characteristics such as bandwidth, input resistance, and output resistance.

20 Filters

20.1 INTRODUCTION

There are many occasions in which we would like to limit certain frequencies or a range of frequencies. In order to do this, we need a filter of frequencies. As we have seen in previous chapters, impedances of inductors and capacitors change with frequencies. Combinations of inductors and capacitors in series and in parallel will then produce a frequency response that will filter some frequencies. The use of passive components, resistors, inductors, and capacitors can be used to create passive filters. We have seen some examples of this behavior with frequency in Chapter 3, where we investigated the response of some RC, CR, and RL series circuits.

In this chapter, we are interested in filters including Op-Amp as an active component to create what is known as an active filter. We can have filters that amplify frequencies up to a cutoff frequency and then attenuate frequencies larger than this cutoff frequency. These filters are called *low-pass filters*. We can have *high-pass filters* that amplify a signal only above the frequency cutoff. We can have filters that amplify only a specific range of frequencies; these filters are the *band-pass filters*. We can also have a filter that attenuates a range of frequencies; these filters are called *band-stop filters*.

In Chapter 19 we saw the frequency response of an Op-Amp. It can be said that the Op-Amp itself is a low-pass filter, where the frequencies above its cutoff frequency f_c are attenuated and frequencies from dc to f_c are amplified.

It is possible to use the Op-Amp to create a filter. As the gain depends on the external components and some components are frequency dependent, we can use an arrangement of external components to adapt a frequency response of an electronic system. An example of a filter can use an inverter amplifier containing an Op-Amp circuit as shown in Figure 20.1. In this amplifier we had replaced the resistor R_{in} and R_f for general impedance Z_{in} and Z_f as the external components. Z_{in} and Z_f can contain capacitors, inductors, and resistors to select the frequency response of the filter.

In this case, the voltage gain can be expressed in terms of the impedances as

$$Gv = \frac{V_O}{V_{in}} = -\frac{Z_f}{Z_{in}} \tag{20.1}$$

Now, if Z_f is an inductor, L, and Z_{in} a resistor, R, then the gain will be

$$Gv = -\frac{R}{j\omega L} = -\frac{1}{j\omega L/R} \tag{20.2}$$

giving a frequency response with a decreasing voltage gain at a rate -20 dB/decade slope passing to 0 dB at $\omega = R/L$. Inductors are not usually used because they can be

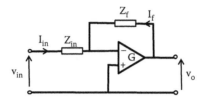

FIGURE 20.1 Circuit diagram of an inverting amplifier with impedances as external components to form a filter.

bulky and expensive with limited frequency response. As an alternative, we can use a capacitor C instead of the inductance as Z_f and the voltage gain will be

$$Gv = -\frac{R}{1/j\omega C} = -j\omega CR \tag{20.3}$$

which will produce a voltage gain increasing with frequency at a rate of +20 dB/decade passing 0 dB at $\omega = 1/RC$.

We can increase the complexity of Z_f and Z_{in} to achieve a desired frequency response. For example, let us make Z_f a resistor R and make Z_{in} a combination of a capacitance C in series with another resistor R as indicated in Figure 20.2.

Here

$$Z_f = R$$

and

$$Z_{in} = R + \frac{1}{j\omega C} = \frac{1 + j\omega CR}{j\omega C}$$

and the voltage gain is

$$\frac{V_O}{V_{in}} = -\frac{Z_f}{Z_{in}} = -\frac{R}{(1 + j\omega CR)/j\omega C} = -\frac{j\omega CR}{1 + j\omega CR} \tag{20.4}$$

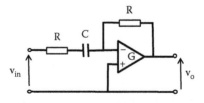

FIGURE 20.2 Circuit diagram of inverting amplifier as high-pass filter.

Plotting the magnitude of this voltage gain against frequency will provide a high-pass filter response with a cutoff at $\omega = 1/RC$.

The voltage gain or transfer function of the filter shown in Figure 20.2 is given by the expression in Equation 20.4. This expression has at the denominator the term $j\omega CR$; plotting the magnitude of term $j\omega CR$ will produce a magnitude that increases in a straight-line with rate +20 dB/decade as the frequency increases. This is plotted in Figure 20.3. The voltage gain for this filter given by the expression in Equation 20.4 has the term $1/(1 + j\omega CR)$ as part of the denominator; this term $1/(1 + j\omega CR)$ will produce a standard pole response with a cutoff at $\omega = 2\pi f_p = 1/RC$. The pole response will have 0 dB gain until f_p, then as the frequency increases the magnitude decreases with a −20 dB/decade rate; as indicated in Figure 20.3. Adding these two responses we can obtain the total response as shown in Figure 20.3; after the cutoff frequency (f_p) the pole response cancel the increase of the $j\omega CR$. The total response is shown in Figure 20.3.

The frequency response then depends on the external components connected to an Op-Amp to produce any filter required. In this example we have a filter that attenuates signals with frequency below f_p and allows signal with frequencies above f_p; then it works as a high-pass filter.

In the following section we look up some standard popular filter circuits commonly used with noninverter amplifiers.

A general active filter using a noninverter amplifier is shown in Figure 20.4. The resistors R_1 and R_2 provide the gain of this filter amplifier and a network, indicated as a block in Figure 20.4, determines the frequency response. The input to the noninverting terminal of the Op-Amp includes input impedance as well as some feedback.

The negative feedback voltage gain for this noninverter amplifier is

$$Gv = 1 + \frac{R_1}{R_2}$$

and the transfer function of the frequency selective network will determine the filter response.

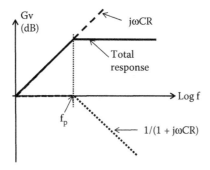

FIGURE 20.3   Frequency response for circuit of Figure 20.2.

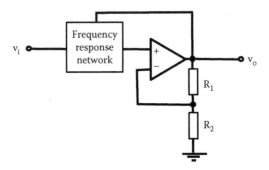

FIGURE 20.4   Generalized active filter using noninverter amplifier.

20.2   LOW-PASS FILTER RESPONSES

Figure 20.5 illustrates low-pass filter response idealized Bode plot curves with several roll-off rates. The different roll-off rates can be obtained using different frequency selective networks.

The real response of a first-order low-pass filter is shown in Figure 20.6, where the cutoff is at 3 dB below the mid-frequency voltage gain.

The difference between this curve and the one indicated in Figure 20.5, for first-order filter, is 3 dB at the corner break of the idealized Bode plot. The bandwidth in this case is defined to be from dc (0 Hz) to the cutoff frequency f_c.

20.2.1   SINGLE FIRST-ORDER LOW-PASS FILTER

In this example of a single first-order low-pass filter, the network to select frequencies is a simple RC circuit as indicated in Figure 20.7.

The RC selective network can be included as part of an active filter. Figure 20.8 shows this selective frequency network as part of a low-pass filter.

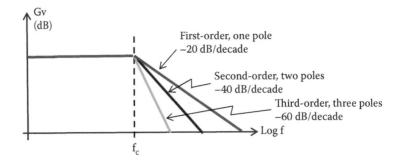

FIGURE 20.5   Low-pass filter response for idealized Bode plot curves with several roll-off rates.

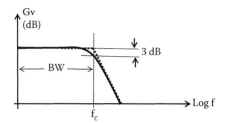

FIGURE 20.6 Low-pass filter response of a single-pole circuit with −20 dB/decade roll-off rate.

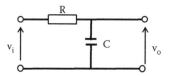

FIGURE 20.7 RC selective frequency network to create a low-pass filter.

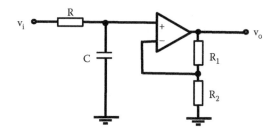

FIGURE 20.8 Circuit diagram of a single first-order low-pass RC filter.

This network was analyzed in Chapter 16. It was found that the transfer function for this RC network (Equation 16.3) is

$$\text{T.F.} = \frac{1}{1 + j\omega RC}$$

with a cutoff frequency of

$$f_c = \frac{1}{2\pi RC}$$

and roll-off of −20 dB/decade.

The frequency response of this network is the one we have seen for a first-order low-pass filter response in Chapter 16; it is indicated in Figure 20.9.

This low-pass filter will reject unwanted frequencies above f_c.

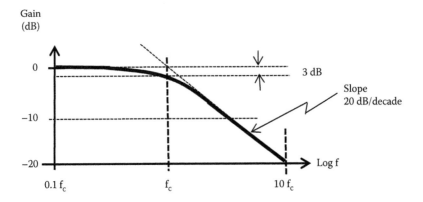

FIGURE 20.9   Frequency response of a single first-order low-pass filter.

20.2.2   SECOND-ORDER LOW-PASS FILTERS

It is possible to include reactive components in the input as well as in the feedback path to improve the frequency response. Some designers have developed selective networks to create filters that have become accepted as standard filters. One commonly used filter that uses capacitors and resistances in the frequency selective network of a noninverter amplifier is the *Sallen and Key filter* (Sallen and Key 1955). With two capacitors in the frequency selective network it is possible to create a filter that has a faster roll-off frequency response, at a rate of 40 dB/decade second. This filter is known as a second-order filter. It is also known as a two-pole filter.

20.2.3   SALLEN AND KEY LOW-PASS FILTER

Figure 20.10 shows a commonly used Sallen and Key second-order low-pass filter.
   This Sallen and Key filter has the same gain determined by R_1 and R_2

$$Gv = 1 + \frac{R_1}{R_2}$$

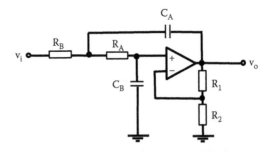

FIGURE 20.10   Sallen and Key second-order low-pass active filter.

and a cutoff frequency

$$f_c = \frac{1}{2\pi\sqrt{R_A R_B C_A C_B}} \qquad (20.5)$$

If the capacitors are chosen to be the same $C_A = C_B = C$ and the resistors are also made the same $R_A = R_B = R$ then the cutoff frequency is the same as the first-order low-pass filter

$$f_c = \frac{1}{2\pi RC}$$

But the roll-off will be in this case −40 dB/decade.

20.3 HIGH-PASS FILTER RESPONSE

Figure 20.11 illustrates a high-pass filter response idealized Bode plot curves with several roll-off rates. The different roll-off rates can be obtained using different frequency selecting networks.

The pass band in this case is defined to be from the cutoff frequency f_c ideally to infinite

20.3.1 First-Order High-Pass Filter

An example of a single-pole high-pass filter network to select frequencies is a simple CR circuit as indicated in Figure 20.12. This circuit is similar to the low-pass filter, but the components are swapped to become a CR network.

This CR selective frequency network is shown in Figure 20.13 as part of a high-pass filter.

This network was analyzed in Chapter 16, the transfer function of this selective frequency CR network is

$$\text{T.F.} = \frac{1}{1 + 1/j\omega CR}$$

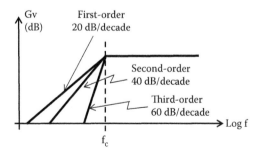

FIGURE 20.11 High-pass filter response idealized Bode plot curves with several roll-off rates.

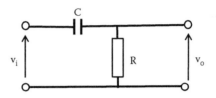

FIGURE 20.12   CR selective frequency network.

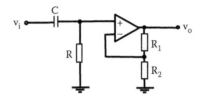

FIGURE 20.13   Circuit diagram of a single-pole high-pass RC filter.

This provides a high-pass filter response with a cutoff frequency equal to

$$f_c = \frac{1}{2\pi RC}$$

with a roll-off of +20 dB/decade.

The frequency response of this network is the one we investigated for one-pole circuit in Chapter 16; it is indicated in Figure 20.14.

This high-pass filter will reject unwanted frequencies below f_c.

20.3.2   SECOND-ORDER FILTERS

By adding reactance to the frequency-selecting network it is possible to create a second-order filter. Alternatively, it is possible to use two identical first-order filters to create a second-order filter as indicated in Figure 20.15.

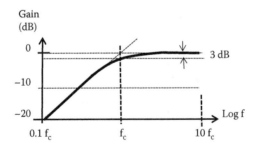

FIGURE 20.14   Frequency response of a first-order high-pass filter.

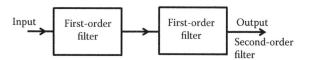

FIGURE 20.15   Block diagram of two identical first-order filters to create a second-order filter.

20.3.3   Sallen and Key High-Pass Filter

The circuit illustrated in Figure 20.16 shows a version of the Sallen and Key second-order high-pass filter.

It is similar to the Sallen and Key low-pass filter, but with the resistors and capacitances swapped. This Sallen and Key filter has the same gain determined by R_1 and R_2.

$$Gv = 1 + \frac{R_1}{R_2}$$

and a cutoff frequency (same as for a low-pass filter)

$$f_c = \frac{1}{2\pi\sqrt{R_A R_B C_A C_B}}$$

As in the case of a low-pass filter if the capacitors are made the same $C_A = C_B = C$ and the resistors are also made the same $R_A = R_B = R$, then the cutoff frequency has the same expression as the low-pass filter (or the first-order filter)

$$f_c = \frac{1}{2\pi RC}$$

but in this case the roll-off is +40 dB/decade.

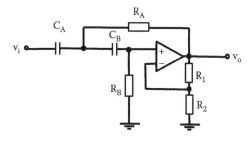

FIGURE 20.16   Sallen and Key second-order high-pass filter.

20.4 BAND-PASS FILTER RESPONSE

The band-pass filter will amplify frequencies in a range between a low-frequency cutoff f_{c1} and a high-frequency cutoff f_{c2}. The frequency response of a band-pass filter is shown in Figure 20.17.

The bandwidth of a band-pass filter is given as the difference between the high- and low-frequency cutoffs. In a band-pass filter is it usual to specify the middle frequency of the filter, f_o, and a quality factor of a band pass is defined in terms of f_o as

$$\text{Quality factor} = Q = \frac{f_o}{BW} \qquad (20.6)$$

The quality factor provides an indication of the selectivity of the filter; it is classified as narrowband or wideband according to the value of Q. Then a narrowband will have a Q > 10 and a wideband will have a Q < 10.

A band-pass filter can be created by cascading a low-pass filter and a high-pass filter, as indicated in Figure 20.18.

Components are selected to provide the cutoffs and create the bandwidth required.

20.5 BAND-STOP RESPONSE

A band-stop filter will reduce the voltage gain of a range of frequencies and allow the frequencies outside this range. Figure 20.19 shows the frequency response of a band-stop filter.

The band-stop filter is also known as notch, band-reject, or band-elimination filter. As with a band-pass filter it is possible to use a low- and a high-pass filter to

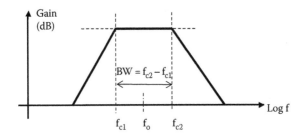

FIGURE 20.17 Frequency response of a band-pass filter.

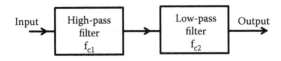

FIGURE 20.18 Block diagram of a low-pass filter and a high-pass filter to create a band-pass filter.

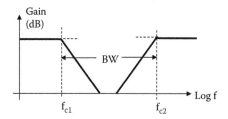

FIGURE 20.19   Frequency response of a band-stop filter.

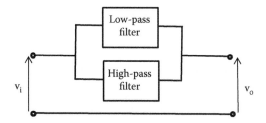

FIGURE 20.20   Block diagram of a band-stop filter using low- and high-pass filters.

create a band-stop filter. These filters can be connected in parallel as indicated in Figure 20.20.

As in the case of a band-pass filter, it is possible to define a bandwidth for this filter. The bandwidth for this filter will be the difference in frequency of the two cutoff frequencies as

$$BW = f_{c2} - f_{c1}$$

20.6   FOURTH-ORDER RESPONSE

We can improve the frequency response of a filter by making the roll-off faster. A fourth-order network can be obtained by cascading two second-order filters as indicated in Figure 20.21.

If the filters that are cascaded have the same frequency cutoff, then this network will create a fourth-order filter with a roll-off of 80 dB/decade. If the frequency cutoffs are different, then they can be combined to create a band pass fourth-order filters.

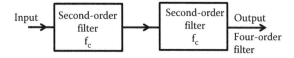

FIGURE 20.21   Two second-order filters cascaded to create a fourth-order filter.

20.7   FILTER RESPONSE CHARACTERISTICS

A filter can be designed to have various response characteristics. Figure 20.22 shows three general types of response characteristics.

The Chebyshev response gives oscillatory changes in amplitude as the frequencies get closer to the cutoff frequency. If the ripple is small the roll-off can be very fast. The Bessel response provides a more steady slow response around and beyond the frequency cutoff. They have a linear phase response. The Butterworth response provides a frequency response closer to one-pole response. They provide an almost constant gain voltage before the cutoff frequency for the low-pass filter or above the cutoff frequency for the high-pass filter. They present a roll-off frequency response of 20 dB/decade/order. The roll-off response is negative for the low-pass filter and positive for the high-pass filter.

20.8   FILTER DESIGN USING STANDARD TABLES

A designed network to select frequencies in a filter can have one of the characteristics indicated in Section 20.7. By changing values of the components in the frequency discrimination network and the amplifier gain, it is possible to obtain a Bessel, Butterworth, or Chebyshev response. Based on practical experience some manufacturing companies and designers have produced tables indicating the different voltage gain required by different cascaded sections, including correction factors to obtain Bessel, Butterworth, or Chebyshev response (Karki 2002). The intention is to write a standard form of the transfer function for a low or high filter. This transfer function for the filter can be expressed as a series of polynomial terms, which includes complex zeros where a polynomial is zero and complex poles where a polynomial term tends to an infinitive value. Depending on the architecture of the filter coefficient can be found to satisfy its response. As complex zero contains real and imaginary parts, coefficient can be determined in terms of the real and imaginary parts of the complex zeros. The Butterworth, Bessel, and Chevyshev filters transfer functions of even orders contain all poles in their expressions. Tables can be constructed by determining an expression of the transfer function of the network that selects the frequency, determining the gain required in the different stages, and determining correction factors for the desired response (Karki 2002). Table 20.1 shows a small sample table based on second-order Sallen and Key filter architecture for Butterworth, Bessel,

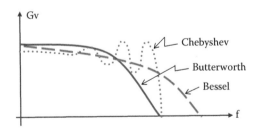

FIGURE 20.22   Chebyshev, Bessel, and Butterworth filter response.

TABLE 20.1

Sample of Sallen and Key Low-Pass Filter Design Parameter Table

Filter Order	Stage	Butterworth		Bessel			Chebyshev (Ripple Steps 2 dB)		
		Gv	Q	Gv	Kc	Q	Gv	Kc	Q
2	First	1.5858	0.7071	1.2677	1.2736	0.5773	2.1138	0.907	1.1284
4	First	1.0839	0.5219	1.0839	1.4192	0.5219	1.9241	0.471	0.9295
	Second	2.2346	1.3065	1.7585	1.5912	0.8055	2.7817	0.964	4.5809

and Chevyshev filters. In this table, Kc is the frequency cutoff correction factor and Q is the quality factor for the filter. The table is for second-order Sallen and Key filters that can be cascaded for a fourth-order filter. It is possible to find tables that provide extensive lists of parameters for different numbers of poles, different values of ripple increment in steps and correction factors, and voltage gains in different stages.

As we cascade filters to create higher orders, the voltage gain in filters will be different. The Kc value in Table 20.1 is a correction factor for the frequency cutoff.

We can illustrate the use of Table 20.1 to design a Sallen and Key filter with an example. The exercise will consist of designing three four-pole low-pass filters with a cutoff frequency equal to 600 Hz to provide Butterworth, Chebyshev, and Bessel responses.

The circuit to be used is the Sallen and Key filter where we have cascaded two second-order filters to create a fourth-order Sallen and Key filter. To make this easier let us assume that capacitors and resistors of the selecting frequency network are the same in each stage, that is, $C_{A1} = C_{B1} = C_1$, $R_{A1} = R_{B1} = R_1$, $C_{A2} = C_{B2}, = C_2$, $R_{A2} = R_{B2} = R_2$. Then, this four-pole filter will have the circuit diagram as indicated in Figure 20.23.

In general, for the Sallen and Key filter using the same capacitors and resistors in each stage the low-frequency cutoff is given by

$$f_c = \frac{1}{2\pi RC}$$

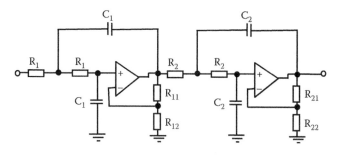

FIGURE 20.23 Sallen and Key fourth-order low-pass filter.

and the gain for the noninverter amplifier, in each stage, is given by

$$Gv = 1 + \frac{R_1}{R_2}$$

Depending on the characteristics of the design some filters will need different gain in different stages. Some of them will also require a correction factor in different stages. A solution to implement this design is to design two second-order sections. Let us design these three filters.

20.8.1 BESSEL

20.8.1.1 First Second-Order Filter
For the Bessel filter from Table 20.1, we see that this filter requires different gain and correction factors in each section.

For the first stage of this Bessel filter there is a correction factor Kc = 1.4192 from Table 20.1, then the cutoff frequency including Kc is

$$f_{c1} = \frac{1}{2\pi R_1 C_1 (Kc)} = 600 \text{ Hz} \tag{20.7}$$

In designing this section we can decide to choose a value of resistance and calculate the capacitor or vice versa. Let us make R_1 = 1 kΩ, then the capacitor will be equal to

$$C_1 = \frac{1}{2\pi R_1 (Kc) f_{c1}} = \frac{1}{2\pi \times 1 \times 10^3 (1.4192)600} \approx 187 \text{ nF}$$

The voltage gain of the first stage according to Table 20.1 is 1.084

$$Gv_1 = 1 + \frac{R_{11}}{R_{12}} = 1.0839$$

If we make R_{12} = 20 kΩ then we can calculate R_{11}, which gives R_{11} = 1.678 kΩ.

20.8.1.2 Second Second-Order Filter
We can repeat the calculations for the second section.

For the second stage Bessel filter there is a correction factor Kc = 1.5912 from Table 20.1, then the cutoff frequency including Kc is

$$f_{c2} = \frac{1}{2\pi R_2 C_2 (Kc)} = 600 \text{ Hz} \tag{20.8}$$

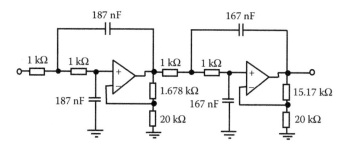

FIGURE 20.24 Fourth-order low-pass Sallen and Key filter design with calculated values for a Bessel response.

Let us make $R_2 = 1$ kΩ then the capacitor C_2 will be equal to

$$C_2 = \frac{1}{2\pi R_2 (Kc) f_{c2}} = \frac{1}{2\pi \times 1 \times 10^3 (1.5912)600} \approx 167 \text{ nF}$$

The voltage gain of the second stage according to Table 20.1 is

$$Gv_2 = 1 + \frac{R_{21}}{R_{22}} = 1.7585$$

If we make $R_{22} = 20$ kΩ then R_{21} can be calculated giving $R_{21} = 15.17$ kΩ. The four-pole low-pass Bessel filter designed is shown in Figure 20.24.

20.8.2 BUTTERWORTH

For the Butterworth filter the cutoff frequency does not have a correction factor in any section and the gain varies according to Table 20.1.

For the first- and second-stage Butterworth filter the cutoff frequency is

$$f_{c1} = \frac{1}{2\pi R_1 C_1} = 600 \text{ Hz}$$

Let us make $R_1 = R_2 = 1$ kΩ. Then the capacitor will be equal to

$$C_1 = C_2 = \frac{1}{2\pi R_1 f_{c1}} = \frac{1}{2\pi \times 1 \times 10^3 600} \approx 265 \text{ nF}$$

The voltage gain of the first stage according to Table 20.1 is

$$Gv_1 = 1 + \frac{R_{11}}{R_{12}} = 1.0839$$

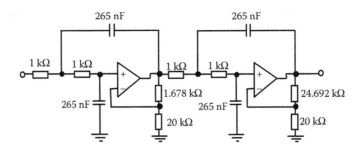

FIGURE 20.25 Fourth-order low-pass Sallen and Key filter design with calculated values for a Butterworth response.

If we make $R_{12} = 20$ kΩ then $R_{11} = 1.678$ kΩ.

Gain second second-order filter. The voltage gain of the second stage according to Table 20.1 is

$$Gv_2 = 1 + \frac{R_{21}}{R_{22}} = 2.2346$$

If $R_{22} = 20$ kΩ then $R_{21} = 24.692$ kΩ.

The fourth-order low-pass Butterworth filter designed is shown in Figure 20.25.

20.8.3 Chebyshev

For the Chebyshev filter, the ripple presents an additional variable. There are tables that contain more details of ripple including increments of 0.1 dB. Table 20.1 includes ripple of 2 dB only for our design let assume that a ripple of 2 dB is adequate.

20.8.3.1 First Second-Order Filter

For the Chebyshev filter, we require different gain and correction factors in each section.

For the first-stage Chebyshev filter there is a correction factor $Kc = 0.471$ at 2 dB ripple from Table 20.1, then the cutoff frequency including Kc is

$$f_{c1} = \frac{1}{2\pi R_1 C_1 (Kc)} = 600 \text{ Hz}$$

Let us make $R_1 = 1$ kΩ then the capacitor C_1 will be equal to

$$C_1 = \frac{1}{2\pi R1(Kc)f_{c1}} = \frac{1}{2\pi \times 1 \times 10^3 (0.471)600} \approx 563 \text{ nF}$$

The voltage gain of the first stage according to Table 20.1 is

$$Gv_1 = 1 + \frac{R_{11}}{R_{12}} = 1.924$$

If we make $R_{12} = 20\ k\Omega$ then $R_{11} = 18480\ \Omega$.

20.8.3.2   Second Second-Order Filter

We can repeat the calculations for the second section.

For the second-stage Chebyshev filter there is a correction factor $Kc = 0.964$ from Table 20.1, then the cutoff frequency including Kc is

$$f_{c2} = \frac{1}{2\pi R_2 C_2 (Kc)} = 600\ Hz$$

Let us make $R_2 = 1\ k\Omega$ then the capacitor will be equal to

$$C_2 = \frac{1}{2\pi R_2 (Kc) f_{c2}} = \frac{1}{2\pi \times 1 \times 10^3 (0.964) 600} \approx 275\ nF$$

The voltage gain of the second stage according to Table 20.1 is

$$Gv_2 = 1 + \frac{R_{21}}{R_{22}} = 2.782$$

If we make $R_{22} = 20\ k\Omega$ then $R_{21} = 35640\ \Omega$.

The fourth-order low-pass Chebyshev filter designed is shown in Figure 20.26.

These three design filters can be simulated to check that the desired response is achieved or if they need to be modified to obtain a better response.

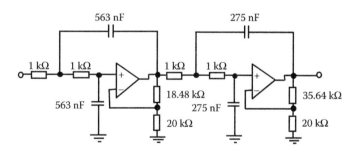

FIGURE 20.26   Fourth-order low-pass Sallen and Key filter design with calculated values for a Chebyshev response.

20.9   KEY POINTS

- There are many occasions in which it is necessary to limit certain frequencies or a range of frequencies. An electronic filter will provide the required frequency response of an electronic system.
- Impedances of inductors and capacitors change with frequencies. Combinations of inductors and capacitors will produce a frequency response that will filter some frequencies. The use of passive components, resistor, inductors, and capacitors can be used to create passive filters.
- Using passive components complemented with Op-Amp as active components creates what is known as an active filter.
- Low-pass filters amplify frequencies up to a cutoff frequency and then attenuate frequencies larger than this cutoff frequency.
- High-pass filters amplify a signal only above the frequency cutoff.
- Band-pass filters amplify only certain ranges of frequencies.
- Band-stop filters attenuate a range of frequencies and let the other frequencies outside the range pass through.
- In an active filter the frequency response depends on the external components connected to an amplifier to produce any filter required. It is possible to include reactive components in the input as well as in the feedback path to improve the frequency response.
- One commonly used filter that uses capacitors and resistors as feedback in a noninverter amplifier is the Sallen and Key filter.
- Low- and high-pass filters can be cascaded to produce a band-pass filter.
- If the filter in the cascade has the same frequency cutoff then this network will create a higher order filter.
- Filters with a roll-off of 20 dB/decade are known as first-order or one-pole filters. Second-order filters have a roll-off of 40 dB/decade. Fourth-order filters will produce a roll-off of 80 dB/decade.
- The Chebyshev response provides oscillatory changes in amplitude as the frequencies get closer to the cutoff frequency. If the ripple is small the roll-off can be very fast. The Bessel response provides a more steady slow response around and beyond the frequency cutoff. The Butterworth response provides a frequency response closer to one-pole response.
- Tables can be used to design filters. Some tables include an extensive list of parameters for different order of filters, different values of ripple increment in steps and correction factors, and gains in different stages.

REFERENCES

Karki, J. 2002. Active low pass filter design. Texas Instrument Application Report SLOA049B. September 2002.

Sallen, R. and Key, E. 1955. A practical method of designing RC active filters. *IRE Transactions on Circuit Theory*, CT-2, 74–85.

21 Applications of Analog Electronics

21.1 INTRODUCTION

There are many applications of analog systems. This chapter presents some simple and common applications of analog systems. Simulation, building, and getting an application to work for general design helps in the understanding of analog electronics. In digital electronics, the realization of an application is based on connecting regular devices in a structured way. In analog electronic applications, circuits may have many choices and experience can play an important part in the success of a project. A successful application might depend on the type of components and on the placement and organization of components. Experienced and competent analog engineers are few and the industry across the world needs them. Before committing a circuit to production a good simulation is required to check all possible cases. Circuits may not work at first attempt and require problem solving. Troubleshooting can be made easier based on a good simulation where expected values of voltages, currents, and electrical powers in different components and sections of a system are known.

21.2 SIMULATION IN CIRCUIT APPLICATIONS

Computer-aided design (CAD) software to simulate circuits can make the analysis of circuits easier, but it is always essential to have a comprehensive understanding of the fundamental principles that govern currents and voltages in circuits. The circuit designer ought to know what to expect from the simulator, and should always be able to predict the signal response. The simulation package is then used to investigate the operation in more detail for different component values or different variations of the input. A circuit simulator package can investigate a worst-case analysis in which all the components are varied according to their tolerances.

The performance of a network can be examined by the application of Kirchhoff's voltage and current laws in order to determine currents in the different nodes and voltages on different closed circuits. Applying Kirchhoff's laws it is possible to express, in any circuit, the relationship of voltages and current as a general impedance matrix as follows:

$$V_1 = Z_{11}I_1 + Z_{12}I_2 + Z_{13}I_3 + Z_{14}I_4 + \cdots + Z_{1n}I_{1n}$$

$$V_2 = Z_{21}I_1 + Z_{22}I_2 + Z_{23}I_3 + Z_{24}I_4 + \cdots + Z_{2n}I_{2n}$$

$$V_3 = Z_{31}I_1 + Z_{32}I_1 + Z_{33}I_3 + Z_{34}I_4 + \cdots + Z_{3n}I_{3n}$$

$$\cdot = \quad \cdot \quad \cdot \quad \cdot \quad \cdot \quad \cdot \quad \cdots \quad \cdot$$

$$\cdot = \quad \cdot \quad \cdot \quad \cdot \quad \cdot \quad \cdot \quad \cdots \quad \cdot$$

$$\cdot = \quad \cdot \quad \cdot \quad \cdot \quad \cdot \quad \cdot \quad \cdots \quad \cdot \tag{21.1}$$

$$V_n = Z_{n1}I_1 + Z_{n2}I_1 + Z_{n3}I_2 + Z_{n4}I_4 + \cdots + Z_{nn}I_{nn}$$

Many computer-aided programs for circuit simulation are based on Kirchhoff's laws. A matrix of the form indicated in Equation 21.1 can be created for any network by following a set of rules based on Kirchhoff's laws; these rules can be written as a piece of software. The voltages and currents at different components or sections of the network are then obtained by a piece of software that solves the matrix for voltages and currents applying matrix algebra. Repetitive calculations are well suited to computers, and a complete accurate response for a complex circuit can be obtained in a matter of seconds.

The first computer simulator to solve circuits, known as *Spice*, was developed in the early 1970s to examine the potential complexity of circuits in integrated form. Spice was run in mainframe computers. As the personal computer (PC) became popular, a version known as Pspice (Pspice is registered as a trademark of MicroSim Corporation) was implemented for PC use. During the years, Pspice has undergone many changes and improvements and now many other similar packages are available.

Circuit analyzer packages usually consist of various subprograms including:

- A graphical circuit editor that captures the *schematics* of the circuit
- An analog and digital *stimulus* generation tool
- A utility to create semiconductor device models and subcircuit *parts* definitions
- An analog, digital, and mixed-circuit *simulator*
- A graphical waveform analyzer used to view and manipulate simulation results

The packages offer different investigations including dc, ac, and transient analysis.

21.3 SELECTION OF COMPONENTS AND CIRCUIT ELEMENTS IN AN APPLICATION

Components are selected according to the different applications. Apart from the defined value and characteristic of a component it is necessary to consider maximum rating, size, and material used in fabrication. In terms of packaging, surface mount (SM) components are small versions of pin-through-hole components. SM components lend themselves for automatic population on printed circuit boards (PCBs). SM components have short legs and are thin compared with pin-through-hole components. Figure 21.1 shows standard and SM components.

FIGURE 21.1   Standard (a) and SM (b) components.

21.3.1 TRANSISTORS

Transistors are selected making sure that the operating point is within the safe field below their maximum values of voltage, current, and power. Frequency and temperature range to be used are also important factors to be considered. The physical size of the transistor is usually related to the respective power dissipation. Manufacturer's data for any transistor usually clearly provides the following

- The safe maximum ratings allowed for the device
- The normal operating voltages, currents, and gains
- Some parameters for a model of the transistor
- Typical graphical characteristics

As the power dissipation is increased, the device heats up. Internal junctions of a transistor, or any semiconductor device, can only undergo certain upper limit temperatures before permanent damage occurs. The data provided by manufacturers is normally for 25°C operation. High-power transistors may require a heat sink to be added if the power dissipation is high. Manufacturers usually provide values for the thermal resistances of junction-to-case and junction-to-ambient to determine the thermal resistance of a heat sink if required.

21.3.2 RESISTORS

Resistors vary in size according to their power rating instead of their value; large power resistance is generally physically large. There are considerable variations of resistors that can be specified by

- The value of resistance
- The power rating
- The tolerance of the resistor
- The type of construction

Resistors are only manufactured in certain npv. Specific values not included in this npv can be obtained by a manufacturer at extra cost. Alternative resistors can be combined to obtain the required value or a variable resistor can be used.

21.3.3 CAPACITORS

Capacitors vary in size according to their value. Large-value capacitors are generally physically large. There are considerable variations of capacitors that can be specified by

- The value of capacitance
- The voltage rating
- Tolerance of the capacitor and
- The type of construction

The voltage rating determines the thickness of the dielectric between capacitor electrodes. The voltage rating will affect the size and the cost. Precision high-quality capacitors are usually based on glass, plastic, or mica dielectrics that have relatively low dielectric constant. Usually, high-frequency circuits do not require large-value capacitors, and capacitors with plastic, mica, or glass can be used. Electrolytic capacitors require the right polarity of voltage to be applied, other capacitors are polarity independent. As in the case of resistors, manufacturers provide tables for npv component values and make them available.

21.3.4 INDUCTORS

Inductors are fabricated on a coil of wire, usually wrapped round a ferromagnetic material. Typical values of inductors are in the range of microhenrys and millihenrys. Inductors are usually used in telecommunication circuits, where they provide frequency selectivity, or in oscillators. In integrated form a device known as gyrator based on Op-Amps, resistors, and capacitors can behave like an inductor, avoiding the fabricating of an inductor in IC form.

21.3.5 NOMINAL PREFERRED VALUES

Resistors and capacitors can be calculated to any value, but when manufactured in quantities for ease of manufacture the values follow clearly defined progressions known *as nominal preferred values (npv)*. In many cases, values of components obtained from a particular design calculation will not correspond with any *npv*. A higher tolerance can be used in order to obtain the required value, or use a variable component, but this will increase the production cost. In many cases, the nearest npv could be used, but then it is necessary to verify whether the design specification can still be satisfied. If a circuit is to be manufactured in large numbers, worst-case analysis should be performed to ensure that design specifications are satisfied for all values of each of the components.

Tables from manufacturers are easily accessible for npv components fabricated at different tolerances (e.g., EIA 2010, Electrical Engineering Community 2015). Various manufacturers of semiconductor components provide all the characteristics of a component and make the information freely available (e.g., Datasheet 2015).

21.4 BUILDING AND REALIZATION OF A CIRCUIT

Once the application has been designed, simulated, and components selected, it can be built. There are "breadboards," which consist of a two-dimensional array of connectors where the connectors in columns are connected to each other. To make a connection between columns a wire can be used as bridge. Components can be plugged into the breadboard without soldering them allowing a quick realization. Figure 21.2 shows an example of breadboard.

As the components are not soldered, these boards are useful only for a simple circuit to quickly check the functioning of an application.

For a more permanent application, components can be soldered in "veroboard." These veroboards are similar to breadboard, but the components are soldered in the board. Figure 21.3 shows a veroboard.

An application built on veroboard is more portable and can be boxed as a complete unit.

For a better finish, reduced area, or for a large number of units to be built a printed circuit board (PCB) can be used. PCBs have only the connection required by the components as conducting material, usually copper, as a layout. Figure 21.4 shows an example of PCB.

Components have to be soldered on the PCB. PCB lends itself for automatic population of components and automatic soldering required for mass production of applications.

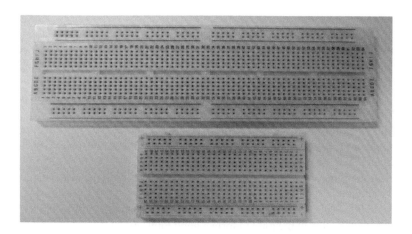

FIGURE 21.2   Sample of breadboard.

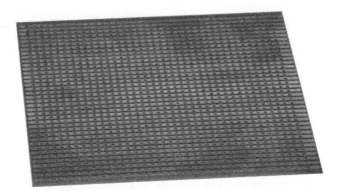

FIGURE 21.3 Sample of a veroboard.

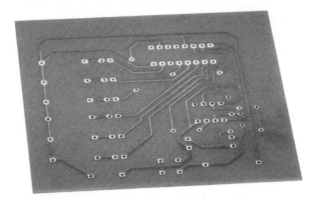

FIGURE 21.4 Example of PCB.

Soldering can be a cause for an application to fail. A bad solder can be unreliable and can deteriorate with time or it may not even make a proper electrical connection. To make a good solder requires practice. A summary of steps to follow in order to solder components can be

- Make sure all parts are clean and free of any grease.
- Make sure the board is firm and place component in the right place securely.
- Apply a fairly small amount of solder on the tip of the hot solder iron and clean the tip on a damp sponge.
- Heat the joint of the board and the component lead with the soldering iron for about two seconds.
- Keep the soldering iron on the joint to continue heating the joint and apply sufficient solder on the joint to form a good joint. Avoid the use of too much solder. It should not take more than 4 s to solder a standard joint.

21.5 TESTING–TROUBLESHOOTING

21.5.1 Testing

After the particular application, or a prototype, is constructed a detailed and complete testing of the whole electronic system is required. The testing is done under various conditions to make sure the application performs as expected. Testing can be done to see performance under different voltages, current, humidity, temperature, etc.

21.5.2 Troubleshooting

If the circuit works as expected there is no need for troubleshooting, but if the output is not the expected one, then it is necessary to determine the source of error and correct it. The initial action will be to identify the symptom. In analog electronics, experience is important; a particular symptom may provide an idea as to where a fault is most likely to have occurred. Electronic engineers use different techniques of troubleshooting depending on their particular experience in analog electronic applications, on the specification and complexity of the circuit, and the nature of the symptom.

The obvious thing to check first is whether the circuit has been energized; power supply has been connected properly, making sure the correct voltage required by the application is applied and working; and fuses are not burned out.

After the power check always carry out a sensory inspection. A good visual examination will provide information on obvious defects such as burned component, broken wires, poor solder joints, or burnt-out fuses. In many cases, a failure can cause large current producing overheating components. Sometimes it is possible to disconnect the circuit and check for overheated areas or components. In some cases is possible to perceive a smell of burn or smoke in damaged components.

Based on experience and knowledge of the circuit purpose, certain symptoms may indicate a particular component being defective. Replacing the most likely component to be faulty may solve the problem. Replacing components may fix the problem, but if many components need to be replaced it can be time consuming. This method is suggested only for systems where the symptoms are evident.

A technique that is very effective is to trace the signal through the circuit. The idea is to follow the signal until an incorrect signal appears. In this way a test point, where the measure voltage or current is wrong can be identified. The preceding process of design and simulation make available details of how the signal is predicted to change as it progresses through the circuit; then voltages and current are known at different test points.

To trace a fault, the circuit can be partitioned in different ways.

One way will be to start at the input and work toward the output. When an *incorrect* signal is found, the area where the problem is located has been isolated. The problem is in the portion of the circuit between the examination point where the error was found and the previous examination point.

An alternative way of tracing is to start at the output of a circuit and work toward the input. This time we check for voltages at each point until a correct measurement is obtained. When a *correct* signal is found the area where the problem is located has been isolated. The problem is in the portion of the circuit between the examination point where the correct signal was found and the previous examination point.

A third way will be to split the circuit in the middle. If the test point at the middle of the circuit has a *correct* signal, then you carry on testing toward the output until the fault is located. If the test point at the middle of the circuit has an *incorrect* signal, then you carry on testing toward the input until the fault is located.

21.6　ANALOG ELECTRONIC APPLICATIONS EXAMPLES

Practical projects help to explain and understand the theory behind analog electronics, as well as providing a good knowledge of the theory that result in the implementation of good applications.

21.6.1　DC Power Supply

21.6.1.1　Description

This application will provide a regulated dc power supply from an ac voltage supply. It will provide a 9 V dc and a variable dc supply up to 15 V. The ac supply is 120 V_{RMS} at 60 Hz.

A power supply will usually require the four stages before the load indicated in the blocks diagram in Figure 21.5.

21.6.1.2　The Transformer Stage

The transformer supplies the required ac voltage value for the rectifier. It also isolates the system from the ac power supply. In this case, we provide a dc power supply up to 15 V_{dc}. Some voltage drop will occur in the next stages. When diodes of the rectifier are conducting there is a small voltage drop in them also the filter and regulator will provide extra voltage drop. In general, in order to compensate for all these drops, we can estimate the voltage required at the output of the transformer as about 2–5 V above the dc output. In our application, let us assume that an extra 2 V is needed at the output of the transformer, and then we require an output of the transformer with a peak voltage of 17 V. The ac supply is 120 V_{RMS}, which provides peak voltage equal to

$$V_p = 120 \times \sqrt{2} \approx 169.7 \text{ V}$$

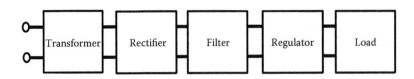

FIGURE 21.5　Block diagram of simple dc power supply.

Then, the transformation ratio required is

$$\text{Transformation ratio} = \frac{169.7}{17} \approx 9.983 \approx 10$$

The transformer needs to provide large peak current to charge capacitors used in the filter smoothing stage. Let us use a current of 3 A for the transformer. Figure 21.6 shows the transformer stage.

21.6.1.3 Rectifier Stage

For full wave rectification we can use a bridge rectifier consisting of four diodes connected as indicated in Figure 21.7. The diodes rating must be higher than the peak forward current and the peak reverse voltage expected in the circuit. Bridges rectifiers come as complete components and available for different current ratios.

21.6.1.4 Filter Smoothing Stage

Usually a capacitor will adequately smooth a rectified voltage. The capacitor can be estimated from the small variable voltage after smoothing, that is, the value of voltage ripple.

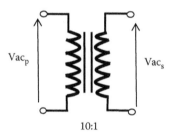

FIGURE 21.6   Transformer stage.

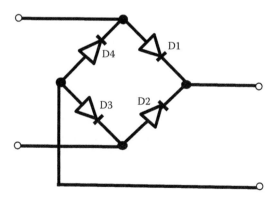

FIGURE 21.7   Full-wave bridge rectifier.

The ripple voltage across the capacitor $v_r(t)$ can be expressed as (Equation 3.14)

$$v_r(t) = \frac{1}{C}\int idt = \frac{1}{C}Q(t) \tag{21.2}$$

The calculation of the smoothing capacitor, C, will require knowledge of how the current changes as a function of time to solve this equation. We would like to have an estimation of the smoothing capacitor, C_s, as the tolerance of a capacitor is usually high (20%, 40% are not uncommon). We can then approximate the charge $Q(t)$ as the product of the dc current, I_C, times the period time T, then

$$v_r \approx \frac{1}{C_s}I_C \times T \tag{21.3}$$

and an expression for the capacitor can be approximated to

$$C_s \approx \frac{I_{dc}T}{V_r} \tag{21.4}$$

where T is the period of the waveform after rectification, I_{dc} the dc current, and V_r the peak-to-peak ripple voltage.

Note that T is the period of the waveform after the rectification, that is, for a half-wave rectification the period will be the same as the ac input power supply and for a full-wave rectification the period will be half the value of the period of the input ac power supply (double the frequency).

In our case, for full-wave rectification, we want a dc current of 3 A. Frequency of the waveform after rectification is 120 Hz (period T = 8.3 ms; double of the ac supply) and we can assume a ripple of 2 V, this gives

$$C_S \approx \frac{I_{dc}T}{2V_r} \approx \frac{3 \times 8.3 \times 10^{-3}}{2 \times 3} \approx 4150\ \mu F$$

This is estimation only; capacitors usually have large tolerances. Let us use an npv of 4700 μF. This is a large capacitance so an electrolytic capacitor can be used with the right polarity.

21.6.1.5 Regulator Stage

It is possible to design a regulator using discrete components that will keep the dc voltage constant for fluctuations of line voltages or load currents. In practice, complete voltage regulators are available in IC form. IC regulators are available from several manufactures. For example, the 78Lnn series provide nominal voltages of 2.6, 5, 6.2, 8, 9, 10, 12, 15, 18, and 24 V. The last two digits of the part number indicate the voltage.

In our case, for the 9 V dc output we can use the 78L09.

For the variable voltage we can use the LM317 regulator. The LM317 can provide voltages between 1.25 and 37 V. Figure 21.8 shows a connection for the LM317 to provide a variable dc output, the regulator component sets a voltage $V_{ref} = 1.25$ V between the common and output terminals; a small current $I_{com} = 50$ μA is provided from the common terminal according to the manufacture.

According to KVL the output voltage is given by

$$V_O = V_{R_1} + V_{R_2} \tag{21.5}$$

and KCL relate the currents as

$$I_2 = I_{com} + I_{ref} \tag{21.6}$$

Then, the output voltage is

$$V_O = V_{R_1} + V_{R_2} = R_1 I_{ref} + R_2(I_{com} + I_{ref})$$

For example, if the variable resistor R_2 is set to 1 kΩ and the resistor R_1 to 180 Ω, I_{ref} will be

$$I_{ref} = \frac{V_{ref}}{R_1} = \frac{1.25}{180} = 6.944 \text{ mA}$$

and the output voltage will be $V_o = R_1 \times I_{ref} + R_2 \times (I_{com} + I_{ref}) = 180 \times 6.944 \times 10^{-3} + 1 \times 10^3(50 \times 10^{-6} + 6.944 \times 10^{-3}) = 8.244$ V.

By changing the value of resistance in the variable resistor the dc output can be varied, the higher voltage will be limited by the dc input voltage to the regulator. Regulator manufacturers recommend adding a capacitor of 1 μF in parallel with the output. The whole dc power supply is shown in Figure 21.9.

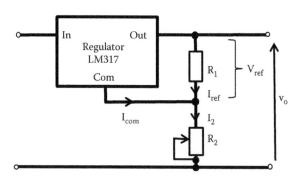

FIGURE 21.8   Circuit for the LM317 regulator to provide a variable dc voltage.

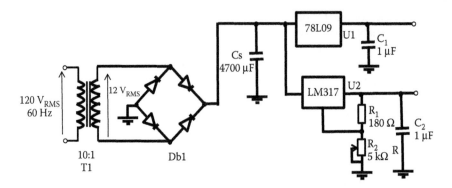

FIGURE 21.9 Dc power supply with a fix 9 V output and a variable output. Components: T1 = transformer 10:1, 5 A; Db1 = rectifier bridge 5 A; Cs = 4700 µF, 25 V; $C_1 = C_2 = 1$ µF; $R_1 = 180$ Ω; $R_2 = 5$ kΩ variable resistor; U1 = voltage regular 78L09; U2 = voltage regular 78L317.

21.6.1.5.1 How Does It Work?

The 10:1 transformer changes the ac voltage (120 V_{RMS}) to a lower voltage (12 V_{RMS}) in this case. The bridge rectifier rectifies the voltage to a dc voltage with a ripple; capacitor Cs smooth the voltage close to 17 V_{dc}, and the regulator fixes the voltage to the desired dc level.

21.6.1.5.2 Activities/Comments

Build this circuit and check its behavior. The application can include other output voltages by adding extra regulator specifics to the dc voltage required. If the ac line voltage is different, then we can use a different transformer with the right transformation ratio. It is possible to obtain negative voltages by connecting the filter and regulator stages to the opposite terminal of the bridge rectifier and using a center tap transformer.

21.6.2 AUDIO AMPLIFIER

21.6.2.1 Description

The circuit will amplify a signal in the 60–50 kHz frequency range. You are asked to design a class A audio amplifier that uses one BJT pin-through-hole transistor and one power supply of 16 V_{dc}. This amplifier feeds a load of 50 kΩ, the minimum BW required is 40 kHz, with a low-frequency cutoff of 60 Hz. With an input ac voltage of 15 m V_{RMS} a voltage gain of 150 is required. The collector current at the operating point is 12 mA and the amplifier should work from −20°C to 144°C. Only components standard npv are available with tolerances of 5% and 10%.

21.6.2.2 Design of a Class A Amplifier

First we write the specifications as we will be designing to specification.

21.6.2.2.1 Specification

One BJT pin-through-hole transistor
One power supply 16 V_{dc}
Load of 50 kΩ
BW minimum 50 kHz
Low-frequency cutoff 60 Hz
Gv = 150 at medium frequencies with v_{in} = 15 mV$_{RMS}$, I_{cQ} = 12 mA
Temp –20 to 144°C
Components: standard npv 5%, 10% tolerance
Class A amplifier

21.6.2.2.2 Design

To obtain a class A amplifier, the operation point (Q) is always in the linear region of the transistor. The collector current is already known from the specifications: 12 mA. We need to determine the value of Vce$_Q$ required by the operation point.

21.6.2.2.3 Select an Operation Point Q

In order to make sure that Q is not near the nonlinear saturation or cutoff regions of the transistor used, we can calculate how much the voltage between collector and emitter will change with an ac signal. At medium frequency with an input voltage of 15 mV$_{RMS}$, the voltage gain is 150 then the output voltage swing can be calculated from the voltage gain

$$G_V = \frac{V_O}{V_{in}} = 150$$

and

$$V_{ORMS} = 150 \ V_{in} = 150 \times 15 \ mV_{RMS} = 2.25 \ V_{RMS}$$

Then

$$V_{Opeak} = 2.25 \times \sqrt{2} \approx 3.182 \ V$$

The collector–emitter voltage will swing by ±3.1820 V around the operating point. The operating point voltage should be selected in such way that the V_{CE} will not be near saturation or cutoff regions. The load line at collector current equal to zero will be equal to Vcc: 16 V. If we assume that the saturation and cutoff regions are around 2 V, to play safe (once the transistor is selected then this could be revisited) we have the range for V_{CE} from 2 to 14 V avoiding saturation and cutoff regions. Figure 21.10 shows the output characteristics of a BJT indicating V_{CE} outside cutoff and saturation regions, from 2 to 14 V.

Now, considering the V_{CE} swing of ±3.1820 V away from the saturation and cutoff region, this leaves a region between (2 V + 3.1820 V) 5.1820 V and

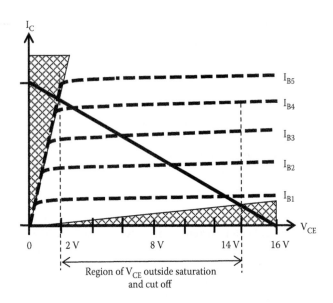

FIGURE 21.10 Output characteristics of a BJT transistor indicating saturation and cutoff regions.

(16 V − 2 V − 3.1820 V) 10.818 V to place V_{CE}. In other words, we can select any value of V_{CE} between about 6 and 11 V to make sure the operating point is in the linear region of the transistor and that the amplifier behaves as a class A amplifier.

Let us select $V_{CE} = 9$ V.

21.6.2.2.4 Choose a Circuit

Now according to specifications, there is a source and one transistor. We could use a fixed bias or an auto-bias circuit to satisfy specifications. We saw in Chapter 7 that auto-bias provides a more stable solution, and then we use an auto-bias circuit. Figure 21.11 shows an auto-bias bias circuit that can be used for this application.

According to KVL

$$Vcc = V_{R_E} + V_{R_C} + V_{CE} \qquad (21.7)$$

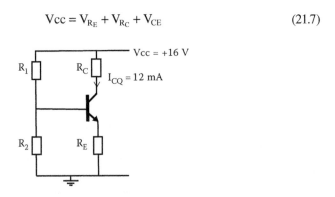

FIGURE 21.11 Auto-bias circuit to be used in this application.

To avoid dependency of β, as seen in Chapter 7, the voltage at the emitter can be made lower than the voltage at the collector.

Let us make the voltage across resistor R_E, $V_{RE} = 2$ V.

Then, from Equation 21.7 we have that the voltage across R_C is

$$V_{R_C} = V_{CC} - V_{CE} - V_{R_E} = 16 - 9 - 2 = 5\,V$$

Since we know the I_C ($\approx I_E$) at Q, we can calculate R_E and R_C as

$$R_C = \frac{V_{R_C}}{I_C} = \frac{5}{12 \times 10^{-3}} = 416.67\,\Omega \quad \text{(not available)}$$

A resistor of 416.67 Ω is not available as standard resistance, then from table for npv resistances we can choose the nearest value of 430 Ω.

npv (tolerance 5%) $R_C = 430\,\Omega$ (A new calculation of V_{Rc} may be required.)

and R_E can be calculated as

$$R_E = \frac{V_{RE}}{I_E} = \frac{2}{12 \times 10^{-3}} = 166.67\,\Omega \quad \text{(not available)}$$

Use npv $R_E = 160\,\Omega$.

21.6.2.2.5 Decide on a Transistor

To select a transistor we go to the information known from specifications, calculations, and decisions already made from specifications:

- One BJT pin-through-hole transistor
- Frequency response; BW of 50 kHz between 60 and 50,060
- I_C maximum larger than 12 mA
- Power maximum larger than 108 mW ($I_C \times V_{CE}$)
- V_{CE} maximum larger than 9 V
- Temperature range within $-20°C$ to $144°C$
- Model for simulation

Almost any general purpose transistor will be appropriate: Say BC546A.

BC546A data from manufacturer: includes I_C max = 100 mA, power = 500 mW, V_{CE} max = 65 V, h_{fe} (β) (at $I_C = 2$ mA, $V_{CE} = 5$ V) min 110–max 220, $Ft_{max} =$ 100 MHz, operating ambient temperature $-65°C$ to $+150°C$ (Data Sheet Catalog 2015). Then, the BC546 transistor is ok.

By using the npv for R_E and R_C the voltages across these resistances will be slightly different. The new values will be

$$V_{R_E} = R_E \times I_E = 160 \times 12 \times 10^{-3} = 1.92\,V$$

$$V_{R_C} = R_C \times I_C = 430 \times 12 \times 10^{-3} = 5.16\,V$$

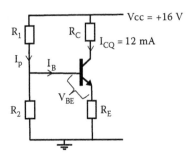

FIGURE 21.12 Auto-bias circuit showing the biasing currents.

21.6.2.2.6 Calculate R_1 and R_2

The current passing through the biasing resistance R_1, I_p in Figure 21.12, should be much larger than the base current, that is, I_B can be neglected with respect to I_p.

21.6.2.2.7 Calculate I_B

We know the value of I_C from the specification, the current gain parameter of the transistor, β, relate I_C and I_B as

$$\beta_{dc} = \frac{I_C}{I_B} \Rightarrow I_B = \frac{I_C}{\beta_{dc}}$$

For the value β we refer to the manufacturer data. The BC546A β (h_{fe}) has values between 110 and 220; by taking the lowest value of $\beta = 110$ will guarantee the largest current for I_B.

Then,

$$I_B = \frac{12 \times 10^{-3}}{110} = 109\ \mu A$$

If I_B is neglected with respect to I_p, in that case the current passing through the potential divider R_1 and R_2 is the same.

We can make I_p larger than I_B by say 20 times. Then

$$I_P = 20 \times I_B = 20 \times 109 \times 10^{-6} = 2.18\ mA$$

Now the voltage across R_2 will be

$$V_{R_2} = V_{R_E} + V_{BE}$$

The BC546A transistor is fabricated in silicon and we can assume the base–emitter junction voltage, V_{BE}, is approximated to be 0.65 V when forward biased. Then

$$V_{R_2} = V_{R_E} + V_{BE} = 1.92 + 0.65 = 2.57\ V$$

and R_2 can be calculated as

$$R_2 = \frac{V_{R2}}{I_p} = \frac{1.98}{2.18 \times 10^{-3}} = 908.26 \; \Omega$$

Use npv $R_2 = 910 \; \Omega$.
The voltage across R_1, V_{R1}, will be

$$V_{R_1} = Vcc - V_{R_2} = 16 - 2.57 = 13.43 \; V$$

and R_1 can be calculated as

$$R_1 = \frac{V_{R1}}{I_p} = \frac{13.43}{2.18 \times 10^{-3}} = 6.16 \; k\Omega$$

Use npv $R_1 = 6.2 \; k\Omega$.

21.6.2.2.8 Calculate C_{c1} and C_{c2}

It is possible to perform a detailed calculation of capacitance required considering the different cutoff frequencies of the different capacitance involved. The error involved in using a rough estimation may be smaller than the tolerance of components used. Usually a rough estimation is good enough to satisfy specifications. Figure 21.13 shows the required coupling capacitors.

We want small impedance for the coupling capacitors; say 20 Ω, at the lowest frequency, then from the reactance expression the coupling capacitors can be estimated as

$$X_{C1,2} = \frac{1}{2\pi f C} = \frac{1}{2\pi 60 \times C_{1,2}} = 20 \; \Omega$$

$$\Rightarrow C_{1,2} = \frac{1}{2\pi f X_{C1,2}} = \frac{1}{2\pi 60 \times 20} = 133 \; \mu F$$

Use npv $C_1 = C_2 = 100 \; \mu F$.

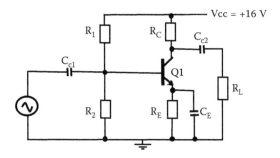

FIGURE 21.13 Auto-bias circuit showing the coupling capacitors.

For the capacitor across the emitter, the reactance of C_E should be much smaller than the emitter resistance.

Say $X_{C_E} \leq R_E/20$

Then,

$$X_{C_E} = \frac{R_E}{20} = \frac{160}{20} = 8\ \Omega = \frac{1}{2\pi f C_E}$$

and C_E can then be estimated as at the low-frequency cutoff as

$$C_E = \frac{1}{2\pi f X_{C_E}} = \frac{1}{2\pi 60 \times 8} = 331.6\ \mu F$$

Use npv $C_E = 330\ \mu F$.

Figure 21.14 shows the final circuit with value for all components.

21.6.2.2.9   How Does It Work?

The power supply Vcc and the biasing resistors R_E, R_C, R_1, and R_2 will fix an operator point on the transistor. As a signal is inputted at the base, an amplify signal will appear at the collector with the same shape as the input signal. The capacitor will block dc voltage and will allow ac signal to be amplified.

21.6.2.2.10   Activities/Comments

Simulate the circuit to check gain and cutoff frequencies. The emitter resistor R_E could be split to reduce the gain to the required value. Capacitances could be varied to obtain the bandwidth required. The upper cutoff frequency required is 50,060 Hz as minimum. Large values will satisfy specification. High-frequency cutoff can be reduced by adding parallel capacitance with the output. Build this circuit and check the behavior.

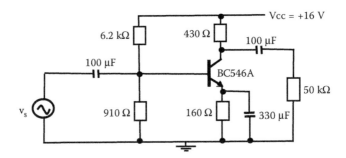

FIGURE 21.14   Circuit diagram for a class A amplifier circuit with calculated value of components. Components: $R_1 = 6.2\ k\Omega$, $R_2 = 910$, $R_C = 430$, $R_E = 160\ \Omega$, $R_L = 50\ k\Omega$, Q1 = BJT BC546A, $C_{c1} = C_{c2} = 100\ \mu F$, and $C_E = 330\ \mu F$.

21.6.3 Sallen and Key Second-Order Butterworth Low-Pass Filter

21.6.3.1 Description

This circuit will allow amplifying signal with frequencies below 7 kHz and block signals with frequencies above 7 kHz. Figure 21.15 shows a circuit for this application.

The frequency cutoff required by this filter is 7 kHz; the expression for the cutoff frequency is given by

$$f_c = \frac{1}{2\pi\sqrt{R_a R_b C_a C_b}} \tag{21.8}$$

For simplicity we can select $C_a = C_b = C$ and $R_a = R_b = R$. Then Equation 21.3 becomes

$$f_c = \frac{1}{2\pi RC} \tag{21.9}$$

In this case, f_c is 7 kHz; we can choose a value for R and calculate C. Let us make $R = 10 \text{ k}\Omega$ and the capacitor C can be obtained as

$$C = \frac{1}{2\pi R f_c} = \frac{1}{2\pi \times 10,000 \times 7000} \approx 2.27 \text{ nF}$$

Now, for this Butterworth filter the voltage gain of the noninverter amplifier is equal to 1.586. Then

$$G_v = \frac{R_1}{R_2} + 1 = 1.586 \tag{21.10}$$

we can choose a value for R_1 and calculate R_2. Let us make $R_1 = 10 \text{ k}\Omega$, then R_2 can be obtained as

$$R_2 = \frac{R_1}{G_v - 1} = \frac{10,000}{1.586 - 1} \approx 17.065 \text{ k}\Omega$$

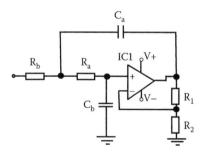

FIGURE 21.15 Circuit for Sallen and Key second-order Butterworth low-pass filter.

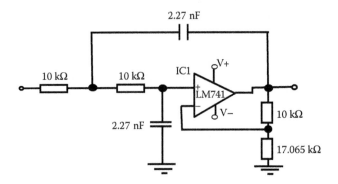

FIGURE 21.16 Circuit for Sallen and Key second-order Butterworth low-pass filter with component designed values included. Components: $R_a = R_b = R_1 = 10$ kΩ, $C_a = C_b = 2.27$ nF, $R_2 = 17.065$ kΩ, and IC1 = LM741 Op-Amp.

The completed circuit including all calculated components is shown in Figure 21.16.

21.6.3.1.1   How Does Work?

The combination of C_a, C_b with R_a, R_b select the frequencies to filter. In this case, block frequencies above 7 kHz, R_1 and R_2 provide the gain for the noninverting amplifier.

21.6.3.1.2   Activities/Comments

Simulate the circuit to check gain and cutoff frequencies as well as the roll-off response. Build this circuit and check its filter behavior.

21.6.4   AUTOMATIC SWITCH ON OF LAMP IN THE DARK USING A BJT

21.6.4.1   Description

This circuit will switch on a lamp when light is not present. This application can be used to illuminate certain site as it gets dark. Figure 21.17 shows a circuit for this application.

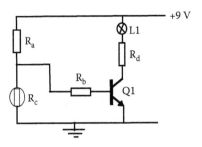

FIGURE 21.17 Circuit for application that switch on a light in the dark. Components: $R_a = 110$ kΩ, $R_b = 2.4$ kΩ, R_c = Light-dependent resistor (LDR), $R_d = 300$ Ω, Q1 = BJT BC546, and L1 = low power lamp 3 V or LED.

LDR (light-dependent resistor) or photoresistor is a resistor that changes its resistance as light is applied.

21.6.4.1.1 How Does Work?

When dark, the resistance of LDR is high and almost all the 9 V will drop across the LDR, the voltage across R_a will be very small, enough current will circulate across R_b to switch the transistor on and the lamp will be on.

When light is applied to LDR, its resistance will decrease. Almost all the 9 V will drop across R_a. The voltage across LDR will be very small and not enough current will circulate through R_b to switch the transistor on, so the lamp will remain off.

21.6.4.1.2 Activities/Comments

Build this circuit and check the behavior. The lamp can be replaced by an LED.

21.6.5 AUTOMATIC SWITCH ON OF A LAMP IN THE DARK USING OP-AMP

21.6.5.1 Description

This circuit works in the same way as application given in Section 21.6.4. It will switch on a lamp when light is not present. This time an Op-Amp is used, which provides larger current and a larger power lamp can be used. Figure 21.18 shows a circuit for this application.

Note: The Op-Amp needs to be biased. In some circuit diagrams, the biasing connections are not indicated. For biasing Op-Amp use pin 4 bias negative voltage, pin 7 bias positive voltage, and pin 8 ground.

21.6.5.1.1 How Does It Work?

The variable resistor R_v works as a potential divider. It sets the level of voltage at terminal (−) on the Op-Amp. When dark, the resistance of LDR is high and the voltage drop across the LDR will be high as well. At a certain amount of light the voltage across LDR, that fixes the (+) input to the Op-Amp, will be higher than input (−). Then the Op-Amp output will be high leading the transistor to saturation. The voltage across the lamp will be almost 9 V and the lamp switches on.

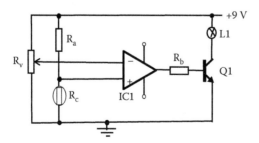

FIGURE 21.18 Circuit for application that switches on a light in the dark using an Op-Amp. Components: $R_a = R_b = 20$ kΩ, $R_v =$ variable resistor 20 kΩ, $R_c =$ LDR, IC1 = LM741, Q1 = BJT BD135, and L1 = lamp 9 V.

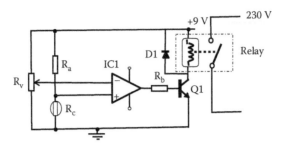

FIGURE 21.19 Circuit diagram for application to switch on a light using relay circuit.

When the light increases, the resistance of the LDR decreases and the voltage across LDR, input (+) of the Op-Amp, will be lower than input (−). Then, the Op-Amp output will be close to 0 V as the Op-Amp biasing use one positive voltage. The transistor is off and the lamp is off.

21.6.5.1.2   Activities/Comments

Build this circuit and check its behavior. Adjusting the value in variable resistor the lamp can be switched on at different levels of illumination. The advantage of this Op-Amp version is that this circuit can draw more current and a lamp of higher wattage can be used. This application can be very useful to switch on and control more powerful systems, large voltage lighting systems, pumps, motors, etc. A device that will permit this is a relay. A relay is an electromechanic switch, where a low electric current switch can control a mechanic switch that can be part of another large system. The lamp L1 shown in Figure 21.18 can be replaced by a relay in parallel with a diode as indicated in Figure 21.19.

21.6.6   AUTOMATIC SWITCH-ON OF LAMP IN THE PRESENCE OF LIGHT USING A BJT

21.6.6.1   Description

This circuit will switch on a lamp when light is present. This application can be used to illuminate certain sites during daytime. This is a variation of the application given in Section 21.6.4, where the resistor R_a and the LDR have been swapped. Figure 21.20 shows a circuit for this application.

21.6.6.1.1   How Does It Work?

When dark, the resistance of LDR is high and almost all the 9 V will drop across the LDR. The voltage across R_a will be very small, not enough current will circulate across R_b to switch the transistor on and the lamp will be off.

When light is applied to LDR its resistance will decrease and some voltage across R_a will produce enough current through R_b to switch the transistor on and the lamp L1 is on.

21.6.6.1.2   Activities/Comments

Build this circuit and check the behavior. The lamp can be replaced by an LED.

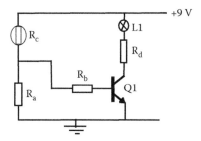

FIGURE 21.20 Circuit for application that switches on a light in the presence of light using a BJT. Components: $R_a = 900\ \Omega$, $R_b = 2.4\ k\Omega$, $R_c = LDR$, $R_d = 300\ \Omega$, Q1 = BJT BC546, and L1 = lamp 3 V.

21.6.7 Automatic Switch-On of Lamp in the Presence of Light Using an Op-Amp

21.6.7.1 Description

This circuit will switch on a lamp when light is present. This application can be used in the same way as application given in Section 21.6.6. This is a variation of the application presented in Section 21.6.5, where the resistor R_a and the LDR have been swapped. Figure 21.21 shows a circuit for this application.

21.6.7.1.1 How Does It Work?

The variable resistor R_v works as a potential divider. It sets the level of voltage at terminal (−) on the Op-Amp.

In darkness, the resistance of LDR is high and the voltage across LDR will be high, making voltage across R_a small. Then the voltage at input (+) is lower than the input (−) of the Op-Amp from the potential divider R_v. As (+) input of the Op-Amp is lower than input (−) therefore output of the Op-Amp will be close to 0 V when the transistor is off and the lamp is off.

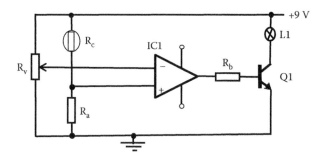

FIGURE 21.21 Circuit for application that switches on a lamp in the presence of light using an Op-Amp. Components: $R_a = 1\ k\Omega$, $R_b = 10\ k\Omega$, $R_c = LDR$, $R_v = $ variable resistor 10 kΩ, Q1 = BJT BD135, L1 = lamp 9 V, and IC1 = LM471.

As the light increases, the resistance of the LDR decreases and the voltage across R_a increases making the Op-Amp input (+) larger than input (−). Therefore, output of the Op-Amp will be close to 9 V and the transistor Q1 goes to saturation switching on the lamp L1.

21.6.7.1.2 Activities/Comments

Build this circuit and check the behavior. By adjusting the value in the variable resistor the lamp can be switched on at different levels of illumination. By replacing the LDR by an element that changes the value of resistance with temperature known as negative temperature coefficient (NTC), this circuit can control the switch on a lamp as the temperature changes. As the temperature increases the resistance of the NTC decreases and changes the voltage level at the (+) input of the Op-Amp. As in the application given in Section 21.6.6 the lamp can be replaced by a combination of relays in parallel with a diode to control larger systems.

21.6.8 HUMIDITY DETECTOR

21.6.8.1 Description

This circuit will detect humidity in a material when the probes are in contact with the material. In the presence of humidity, an LED will light. Figure 21.22 shows a circuit for this application.

21.6.8.1.1 How Does It Work?

When the probe leads are not connected to a material, the input will see a very high resistance of the air. As the probes are connected to a humid material such as wet sand, ground, skin, the resistance is considerably reduced. Small currents will reach the base of Q1; the output current from the emitter of Q1 will provide a current to the base of Q2 and transistor Q2 will be on to saturation and the LED will light up.

21.6.8.1.2 Activities/Comments

Build this circuit and check the behavior. The LED can be replaced by a lamp. As in previous applications, the lamp can be replaced by a relay to control large systems.

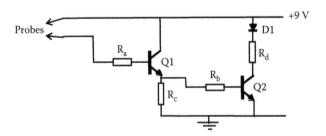

FIGURE 21.22 Circuit for a humidity detector application. Components: $R_a = 2.4\ k\Omega$, $R_b = 2.4\ k\Omega$, $R_c = 240\ \Omega$, $R_d = 330\ \Omega$, Q1 = BJT BC546, Q2 = BJT BD135, and D1 = LED.

21.6.9 DELAY SWITCH

21.6.9.1 Description

This circuit will keep a light on for a short time after it is switched off.

Figure 21.23 shows a circuit for this application.

21.6.9.1.1 How Does It Work?

As the switch is closed, a current to the transistor base will circulate making the collector current increase close to saturation and the lamp is switched on. Also, the capacitor C_a will be charged. As the switch is open the capacitor will discharge producing a current in the base of Q1 and the lamp will remain on until the capacitor is discharged.

21.6.9.1.2 Activities/Comments

Build this circuit and check its performance. The capacitor can be replaced by a different capacitance that will change the time the lamp remains on.

In any case, in this circuit, the lamp will not stay on for a long time after the switch is off because the current from the capacitor is small. This circuit can be improved by adding another general purpose power transistor (BD135) as indicated in Figure 21.24. The current is increased with the addition of this transistor; the emitter current from Q1 goes into the base of the second transistor Q2. The combination of these two transistors is known as a *Darlington* pair. This combination of transistors will increase the time that the lamp remains on. Also, this Darlington

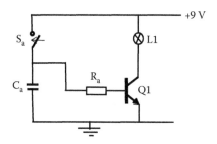

FIGURE 21.23 Circuit for a delay switch. Components: $R_a = 2.4\ k\Omega$, $C_a = 3300\ \mu F$ electrolytic 16 V or more, Q1 = BJT BC546, S_a = switch, L1 = lamp 9 V.

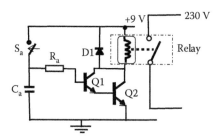

FIGURE 21.24 Circuit for a delay switch with a Darlington pair.

combination will allow to drive load with large current with the inclusion of a relay as indicated in Figure 21.24.

21.6.10   SMOKE DETECTOR

21.6.10.1   Description

This circuit will detect and initiate an alarm in the presence of smoke. Figure 21.25 shows a circuit for this application.

21.6.10.1.1   How Does it Work?

The circuit is similar to application given in Section 21.6.7, light detector, but this time the light comes from the LED. When no smoke is present the light from the LED is applied to the LDR and its resistance decreases and a large voltage is applied to input terminal (−) of the Op-Amp. The output of the Op-Amp is zero and the transitor Q1 is off and the siren is not activated. When smoke is present, the illumination on LRD is reduced and its resistance changes as well. The output of the Op-Amp increases saturating the Q1 and the siren is then activated.

21.6.10.1.2   Activities/Comments

Build this circuit and check its behavior. As in previous applications, the siren can be replaced by a combination of relay in parallel with a diode to control larger systems.

21.6.11   MULTIVIBRATORS AND THE 555 TIMER

Inside digital gates and digital components analog circuits will produce the required behavior. Usually a transistor driven to saturation will produce a "zero" and a transistor driven to saturation will produce a "one" output. There are dedicated ICs especially designed to accurately produce a required digital function packaged as a cell block.

21.6.11.1   Astable Multivibrator

21.6.11.1.1   Description

The stable multivibrator consists of a circuit with two switching transistors, a cross-coupled feedback network, and two time delay capacitors, which allow oscillation

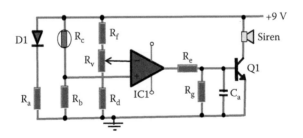

FIGURE 21.25 Circuit for a smoke detector using an Op-Amp. Components: $R_a = R_d = R_e = 390\ \Omega$, $R_b = R_f = 11\ k\Omega$, $R_c = $ LDR, $R_g = 1.1\ k\Omega$, $C_a = 330\ \mu F$ electrolytic 12 V or more, Q1 = BJT BD135, siren = BZ1 buzzer, D1 = white LED, IC1 = LM741.

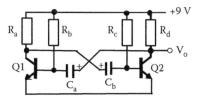

FIGURE 21.26 Circuit for an astable multivibrator. Components: $R_a = R_d = 390\ \Omega$, $R_b = R_c = 18\ k\Omega$, $C_a = C_b = 150\ uF$ electrolytic 12 V or more, and Q1 = Q2 = BJT BC546.

between the two states. Figure 21.26 shows a circuit for this application. Both transistors are biased for linear operation and are operated as common emitter amplifiers with 100% positive feedback.

21.6.11.1.2 How Does It Work?

At the connection of the dc power supply let us assume that transistor Q1 is *on* and transistor Q2 is *off*, nevertheless a small current will circulate through R_d, which allows charging C_a; when C_a is fully charged it will block the current to the base of transistor Q1 and the transistor Q1 is off. As the transistor Q1 is switched off the voltage at the collector of transistor Q1 goes to about 9 V and the transistor Q2 is switched on. Now a current will circulate through R_a and allow the charging of C_b. When C_b is fully charged it will block the current to the base of transistor Q2 and the transistor Q2 is off. When C_b is charging, C_a is discharging through R_c. The whole sequence starts again producing an output similar to a square wave signal.

21.6.11.1.3 Activities/Comments

Build this circuit and check its behavior. You can change the values of resistors and capacitors to change the frequency of the output voltage.

Multivibrators can be modified to include the application of an external trigger voltage pulse to force the change from one stable state to the other. For example, a multivibrator with two external trigger circuits will create a bistable multivibrator, commonly known as *flip-flop*. In each of the two states, one of the transistors is *cut off* while the other transistor is in *saturation*. This means that the bistable circuit is capable of remaining indefinitely in either stable state.

21.6.11.1.4 The 555 Timer

Multivibrators can be easily constructed from discrete components to produce basic square wave output waveforms. The "*555 Timer*" is a specially designed IC to accurately produce a required output waveform with the addition of a few external components. The 555 timer is a useful, cheap, and popular oscillator timing device that can generate single pulses or long time delays. Figure 21.27 shows the 555 internal block diagram.

The *555 timer* uses three 5 kΩ resistors as voltage dividers providing reference voltage for two Op-Amps working as comparators. This is why it is known as the "555 timer." The lower comparator connected to the Trigger (pin 2, inverting terminal) has a voltage reference of one-third Vcc (noninverting terminal). The upper

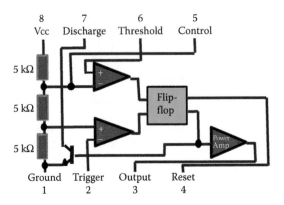

FIGURE 21.27 The 555 timer internal block diagram.

comparator is connected to the Threshold (pin 6, noninverting terminal) and has a voltage reference of two thirds Vcc (inverting terminal). Voltage levels can be connected to Trigger and Threshold to obtain combinations of outputs from the comparators that become inputs to the flip-flop to produce a stable state. The output from the flip-flop is passed through a power amplifier to increase the current drawn from the timer (pin 3). The voltage level applied to Trigger and Threshold terminals of the comparators is usually obtained using resistors and capacitors. Capacitors are charged via external resistance and Vcc and discharged internally using a NPN transistor (pin 7, discharge). The flip-flop can be reset (pin 4) or discharged (pin 7). Figure 21.28 shows the package of a standard 555 timer.

The external connections to the 555 timer are as follows:

- Pin 1: Ground, connects the 555 timer to the negative (0 V) voltage.
- Pin 2: Trigger, the negative input to lower comparator.
- Pin 3: Output, it can draw a current up to 200 mA at an output voltage equal to Vcc − 1.5 V.
- Pin 4: Reset, it resets the flip-flop to control the output. It is usually connected to Vcc to prevent any unwanted resetting.

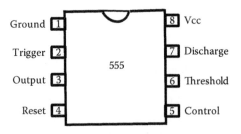

FIGURE 21.28 Standard 555 timer package and pin out.

- Pin 5: Control voltage, it overrides the two thirds Vcc level of the voltage divider input to the upper comparator. A voltage to this pin can control the width of the output signal, independently of an external RC timing network. It is usually connected to the ground through a capacitor when not required to eliminate noise.
- Pin 6: Threshold, the positive input to the upper comparator. A voltage higher than two thirds Vcc will reset the flip-flop changing the output from *high* to *low* state.
- Pin 7: Discharge, it will discharge an external timing capacitor when the output of the timer switches to low.
- Pin 8: Supply, it connects the timer to the dc power supply +Vcc. Usually between 4.5 and 15 V.

21.6.12 Humidity Detector Using 555 Timer

21.6.12.1 Description

This circuit will detect humidity in a material when the probes are in contact with the material. In the presence of humidity an LED will flash an intermittent light. Figure 21.29 shows a circuit for this application.

21.6.12.1.1 How Does It Work?

When the probe leads are not connected to a material, the input will be presented by a very high resistance of air. The capacitor C_1 is discharged. As the probes are connected to a humid material the resistance between the probes is considerably reduced and C_1 starts to charge through R_b and R_a. When the capacitor reaches two-third of applied bias voltage, the output voltage of the 555 timer will be zero and D1 is on. From this moment C_1 is discharged through the internal transistor of the 555 timer. When the voltage of C_1 reaches one-third of the supply voltage the output voltage of the 555 timer will be 9 V and D1 is off. The process is repeated again and the LED light is on and off, indicating humidity. The frequency of the LED flashing will be increased if the material touched by the probes is more humid.

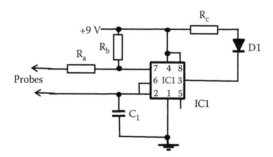

FIGURE 21.29 Circuit for a humidity detector application using a 555 timer. Components: $R_a = 680\ \Omega$, $R_b = 8.2\ k\Omega$, $R_c = 470\ \Omega$, $C_1 = 15$ uF electrolytic 10 V or more, IC1 = NE555, and D1 = LED.

21.6.12.1.2   Activities/Comments

Build this circuit and check its performance. The LED can be replaced by a combination of relays in parallel with a diode to control larger systems such as an automatic irrigation system.

21.6.13   TWO-TONE MUSICAL INSTRUMENT USING 555 TIMERS

21.6.13.1   Description

This circuit will play two tones; the two tones can be varied to create a simple musical instrument. Figure 21.30 shows a circuit for this application.

21.6.13.1.1   How Does It Work?

A particular value on R_{v1} and R_{v2} will create outputs from the two 555 timers with specific frequencies that produce particular tones. By changing the value of the variable resistances a change in frequency is obtained at the output of the 555 timer to create different tones.

21.6.13.1.2   Activities/Comments

Build this circuit and check its behavior. You can make this circuit to produce the musical scale at different values in the variable resistance. The circuit can also be used as alarm, pedestrian zebra crossing, etc.

21.6.14   AUTOMATIC SWITCH-ON OF LAMP IN THE PRESENCE OF LIGHT USING A 555 TIMER

21.6.14.1   Description

This circuit will switch on a lamp when dark. This time a 555 timer is used to control a lamp. Figure 21.31 shows a circuit for this application.

21.6.14.1.1   How Does It Work?

The variable resistor R_v works as a potential divider. It sets the level of voltage at terminal (2) on the 555 timer. This sets the level voltage of the lower comparator

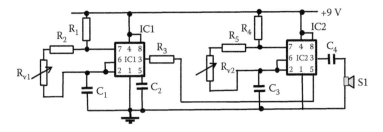

FIGURE 21.30 Circuit for a two tones musical instrument using 555 timers. Components: $R_1 = R_2 = R_3 = R_4 = R_5 = 8.2$ kΩ, $R_{V1} = R_{V2} =$ variable resistor 100 kΩ, $C_1 = 47$ nF, $C_2 = C_3 = 220$ nF, $C_4 = 10$ uF electrolytic 10 V or more, IC1 = IC2 = NE555, and S1 = speaker 8 Ω.

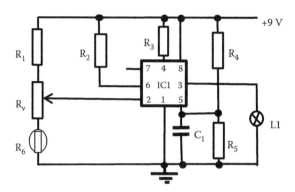

FIGURE 21.31 Circuit for application that switch on a light in the dark using a 555 timer. Components: $R_1 = R_2 = R_3 = R_4 = R_5 = 12\ k\Omega$, $R_6 = $ LDR, $C_1 = 330\ \mu F$, L1 = lamp 9 V. $R_V = $ variable resistor 100 kΩ, and IC1 = NE555.

inside the 555 timer. When light is applied over the LDR its resistance decreases and the output voltage from the timer will be approximately 0 V, and the lamp L1 will be off.

In darkness, the resistance of LDR is high and the output voltage from the timer will be approximately 9 V switching the lamp L1 on.

21.6.14.1.2 Activities/Comments

Build this circuit and check its behavior. Adjusting the value in variable resistors, the lamp can be switched on at different levels of illumination.

This circuit, as other examples in this section, can control a larger power system through a relay. For example, by replacing the lamp L1 for a relay we can switch on a bulb used in domestic lighting at larger voltage. Figure 21.32 shows a modification of this application to include a relay to control higher current.

In realizing any of these applications it is crucial to follow safety consideration to avoid damage equipment and more important to prevent accident to persons. Special care should be taken when soldering components and only experienced practitioners should use voltage over 30 V.

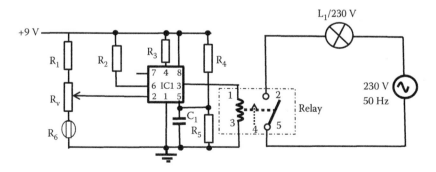

FIGURE 21.32 Modification of circuit to include a relay to switch on a light in the dark.

21.7   KEY POINTS

- Practical projects help us to understand the theory behind analog electronics, and good understanding leads to the implementation of good applications.
- In applications of analog electronics, the experience of a practitioner or designer can add to the success of a project.
- Before committing a circuit to production a good simulation is required to check all possible cases.
- Circuits may not work the first time but troubleshooting will be made easier if based on a good simulation where expected values of voltages and current are known.
- Troubleshooting processes include identifying the symptoms of the fault, carrying out a power check on the circuit, performing a physical check of the circuit, and applying a signal tracing technique to isolate the fault.
- Components are selected according to the different applications. Apart from the defined value and characteristics of a component, it is necessary to consider maximum rating, size, and material used in fabrication, and type of packaging.

REFERENCES

Datasheet 2015 http://www.datasheetcatalog.com
Data Sheet Catalog 2015 http://pdf.datasheetcatalog.com/datasheet/philips/BC546_547.3.pdf
EIA 2010 Electronic Industries Association http://logwell.com/tech/components/resistor_values.htm
Electrical Engineering Community 2015 http://www.eeweb.com/toolbox/resistor-tables

22 Future Trend of Analog Electronics

22.1 INTRODUCTION

Digital electronics have experienced a huge advance within electronic systems. While analog electronics have not developed at the same pace as digital electronics, it is expected that analog will complement and match the advances in digital applications. Analog electronics should, in the future, enjoy similar rates of progress as digital electronics to improve applications in electronic systems and reduce any problems causing bottlenecks in new applications. For example, in operating biological signals at micro-volt level, the digital systems cannot manipulate these signals before an analog system amplifies them to a level in which analog to digital conversion can work on them.

In a digital system a great deal of the design and performance improvement can be done by software-based programs. In contrast, improvement in analog electronics requires a good deal of human intervention because of their time-variable nature; therefore, there is large room for improvement with clever designs by humans. At present, analog systems are relatively expensive and they consume an irregular amount of power compared with digital circuits.

One challenge of analog systems is to reduce the draining power and improve their versatility in applications of consumer electronic systems.

Semiconductor technology advances allow reducing the size of electronic components from micro-level channels transistor to nano-level channels transistors. They offer a bright future in the upcoming years in terms of the size, power consumption, speed, and frequency range. A challenging future of analog electronics will be to perform analog operations in smaller transistor devices. There are areas in analog electronics that can take advantage of new process technologies, such as high-speed data converters and design at high frequencies.

New analog systems are mainly based on using operational amplifiers as building blocks. An alternative approach is to use new analog circuits known as comparator-based switched capacitor (CBSC) circuits. Using CBSC as building block can lead to more user-friendly and power-efficient analog systems. New applications may appear in the future using the advantages of CBSC.

There has been a massive increase and advance in sophisticated digital signal processing in new electronic products. New generation of cellular smart telephone and wireless devices use different frequencies for their signal processing. In the future, the analog function will not only do the information extraction, but will prepare the signal for algorithms and processing required by the digital signal processing unit. Analog function will also provide amplification, analog to digital conversion, digital to analog conversion, and isolation.

Recent developments in the communications industry impose new challenges on the design of wireless terminals. The ever-increasing diversification of services has led to highly complicated equipment, supporting extended functionality. The digital section of wireless terminals greatly benefits from the advantages offered by the rapid advancement of semiconductor technologies. Miniaturization allows an increasing complexity and higher operating frequencies, leading to better performance and more integrated functionality of digital circuits (Cafaro et al. 2007).

The future then is a mixed technology between analog and digital electronics working jointly in complete systems. Analog design is more difficult and requires more experience than digital design. The future will need analog design engineers with a good understanding of the fundamentals of electronic concepts.

Science is always developing new electronic materials, new practices or procedures of transferring information, new and powerful microprocessors, and new processes of storing information. Analog electronics engineers should be ready to satisfy the ever-faster operating speed requirements, fulfil the ever-tighter requirement of power consumption and management, and match the higher level of integration required by new systems.

22.2   RECONFIGURABLE ANALOG CIRCUITRY

Traditionally, the bottleneck in the development of mixed signal ICs has been the analog interface. Next to the increasing digital functionality, analog circuits must support extensive reconfiguration, not only to adapt to the processed signal characteristics, but also to deal with various environmental effects, variable supply voltages, fabrication tolerances, and a wide range of operating temperatures specific to portable equipments (Damico et al. 2004; Kenington 2005).

The reconfiguration of analog circuits in an array of circuits, that can be programmed to fulfill different functions, offers a new attractive way of designing analog systems to make them easy to mix with digital systems. The universal reconfigurable circuits called field programmable analog array (FPAA) have been gaining popularity among analog designers. As a result, it is possible to implement a variety of circuit architectures using FPAA. The new technology of FPAA works in a similar fashion as the well-established uncommitted array of gates in digital electronic design known as the field programmable gate arrays (FPGA). Combination of FPAA and FPGA provides a basis for the development of dynamically reconfigurable electronic mixed systems. This mixed signal circuit is especially suitable for applications in mobile and portable telecommunication systems, in general. This combination of digital and analog arrays provides a better power dissipation management through proper task of hardware processing functions to analog and digital circuits.

An FPAA consists of a regular two-dimensional array of analog identical configurable cells. Each cell can contain devices based on uncommitted operational amplifiers, static and dynamic switches, and banks of programmable switched capacitors.

The static switches are programmed once for a particular application. The dynamic switches are switched intermittently during the circuit operation. Both static and dynamic switches are electronically controlled providing the potential to tailor the functionality of each cell in the array.

A block diagram of a FPAA is shown in Figure 22.1. The crossbar style routing network, inherited from digital FPGA, is a feature specific to most of the FPAA architectures. Switches (Sw) allow routing the signal throughout the entire array and access to any of the configurable analog blocks (CABs). The structure of each CAB depends on the target application mapping a wide range of analog functionality to the FPAA.

Reconfigurable analog systems are gaining increasing significance in all application areas of the semiconductor industry. Modern equipment destined for telecommunications, medicine, or transport must support a high degree of integration in functionality and switching between different features. Changing functionality must be transparent to the end user and while it is achieved by software, it must be supported at hardware level by the underlying circuitry (Kenington 2005).

The optimization of the cost-to-performance ratio and of the equipment portability, are key aspects of reconfigurable circuit design. Outstanding results may be achieved by combining wide range parameter programmability with topological reconfiguration (Csipkes et al. 2012).

Work on analog filters that retain the generality in what concerns supported synthesis methods, filter types, and filter order has been published. These applications often use evolutionary methods to find the filter topology required by the given application (Pankiewicz et al. 2002, Becker et al. 2008).

There is a variety of applications of FPAA; an example of FPAA proposes a filter based on open-loop active elements provides a generalized approach to complex polyphase (Csipkes et al. 2010b).

Global roaming, worldwide equipment compatibility, and the ever-increasing diversity of services supported by providers are the main driving forces of emerging telecommunication technologies, such as the software defined radio (SDR), or the cognitive radio (CR) (Kenington 2005, Cafaro et al. 2007).

One key element of reconfigurable analog front ends in practical SDR realizations is the channel or band select filter. The classical approach to multimode operation uses switched signal paths in order to adapt the receiver to the requirements of

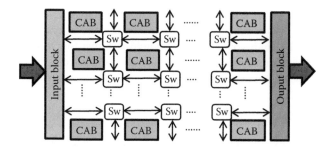

FIGURE 22.1 General architecture of a FPAA. (From Hintea, S. et al. 2010. A variable topology analog filter suitable for multi-mode wireless applications. *Proceedings of the 14th International Conference, KES2010. Knowledge-Based Intelligent Information and Engineering Systems,* Cardiff, UK, 603–612. ISBN-0302-9743. ISBN-13-978-3-642-15383-9. With permission.)

a limited number of supported standards. As the number of supported standards increased, filters evolved into building blocks featuring programmable frequency parameters over a relatively wide range of frequencies (Damico et al. 2004). The next step in the evolution process is to implement fully reconfigurable filters that can adapt to the spectral characteristics of a wide variety of communication standards while maintaining a relatively low hardware overhead, low consumption, and high cost efficiency. This adaptation may be implemented by means of software that dynamically changes the frequency parameters, topology, and even the approximation or synthesis technique.

FPAAs (field programmable analog arrays) have been traditionally associated with rapid prototyping and design cycle shortening in analog design. They have been mostly regarded as a monolithic collection of analog building blocks, user-controlled routing resources, and memory elements (D'Mello et al. 1998). FPAA systems offer advantages in applications where signal frequencies are high and low-power dissipation is required. Examples of FPAA include filter applications in multimode wireless communication equipment (Csipkes et al. 2008, 2009, 2010a,b, Hintea et al. 2008) systems for Software Radio (Csipkes et al. 2008, Hintea et al. 2008) as well as a large number of diverse analog applications (Hintea et al. 2010).

22.3 ANALOG DEVICES AT HIGH POWER

New advances in semiconductor technologies have made it possible to fabricate new electronic devices that can work at high voltages. Operations performed at low or medium voltage can now be executed at high voltage using the new electronic components. This makes it possible for analog electronics to make an impact into the field of power electricity. In general, the separation between electronics and electricity can become indistinguishable for some applications. In terms of generation of electricity new alternative ways of generation have been developed using renewable sources of energy. Analog electronics presents a good source for future applications at high voltages and currents. An example is the use of compound semiconductors such as silicon carbide (SiC), which has permitted the fabrication of devices that can work at voltage in the range of several thousand volts (Planson et al. 2008, Bhalla et al. 2013). These new devices offer the advantage of high breakdown fields and low on-resistance compared to conventional silicon devices; this is significant for high-voltage power switching transistors, creating electrical systems of higher efficiency. Also, SiC semiconductor devices are able to operate at temperatures of several hundred degrees Celsius without major variation in their electrical properties. Research on SiC and other compound semiconductors is ongoing to obtain devices with even higher voltages and lower switching losses (Hatori et al. 2014, Casady 2015).

Silicon carbide devices enable the creation of very efficient and compact power electronic systems supporting entirely new power electronic system structures to make renewable power generation in large plants and power transmission possible. Examples of the use of SiC devices have produced high-voltage dc converters with high power reaching higher efficiency and higher switching frequency compared to conventional silicon devices (Thoma and Kranzer 2015).

Research in devices fabricated using silicon-on-insulator (SOI) (Hiraoka et al. 2001, Huque et al. 2010) technologies can produce the next generation of power semiconductor devices. The next generation of power semiconductor devices will offer important advantages over the existing generation, such as higher breakdown voltage, higher current capabilities, higher frequency range, lower on-state resistance, and lower leakage currents. All these properties create the potential to develop new efficient power systems in IC form.

22.4   FUTURE ADVANCES IN APPLICATIONS OF ANALOG ELECTRONICS

Analog electronics will advance jointly with digital electronic development as they complement each other. There are many areas in which analog electronics will play a large role in implementing new applications. New analog applications will certainly improve the condition of life for many people of our planet.

22.4.1   NANOTECHNOLOGY

Nanotechnology deals with systems designed and manufactured of the order 10^{-9} m size. It is possible to find structures of some decades of nanometers in size. Nanotechnology may bring forward new ways of creating electronic systems (Hamedi-Hagh and Bindal 2008, Ytterdal 2008). It may introduce new circuit materials, new nano-processors, and new methods of transferring information. The near future will take advantage of many nano-technological applications within the electronics industry.

22.4.2   NEW ANALOG BUILDING BLOCK WORKING VOLTAGE UNDER 1 V

New optimized analog circuits that eliminate the use of operational amplifiers as building blocks are being developed (Shaterian et al. 2013). These new circuits use an architecture that relies on circuit blocks based on CBSC circuits (Zamani et al. 2011). These building blocks can handle small voltages of less than 1 V resulting in greater power efficiency. This can give a new dimension to applications in consumer electronics based on analog electronics.

22.4.3   ANALOG ELECTRONICS SEES A REVIVAL IN THE MUSIC INDUSTRY

Analog electronics in music is experiencing a revival. Some old products are being reintroduced and recording studios are incorporating analog elements into previously pure digital processes (D'Angelo and Valimaki 2013). Analog electronics can produce sounds that are fundamentally distinct; a parameter when changed in digital form happens in separate steps while in analog the range of variation is infinite. Some digital software does not work as quickly as analog for certain sound effects, for example, the way the wave envelopes respond. Many recording studios and artists are now using at least some analog technology in their music production. Many acoustic recording artists are introducing analog elements to their recording process. In the future, it is expected that an increasing number of hybrid solutions will be produced that include a variety of analog processing elements incorporated directly into digital systems.

22.4.4 NEW TYPE OF ANALOG COMPUTING

In some applications, the instruction of programming code can accept tolerance for errors. For these types of applications, it is possible to build neural network models to approximate programming that are intended to run in a regular processor. These models can run on special neural processing accelerators (NPUs) (Pietras 2014). Complete systems that use this approximate computing model can be built to run at lower voltages reducing the energy consumption required. These NPUs can be built in analog ICs that will be faster and use much less power than their digital equivalents.

This approach will not be useful for applications that cannot handle some errors. For applications such as bioinformatics, data mining, large-scale machine learning, speech recognition, and image recognition it is possible to use this new way of analog programming (Roska et al. 1990, Esmaeilzadeh et al. 2013). A process that can deliver with some minor inaccuracy in application, such as image processing, can still produce a good solution. For example, in the use of image processing to determine and identify microorganisms in drinking water (Fernández-Canque 2012), using analog systems to extract information and neuro-networks in the identification stage have provided a good solution.

22.4.5 BIOTECHNOLOGY

Many bio signals at micro-level voltage and new bio cells producing electricity may play an important role in future applications. These will be analog in nature and development in bioengineering will have a large input from analog electronics (Chung et al. 1999, Sarpeshkar 2013).

22.5 KEY POINTS

- The future will be a mixed technology with analog and digital electronics working jointly in complete systems.
- Analog electronic design is more intricate and requires more experience. Analog design engineers will need a good understanding of the fundamentals of electronic concepts.
- Analog electronics should in the future enjoy similar rates of progress as digital electronics to improve applications in electronic systems and reduce any problems causing bottlenecks in new applications.
- Semiconductor technology advances allow reducing the size of electronic components from micro-level channels transistors to nano-level channels transistors. They offer a bright future in the upcoming years in terms of the size, power consumption, speed, and frequency range.
- FPAAs for universal reconfigurable circuits have been gaining popularity among analog designers. As a result, it is possible to implement a variety of circuit architectures using FPAA.

- There are many areas where analog electronics will play a large role in implementing new applications such as FPAA, power electronics, nano-technology, music industry, analog computing, and biotechnology.
- Analog electronics engineers should be ready to satisfy the ever-faster operating speed requirements, fulfil the ever-tighter requirements of power consumption and management, and match the higher level of integration required by new systems.

REFERENCES

Becker, J., Henrici, F., Trendelenburg, S., Ortmanns, M., Manoli, Y. 2008. A hexago-nal field programmable analog array consisting of 55 digitally tunable OTAs. *IEEE International Symposium on Circuits and Systems*, Seattle, WA, 2897–2900.

Bhalla, A., Fursin, L., Hostetler, J., Li, X., Fox, M. 2013. Breakthrough SiC GTO's enable efficient & compact next generation high voltage grid conversion. *PCIM Europe 2013 Conference*, Nuremberg, 34–48.

Cafaro, G., Correal, N., Taubenheim, D., Orlando, J. 2007. A 100 MHz-2.5 GHz CMOS transceiver in an experimental cognitive radio system. *SDR Forum Technical Conference*, Honolulu, HI, 189–192.

Casady, J. 2015. New generation 10 kV SiC power MOSFET and diodes for industrial applica-tions. *PCIM Europe 2015*, Nuremberg, 1–8.

Chung, W.-Y., Lin, K.-P., Yen, C.-J., Tsai, C.-L., Chen, T.-S. 1999. Analog processor chip design for the bio-signal readout circuit application. *Engineering in Medicine and Biology. 21st Annual Conference and the 1999 BMES/EMBS Conference*, Vol. 2, IEEE Conference Publications, Atlanta, GA, 24–31. DOI: 10.1109/IEMBS.1999.804036.

Csipkes, G., Csipkes, D., Hintea, S., Fernandez-Canque, H. 2010a. A generalized approach to complex filter design with open loop active cells. *Proceedings of the 12th International Conference. IEEE 2010. Optimization of Electrical and Electronic Equipment OPTIM 2010*, Brasov, Romania, 1027–1033. ISSN 1842-0133, ISBN 978-973-131-080-0.

Csipkes, D., Csipkes, G., Hintea, S., Fernandez-Canque, H. 2010b. An analog array approach to variable topology filters for multi-mode receivers *Journal of Electronics and Electrical Engineering. Kaunas: Technologija*, 9(105), 43–48.

Csipkes, G., Csipkes, D.P., Farago, P., Fernandez-Canque, H., Hintea, S. 2012. An OTA-C programmable analog array for multi-mode filtering applications. *Proceedings of the 13th International Conference. IEEE 2012. Optimization of Electrical and Electronic Equipment OPTIM 2012*, Brasov, Romania, 1027–1033. ISSN 1842-0133, ISBN 978-4673-8/12.

Csipkes, G., Hintea, S., Csipkes, D., Rus, C., Festila, L., Fernandez-Canque, H. 2008. A digitally reconfigurable low pass filter for multi-mode direct conversion receivers. *I Knowledge-Based Intelligent Information and Engineering Systems*. Vol. 5179, Springer, Berlin, 335–342. ISSN0302-9743.

Csipkes, G., Hintea, S., Csipkes, D., Rus, C., Festila, L., Fernandez-Canque, H. 2009. A highly linear low pass filter for low voltage reconfigurable wireless application. *JAMRIS Journal of Automation, Mobile Robotics & Intelligent Systems*, 3, 76–81.

Damico, S., Gianini, V., Baschirotto, A. 2004. A low-power reconfigurable analog filter for UMTS/WLAN receivers. *Proceedings of the Norchip Conference*, Oslo, Norway, 179–182.

D'Angelo, S., Valimaki, V. 2013. An improved virtual analog model of the Moog ladder fil-ter. Acoustics, Speech and Signal Processing (ICASSP). *2013 IEEE International Conference*. IEEE Conference Publications, Vancouver, Canada, 729–733. DOI: 10.1109.

D'Mello, D.R., Gulak, P.G. 1998. Design approaches to field-programmable analog integrated circuits. *Journal of Analog Integrated Circuits and Signal Processing*, 17, 7–34.

Esmaeilzadeh, H., Sampson, A., Ceze, L., Burger, D. 2013. Neural acceleration for general-purpose approximate programs. *Micro, IEEE*, 33(3), 16–27. *IEEE Journals & Magazines*, 16–27. DOI: 10.1109.

Fernández-Canque, H. 2012. *Machine Vision Application to Automatic Detection of Living Cells/Objects. Human-Centric Machine Vision*, Chapter 6. INTECH, Rijeka, Croatia, 99–124. ISBN 978-953-51-0563-3.

Hamedi-Hagh, S., Bindal, A. 2008. Spice modeling of silicon nanowire field-effect transistors for high-speed analog integrated circuits. *Nanotechnology, IEEE Transactions on*, *IEEE Journals & Magazines*, 7(6), 766–775. DOI: 10.1109.

Hatori, K., Ota, K., Sakai, Y., Kitamura, S., Tetsuo Motomiya, T. et al. 2014. *The Next Generation 6.5 kV IGBT Module with High Robustness*. PCIM Europe 2014 conference: PCIM Europe, Nuremberg, 2897–2900.

Hintea, S., Csipkes, G., Rus, C., Csipkes, D., Fernandez-Canque, H. 2008. On the design of a reconfigurable OTA-C filter for software radio. *International Journal of Knowledge-Based Intelligent Information and Engineering Systems*, 12(3), 245–253. IOS Press. ISSN 1327-2314.

Hintea, S., Doris Csipkes, D., Csipkes, G., Fernandez-Canque, H. 2010. A variable topology analog filter suitable for multi-mode wireless applications. *Proceedings of the 14th International Conference, KES2010. Knowledge-Based Intelligent Information and Engineering Systems*, Cardiff, UK., 603–612. ISBN-0302-9743. ISBN-13-978-3-642-15383-9.

Hiraoka, Y., Matsumoto, S., Sakai, T. 2001. Low on-resistance SOI power MOSFET using dynamic threshold (DT) concept for high efficient DC-DC converter. Power Semiconductor Devices and ICs, 2001. ISPSD '01. *Proceedings of the 13th International Symposium on*. IEEE Conference Publications, Osaka, Japan, 267–270. DOI: 10.1109/ISPSD.2001.934606.

Huque, M.A., Tolbert, L.M., Blalock, B.J., Islam, S.K. 2010. Silicon-on-insulator-based high-voltage, high-temperature integrated circuit gate driver for silicon carbide-based power field effect transistors. *Power Electronics, IET*, 3(6), 1001–1009. DOI: 10.1049.

Kenington, P.B. 2005. *RF and Baseband Techniques for Software Defined Radio*. Artech House Mobile Communications Series, Boston, London.

Pankiewicz, B., Wojcikowski, M., Szczepanski, S., Sun, Y. 2002. A field programmable. Analog array for CMOS continuous-time OTA-C filter applications. *IEEE Journal of Solid-State Circuits*, 37, 125–136.

Pietras, M. 2014. Hardware conversion of neural networks simulation models for neural processing accelerator implemented as FPGA-based SoC. Field programmable logic and applications (FPL). *2014 24th International Conference on*. IEEE Conference Publications, Munish, Germany, 1–4. DOI: 10.1109.

Planson, D., Tournier, D., Bevilacqua, P., Dheilly, N., Morel, H. et al. 2008. SiC power semiconductor devices for new applications in power electronics. *Power Electronics and Motion Control Conference, EPE-PEMC*, Poznan, Poland, 2457–2463.

Roska, T., Bartfai, G., Szolgay, P., Sziranyi, T., Radvanyi, A., Kozek, T., Ugray, Zs. 1990. A hardware accelerator board for cellular neural networks: CNN-HAC. Cellular Neural Networks and their Application. *CNNA-90 Proceedings of the 1990 IEEE International Workshop*. IEEE Conference Publications, Budapest, Hungary, 160–168. DOI: 10.1109/CNNA.1990.207520.

Sarpeshkar, R. 2013. Ultra energy efficient systems in biology, engineering, and medicine. *Energy Efficient Electronic Systems (E3S), 2013 Third Berkeley Symposium*. IEEE Conference Publications, Berkeley, CA, 1. DOI: 10.1109/E3S.2013.6705873.

Shaterian, M., Twigg, C.M., Azhari, J. 2013. An MTL-Based configurable block for current-mode nonlinear analog computation. *Circuits and Systems II: Express Briefs, IEEE Transactions on*, 60(9), 587–591. DOI: 10.1109.

Thoma, J., Kranzer. D. 2015. Demonstration of a medium voltage converter with high voltage SiC devices and future fields of application. *PCIM Europe 2015*, Nuremberg, 1–8.

Ytterdal, T. 2008. Nanoscale analog CMOS circuits for medical ultrasound imaging applications. *Solid-State and Integrated-Circuit Technology, 2008. ICSICT 2008. 9th International Conference on*. IEEE Conference Publications, Beijing, China, 1697–1700. DOI: 10.1109.

Zamani, M., Dousti, M., Nohoji, A.H.A. 2011. A10-bit, 20-MS/s, fully differential single transfer phase CBSC pipelined ADC in 0.18 μm CMOS. *Electronic Devices, Systems and Applications (ICEDSA), 2011 International Conference on*. IEEE Conference Publications, Kuaklalumpur, 77–82. DOI 10.1109, ISBN 9781612843889.

23 Computer-Aided Simulation of Practical Assignment

23.1 INTRODUCTION

Pspice is a computer-aided simulation package widely used in industry and academic institutions to solve analog circuits. Pspice is a registered trademark of Microsim Corporation.

It provides various types of circuit analysis. The full version contains several thousands of components from different manufacturers. There are student or evaluation versions of this package that can be obtained at no cost. The product marketing Department of Microsim Corporation can advise on details of how to acquire an evaluation copy for educational purposes. The student and evaluation version of Pspice have a limited number of components and limited size of circuit to be used. This section is intended as a brief introduction for using all the versatility and potential of Pspice. For more advanced use, you are recommended to consult books (such as Keown 1998, Hermiter 1999) on how to use this package. In Chapter 21, Section 21.2 the characteristics of Pspice and its potential were included.

In this section, we attempt to use Pspice to solve some simple circuits. The aim of this chapter is to perform some simulation using Pspice to complement material already covered in other chapters.

Spice and Pspice have been developed through the years to become a powerful package that can investigate analog as well as digital circuits. This package has been developed by different software companies and has acquired distinctive names. A simple introduction to Pspice early version is presented below. If we become familiar with this version, it will be easy to use other versions that are variations of Pspice such as MicroSim Pspice, Cadence Pspice, and Orcad Pspice.

Many CAD programs for circuit simulation are based on Kirchhoff's laws. Spice was developed in the early 1970s in response to the potential complexity in analyzing integrated circuits. Spice stands for "Simulation Program with Integrated Circuit Emphasis." Pspice was written for personal computers (PCs). During the years since its development it has undergone many changes and improvements and it now has many variations.

Pspice consists of the following subprograms:

Schematics is the graphical circuit editor and design manager for Pspice.
Pspice is the analog, digital, and mixed circuit simulator.

Probe is the graphical waveform analyzer used to view and manipulate Pspice
   simulation results.
Stimulus editor is the analog and digital stimulus generation tool.
Parts is the utility for creating semiconductor device models and subcircuit
   definitions.

Pspice is a very powerful package for simulating circuits. It will provide a number of different types of simulations such as dc analysis, ac analysis, transient analysis, and bias point analysis. When you use pspice you usually indicate the type of
analysis required, otherwise the package does the default analysis, which is the bias
point analysis. This analysis will provide particular values of voltages at nodes and
current in each source. Since it is one value of voltage at a node, the package will
not take you to "Probe," the section that provides graphical results. If a different
analysis is performed, where a quantity is varied, Pspice will provide a graph with
the results.

23.2   FIRST INTRODUCTION TO USING PSPICE

This introduction to Pspice includes a tutorial that will assist you in getting familiar
with using a Pspice-based design package. The introduction will be based upon a
simple *hand on* example consisting on a dc source connected to two resistors in
series as shown in Figure 23.1. This example will take you step by step through the
process of creating a new circuit schematic, settingup Pspice analysis, and viewing
the simulation results.

Figure 23.1 shows a simple circuit, which will help in using a schematics diagram.
Step-by-step details of how to create this schematic will be given below. This circuit
will be used to compute the *bias point* calculations.

The example is written in a way that indicates all details of steps required for you
to start with the package. Once you became familiar with the package these steps
will become transparent to you and you can use short cuts to explore and learn the
potential of this package. If you are familiar with this package you can move to the
following section to perform some assignments in this section.

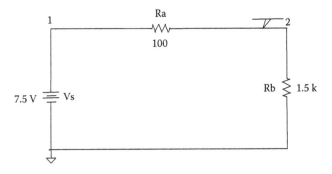

FIGURE 23.1   Circuit diagram to be used as example for Pspice introduction.

23.2.1 CIRCUIT CREATION/SCHEMATICS

Once you have activated this package, invoke *schematic* by double clicking on the Schematics icon. Schematics tool will allow you to draw the circuit (schematics) you want to analyze (in this case the circuit diagram shown in Figure 23.1).

You can begin drawing your schematic directly onto the blank worksheet provided. Pull-down menu selections are accomplished by clicking on the desired menu item. For example, menu items within the *File/Menu* can be selected either by clicking on them, or by typing the letter that is underlined. For example once you have pulled down the *File/Menu*, you can select the *New* command by typing the letter N, since that is the letter that is underlined. (If you were previously working on another schematic, simply select *File/New* to begin a new one.) The steps needed to place the components in the schematic are outlined below.

Let us start drawing the circuit of our example by selecting and placing the components required.

Note: Click means single click on the left button of a computer mouse.

Voltage source:

1. Select *Draw/Get New Part* (or type <Ctrl G>)
2. Type *VDC* in the Add Part dialog
3. Click OK or press <enter>
4. Click to place the source on your schematic page
5. Right click to end placement mode

Resistors:

1. Select *Draw/New Part* (or type <Ctrl G>)
2. Type *R* in the Add Part dialog
3. Click OK or press <enter>
4. Click to place the first resistor (R1) on your schematic page
5. Type <Ctrl R> to rotate the resistor outline
6. Click to place the second resistor (R2)
7. Right click to end placement mode

23.2.2 TO CONNECT THE COMPONENTS

Select *Draw/Wire* (or type <Ctrl W>) to enter wiring mode. Your cursor will change from an arrow to a pencil. Click on the connection point of the first pin that you wish to connect. Double click on the connection point of the second pin to place and end the wire. Your cursor will return to the default arrow. Double click right to resume wiring mode. Your cursor will change back to a pencil. Again, click on the first pin to start the wire, and double click on the second pin to end and place the wire. Continue this sequence until your circuit is appropriately wired. If you make a mistake, you can delete a part of wire by selecting it (single click), then press the <Del> button or select Edit/Cut.

To place the analog ground symbol:

1. Select *Draw/Get New Part* (or type <Ctrl G>)
2. Type *AGND* in the Add Part dialog
3. Click OK or press <enter>
4. Click to place the ground symbol, noting that it is connected by a junction dot
5. Right click to end placement mode

To place the viewpoint symbol:

1. Select *Draw/Get New Part* (or type <Ctrl G>)
2. Type *VIEWPOINT* in the *Add Part* dialog
3. Click OK or press <enter>
4. Move the outline of the viewpoint so that its hotspot (pin end) touches the wire joining R1 and R2
5. Click to place the viewpoint
6. Right click to end placement mode

In all the above cases of selecting and placing parts on the schematic page we typed in the recognized name of the part(s) we wanted to place. If we were unsure of the name, or wanted to see what parts are available to be placed, we could click on the *Browse* button in the *Add Part* dialog to browse the parts in loaded symbol libraries.

23.2.3   NAMES TO ELEMENTS

To assign particular names to the devices in the schematic (like "Vs" for the VDC, and "Ra", "Rb" for the resistors), simply double click on the element then click on the *PKGREF* option (reference designator), and type the new name in the *Edit Reference* designator dialog.

Note: Make sure that the nonchangeable attributes is *not* selected.

1. Double click on the Reference Designator of the *VDC* symbol, V1
2. Type Vs in the *Value* field
3. Click on *Save Attr*
4. Click OK or press <enter>

Do the same thing to change the resistor's name from R1 to Ra and from R2 to Rb. You can change the values of the other components using the same method.

Node numbers: To assign particular names (or labels) to the nodes, as shown in Figure 23.1

1. Double click on the wire connecting Vs to Ra
2. Type 1 in the *Reference Designator* field

3. Click OK or press <enter>
4. Double click on the wire connecting Ra and Rb
5. Type 2 in the *set Attribute Value* dialog
6. Click OK or press <enter>

Component value: Now let us assign a dc value of 7.5 V to the voltage source:

1. Double click on the Vs symbol
2. Double click on the *DC attribute*
3. Type 7.5 V in the Value field of the dialog
4. Click on the *Save Attr* button or press <enter>
5. Click OK to end editing

To save your schematic:

1. Select File/Save
2. Select the place where you want this file to be saved (e.g. select a driver in your computer to save your work space or save it in and external memory stick)
3. Type *example1* as the schematic file name (this creates a file named "example1.sch")
4. Click OK or press. <enter>

23.2.4   RUNNING A SIMULATION

After having created the schematic, you can run Pspice by selecting *Analysis/Simulate*. Upon selecting *Analysis/Simulate* Pspice will check electrical connections and perform a netlist creation. You will see a screen that gives up-to-date information about the simulation progress. Pspice automatically generates the output file (*example1.out*) relevant error and/or warning messages would be written to the output file. Once the simulation is completed you can close this screen and you can browse the simulation output by selecting *Analysis/Examine Output*. Once you click on "*Examine Output*" Pspice will display the information for this circuit. It will indicate the components used as text lines (netlist), their values, connection of components to different nodes and labels. It will also display the results of the simulation, providing the current in a source and voltages at the different nodes with respect to ground.

The *VIEWPOINT* symbol was initially a blank line. Now that the simulation has finished, and bias point voltages are available, the viewpoint will indicate the voltage value with respect to ground at the node where the symbol was connected.

Usually when you use Pspice you first indicate the type of analysis required. If you do not indicate the analysis required the package does the default analysis, which is the *bias point analysis* that we completed above. The *bias point analysis* provides the value of a voltage of a particular node. As it is only one value, the package will not produce a graph result, and the results are indicated as text in the *Examine*

Output display. If a quantity is varied then the package will take you to probe to display results as a graph. Let us illustrate this with a dc analysis, where we change the values of the voltage source to cover a range of values for the source.

23.2.5   DC ANALYSIS

A dc analysis will allow you to vary the voltage and/or current sources and graphically display the wave form results in *probe*. To do this, specify a *DC sweep* analysis. To set up the dc analysis parameters, where we will sweep the Vs from 0 to 12 V increments:

1. Select *Analysis/Setup*, and then click on the *DC Sweep* button. This will bring upon the DC Sweep Analysis Setup dialog.
2. Click in the *Name Field* and type Vs
3. Click in the *Start Value* field and type 0
4. Click in the *End Value* field and type 12
5. Click in the *Increment* field and type 0.5
6. Click on Ok to end dialog
7. Click on the box to the left of the DC Sweep button (putting an "x" in the box)
8. Click OK to end Analysis Setup dialog

Now we are ready to run the simulation.

From menu select *Analysis* and click on *Simulate*. The probe window will appear once the simulation has finished.

Probe is the graphics postprocessor for Pspice. Probe accepts commands through a menu displayed. Most of these commands are self-explanatory. To display a waveform, select the *Trace/Add*. Click on V(1) and V(2), which are listed in the "*Add Traces*" list box. Probe contains various ways of displaying results in graphic form, for instance, V(1,2) will show the voltage across nodes 1 and 2. I(Ra) will show the current through Ra.

The *Add Trace* command also allows you to enter arithmetic expressions of voltages and currents. For instance, V(1,2)*I(Ra) will display the electrical power dissipated by Ra.

V(1)/V(2) will show the voltage ratio of voltage at node 1 to voltage at node 2.

Once the simulation is complete, Pspice will have acquired all values of voltages and current in the circuit and probe allows displaying any required output.

23.3   PRACTICAL ASSIGNMENTS USING PSPICE

The assignments included in this section will allow you to explore and analyze some examples of topics covered in previous chapters. In previous chapters, we have solved some examples for particular results. With the use of a package like Pspice, we can quickly investigate the behavior to variation of components or parameters in an electronic system. Performing these assignments will complement the theory of analog electronic applications as well as getting you more familiar with the use of packages such as Pspice.

23.3.1 ASSIGNMENT 1. DC NETWORKS: BIAS POINT ANALYSIS

Aim: Use of *Bias Point* and DC *Sweet* analysis in Pspice, use of node numbers and proof of Kirchhoff's laws.

23.3.1.1 Experiment 23.1

For the circuit shown in Figure 23.2, draw the schematic for this circuit and perform a *Bias Point Details* analysis.

Note: Use a DC source (VDC). To label a node double click (mouse left button) on the wire and type the node number.

Obtain the results by reading the output file (from menu: *Analysis/Examine Output values* or use the V and I measurement icon from main menu). Identify the right voltages and currents and complete Table 23.1.

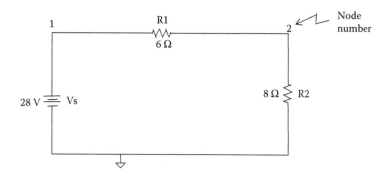

FIGURE 23.2   Circuit diagram for Experiment 23.1.

TABLE 23.1
Bias Point Result for Experiment 23.1

	Value
Voltage across R1	
Voltage across R2	
Current, I	
Voltage across both resistors R1 + R2	
Power dissipated by R1	
Power dissipated by R2	
Power dissipated by R1 + R2	
Total power dissipated Vs*I	

23.3.1.2 Experiment 23.2

For the circuit shown in Figure 23.3, draw schematics and perform a *Bias Point Details* simulation to obtain all the voltages and all currents in the circuit and complete Table 23.2.

The values of the resistors are in ohms.

Using the results from Experiments 23.1 and 23.2:

- Verify that KCL is valid in all nodes
- Verify that KVL is valid in all closed circuits
- Comment on your results

23.3.1.3 Experiment 23.3

For the circuit shown in Figure 23.2, draw schematics and perform a dc analysis (*DC Sweep*) simulation by changing the values of source V1 in the range of 6–16 V.

Use probe to display a graph for voltages across all the resistors, that is, V(1,2) voltage across R1; V(2,3) voltage across R2; V(2,0) voltage across R3.

Draw a graph with *all* voltages in Figure 23.4 as function of the source voltage V1.

Use probe to obtain a display of all currents, that is, I(R1), I(R2), I(R3), etc.

Draw a graph with *all* currents in Figure 23.5 as function of the source voltage V1.

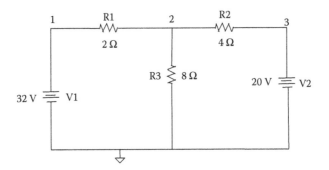

FIGURE 23.3   Circuit diagram for Experiment 23.2.

TABLE 23.2

Bias Point Details Results for Experiment 23.2

	Voltage	Current	Power
R1			
R2			
R3			
V1			
V2			

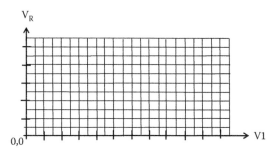

FIGURE 23.4   Graph to include results of all voltages in Experiment 23.3.

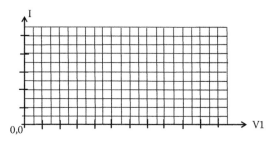

FIGURE 23.5   Graph to include results of all currents in Experiment 23.3.

Use probe to obtain a display of the power dissipated by resistor R3 when V1 varies from 1 to 230 V.

Draw a graph the power dissipated by R3 in Figure 23.6 as function of the source voltage V1.

Note: A negative current or voltage indicates the direction of the current or the polarity of a voltage. To obtain the electrical power you need the magnitude values for V and I, that is, use absolute values. In Pspice use ABS(...).

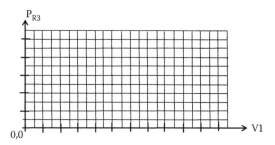

FIGURE 23.6   Graph of power dissipated by R3 as input voltage V1 is changed from 1 to 230 V in Experiment 23.3.

23.3.2 Assignment 2. AC Networks: AC Sweep

Aim: Use of ac sweep/probe to display and obtain *magnitude and phase* of ac voltages and currents.

23.3.2.1 Experiment 23.4

For the circuit shown in Figure 23.7 draw a schematics and perform an ac analysis (AC sweep) simulation to display the voltages across resistors, inductors, and capacitor plus currents in the circuit in the 10–10 kHz frequency range.

Amplitude and phase: Use probe to display the appropriated *amplitude and phase* that allows determining an accurate value for the voltage and current for resistor R1. From the appropriate display, determine the values to complete Table 23.3 accurately.

Impedance: Use probe to obtain and display the total *impedance* versus frequency. Draw a graph of the magnitude of this impedance as function of frequency in Figure 23.8 indicating clearly the *resonance* point (largest impedance). Use appropriate displays with probe to obtain values to complete Table 23.4 at the resonance frequency.

23.3.2.2 Experiment 23.5

For the circuit shown in Figure 23.9 draw schematics and perform an ac analysis (*AC sweep*) simulation for the range of frequencies between 40 and 100 Hz to obtain the voltages (amplitude and phase) across the resistor, inductor, and capacitor plus the current in the circuit.

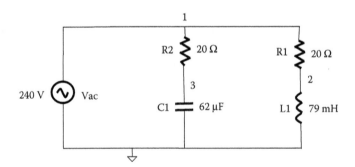

FIGURE 23.7 Circuit diagram to be used in Experiment 23.4.

TABLE 23.3

Result for Voltage, Current, and Power Dissipated by R1 in Experiment 23.4

	100 Hz	1 kHz
Amplitude voltage across R1		
Amplitude current through R1		
Angle between voltage and current in R1		
Powr dissipated by R1		

Magnitude of impedance

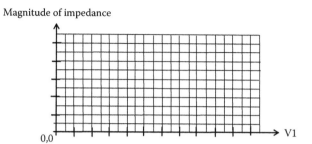

FIGURE 23.8   Graph of impedance versus frequency for Experiment 23.4.

TABLE 23.4
Result for Impedance versus Frequency
at Resonance for Experiment 23.4

	Value
Frequency at resonance	
Magnitude of impedance at resonance	
Magnitude of current at resonance	
Total power dissipated at resonance	

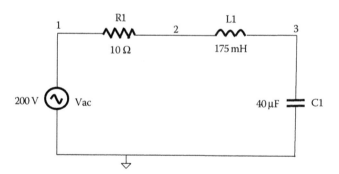

FIGURE 23.9   Circuit diagram for laboratory Experiment 23.5.

Note: Use an ac source, that is, VAC and in Analysis Setup use Total Pts = 500, Start Freq = 40, End Freq = 100, logarithmic scale (decades).

From the displayed graph complete Table 23.5 at a frequency of 50 Hz.

Use probe to display a plot of power dissipated by R1(P_{R1}) against frequency and draw P_{R1} in Figure 23.10.

Draw a *phasor diagram* for all voltages and currents at f = 50 Hz. Clearly indicate scales used for current and voltages.

TABLE 23.5
Result of the RLC Circuit at 50 Hz in Experiment 23.5

	Magnitude	Phase
V_{R1}		
V_{L1}		
V_{C1}		
I		

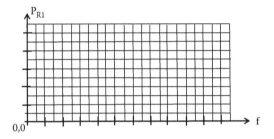

FIGURE 23.10   Graph of power dissipate by resistor R1 versus frequency for Experiment 23.4.

23.3.2.3   Experiment 23.6

For the circuit shown in Figure 23.9, draw schematics and perform an ac analysis (AC sweep) simulation to display the current in the circuit in the range of frequencies between 10 and 500 Hz.

Use probe to display a plot of amplitude and phase of the current. In Figure 23.11, draw the (a) amplitude and (b) phase of the current for this circuit versus frequency and obtain an accurate value for the resonance frequency.

From the displayed graph determine accurate values to complete Table 23.6.

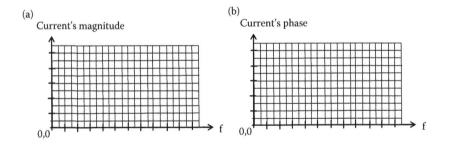

FIGURE 23.11   Graph of magnitude of current (a) and phase (b) versus frequency for simulation of circuit shown in Figure 23.9 of Experiment 23.6.

TABLE 23.6
Results of Current and Frequency at Resonance for Experiment 23.6

	Value
Frequency at resonance	
Magnitude of current at resonance	
Phase of current at resonance	

Impedance's magnitude

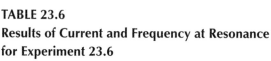

0,0 → f

FIGURE 23.12   Graph of magnitude of impedance versus frequency for Experiment 23.6.

TABLE 23.7
Results of Impedance versus Frequency at Resonance for Experiment 23.6

	Value
Frequency at resonance	
Magnitude of impedance at resonance	
Phase of impedance at resonance	

Use probe to obtain a display of the impedance "seen" by the source versus frequency. In Figure 23.12 include a graph of magnitude of impedance versus frequency for Experiment 23.6. From the displayed graph determine accurate values to complete Table 23.7.

For all experiments in Assignment 2 comment and discuss results.

23.3.3   ASSIGNMENT 3. BJT OPERATING POINT, Q, STABILITY

Aim: The aim of this experiment is to investigate the stability of the operating point of fixed bias, auto-bias, and two sources bias circuits as *temperature* and β varies.

23.3.3.1 Experiment 23.7

Measurement of nominal values: Draw a schematics and run Pspice—*Bias Point Details*—simulation for the circuits shown in Figure 23.13a–c. Once the bias point calculation is completed *examine output file* to obtain the required values for the operation point Q (Ic, Vce).

Procedure: Draw (*schematic*) circuit Figure 23.13a and run the *Bias Point Details* analysis (default analysis for Pspice). Once the analysis is completed then browse output file (i.e., select *Analysis* from menu and then *Examine Output*).

Note: V_{cc} and V_{dd} are dc voltages sources. Use node numbers.

From the *output file* examine the following: voltages in all nodes, current passing through sources (current is taken as positive when circulating anticlockwise), and transistor's parameters including voltages and current in the transistor. Write down values for the operating point Q (Ic, Vce) in Table 23.8. Repeat this experiment for circuit shown in Figure 23.13b and c to complete Table 23.8.

Ic_n and Vce_n are taken as the measured *nominal* values. These values are taken with default parameters, that is, temperature = 27°C, $\beta = 255.9$, Is = 14.34×10^{-15} A, etc. (as for Q2N2222 model).

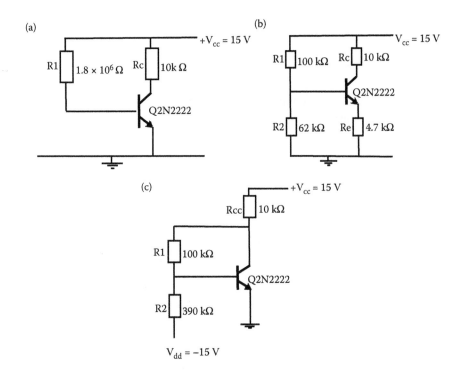

FIGURE 23.13 Circuit diagram for three biasing circuits used in Experiment 23.7. (a) Fixed bias, (b) auto-bias, and (c) two sources biasing.

TABLE 23.8

Results for Experiment 23.7 for T = 27°C and β = 255.9 (*Nominal Values*)

Circuit	Ic_n	Vce_n
23.13 (a)		
23.13 (b)		
23.13 (c)		

Temperature change: From the analysis menu of Pspice change the temperature to include values of T = 40°C and perform a Pspice—*Bias Point Details*—simulation for the circuits shown in Figure 23.13a–c. Once the bias point calculation is completed *examine output file* to obtain the required values for the operation point Q (Ic, Vce) to be included in Table 23.9. Also determine the percentage change in the Q point for the different types of biasing as the temperature changes to complete Table 23.9.

Repeat the experiment changing the temperature value to −40°C to complete Table 23.10.

Beta, β, change: Change the model for the Q2N2222 transistor to include values of β = 50. Keep the temperature at 27°C and perform Pspice—Bias Point Details—simulation for the circuits shown in Figure 23.13a–c. Once the bias point calculation is completed *examine output file* to obtain the required values for the operation point

TABLE 23.9

Results for Experiment 23.7 for T = +40°C and β = 255.9

Circuit	Ic	Vce	% Ic Change $((Ic_n - Ic)/Ic_n) \times 100$	% Vce Change $((Vce_n - Vce)/Vce_n) \times 100$
23.13 (a)				
23.13 (b)				
23.13 (c)				

TABLE 23.10

Results for Experiment 23.7 for T = −40°C and β = 255.9

Circuit	Ic	Vce	% Ic Change $((Ic_n - Ic)/Ic_n) \times 100$	% Vce Change $((Vce_n - Vce)/Vce_n) \times 100$
23.13 (a)				
23.13 (b)				
23.13 (c)				

TABLE 23.11
Results for Experiment 23.7 for T = 27°C and β = 50

Circuit	Ic	Vce	% Ic Change $((Ic_n - Ic)/Ic_n) \times 100$	% Vce Change $((Vce_n - Vce)/Vce_n) \times 100$
23.13 (a)				
23.13 (b)				
23.13 (c)				

TABLE 23.12
Results for Experiment 23.7 for T = 27°C and β = 400

Circuit	Ic	Vce	% Ic Change $((Ic_n - Ic)/Ic_n) \times 100$	% Vce Change $((Vce_n - Vce)/Vce_n) \times 100$
23.13 (a)				
23.13 (b)				
23.13 (c)				

to complete Table 23.11. Determine the percentage of change in Q point for the different biasing to be included in Table 23.11.

Note: Components are modelled for simulation in a circuit. The model of a component or device consists of a list of parameter according to the characteristic of the component/device. It is possible to modify the model used by Pspice by changing parameters in the model.

To edit a model:

Select the component whose parameters are to be modified.
Select *Model* from the *Edit* menu. This will bring up the Edit Model dialogue.
Select *Edit Instance Model (text)* which allows to edit a copy of the model.
Change the required parameter and click Ok.

In a BJT model the parameter for the forward dc current gain, b, is labelled as BF and the capacitance between Base and Collector is labelled as CJC.
Repeat experiment with β = 400 to complete Table 23.12.
For all experiments in this section, *comment and discuss* results in relation to variation of operating point with changes of temperature and β for the fixed bias, auto-bias, and two sources biasing of an amplifier.

23.3.4 ASSIGNMENT 4. BJT AMPLIFIER ANALYSIS

Aim: The aim of this session is to analyze the behavior of an amplifier using BJT and to determine the effects of the different components on the response of the amplifier, in particular the capacitance and the emitter resistance effects on the ac response.

23.3.4.1 Experiment 23.8

23.3.4.1.1 Dc Analysis

This is similar to operating point analysis, except that one or more dc sources can be swept over a specified range in discrete steps.

Draw a schematic and run Pspice—dc *analysis*—for the circuit shown in Figure 23.14. The input voltage V_{in} is a dc source to be swept from 0 to 3 V in steps of 0.03 V. V_{cc} is a dc voltage source of 3 V. Once the circuit is simulated, probe will allow you to plot the output voltage V_o against the input voltage V_{in}. Use your knowledge of basic operation of a transistor to satisfy yourself that the output obtained is correct.

Note: Label all nodes.

Draw a graph in Figure 23.15 with the output voltage, V_o, as the vertical axis and the input voltage, V_{in}, as the horizontal axis.

In your own words explain this result:

Make yourself familiar with probe. (Change range of values, change variables to be displayed, change of axis to be displayed, etc.) Modify the different options available in probe to change the display of results. Probe is a powerful program used to produce results in graphical form. You should be able to display results in the most appropriate form.

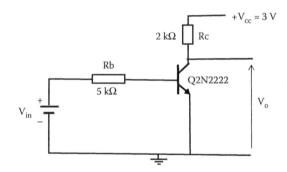

FIGURE 23.14 Circuit diagram for the fixed bias amplifier used in Experiment 23.8.

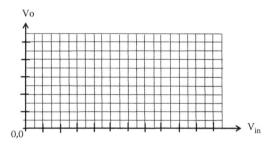

FIGURE 23.15 Graph of magnitude of output voltage versus input voltage for Experiment 23.8.

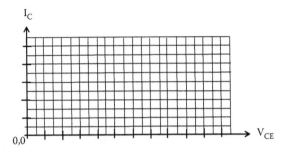

FIGURE 23.16 Graph of the load line for circuit shown in Figure 23.14 for Experiment 23.8.

Load line: In probe, it is possible to control the display of results to plot variables in different forms. Once you are familiar with probe, plot the *load line* for this circuit by changing the vertical axis as the collector current, I_C, and the horizontal axis as the voltage between collector and emitter, V_{CE}. Draw the load line in Figure 23.16.

23.3.4.2 Experiment 23.9

23.3.4.2.1 BJT Amplifier AC Analysis

The amplifier shown in Figure 23.17 was designed to meet the (i) through (v) *specifications* indicated below.

Some of the specification for the amplifier used in Experiment 23.9

  i. Mid-band voltage gain ≈ 10
 ii. Low-frequency cutoff, $f_L < 30$ Hz
iii. High-frequency cutoff, $f_H > 100$ kHz

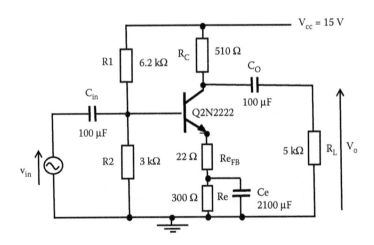

FIGURE 23.17 Circuit diagram for the auto-bias amplifier used in Experiment 23.9.

iv. Class A amplifier
v. The amplifier should be capable of delivering a peak-to-peak voltage to the load without clipping

As part of this experiment it is required to verify that the designed circuit fulfils the above specifications for this circuit.

Verification: Draw a schematic diagram for the circuit shown in Figure 23.17 and use Pspice to verify that the design meets the desired specifications. Use appropriate analysis and appropriate inputs to verify all the specifications from (i) through (v).

Procedure: Run an *ac analysis* with an ac source as input (magnitude less than 50 mV, phase 0°, no dc). Plot the voltage gain against frequency to verify (i)

Measured mid-band voltage gain v_{in}/v_o _____

Use probe to plot the voltage gain (in decibels) against frequency (decades) and obtain the cutoff frequencies to verify (ii) and (iii) (remember definition of cutoff frequency: 3 dB down from mid-band frequency. Select a frequency range to include low- and high-frequency cutoff).

Sketch below voltage gain G_v (in dB) versus freq. (log scale, decade) in Figure 23.18 including low- and high-frequency cutoff points.

Measured low-frequency cutoff, f_L _____
Measured high-frequency cutoff, f_H _____

Run a *transient analysis* with a transient source to verify (iv) and (v). Use a transient source for the transient analysis: *VSIN* attribute/description: *VOFF* = offset voltage = 0, *VAMPL* = amplitude = under 50 mV, *FREQ* = frequency 100, *TD* = time delay = 0, *DF* = damping factor = 0, *PHASE* = phase angle = 0.

Use probe to plot input signal and output signal and compare for distortion and amplitude.

Measured results: Class A amplifier Y/N?_____
Peak-to-peak input voltage_____
Peak-to-peak output voltage_____

Provide comments and analysis on your results:

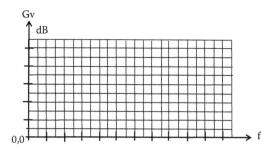

FIGURE 23.18 Sketch voltage gain G_v (in dB) versus frequency for Experiment 23.8.

23.3.4.2.2   Feedback Effect

Examine the effect of feedback resistance Re_{FB} on voltage gain and bandwidth.

Procedure: Change the value of Re_{FB} from 22 to 1 Ω and repeat procedure for verification as above to obtain new values for cutoff frequencies and to check for distortion. Include your values in Table 23.13. Repeat procedure for $Re_{FB} = 44\ \Omega$ and complete Table 23.13.

Provide an analysis and comments on your results:

23.3.4.2.3   Coupling Capacitors Effects

Examine the effect of capacitances C_{in}, C_O, and C_e on voltage gain and cutoff frequencies.

Procedure: Change the value of Re_{FB} to its original value of 22 Ω.

Change the value of capacitances C_{in}, C_O, and C_e and run an *ac analysis* to obtain new values for cutoff frequencies and new voltage gain to complete Tables 23.14 through 23.16.

Provide an analysis and comments on your results:

23.3.4.2.4   Miller Effect (High-Frequency Cutoff)

Examine the effect of an output parallel capacitance, C_P.

TABLE 23.13

Results for Experiment 23.9 for Different Values of Feedback Resistor Re_{FB}

Feedback Resistor Re_{FB}	22 Ω	1 Ω	44 Ω
Measured mid-band voltage gain			
Measured f_L			
Measured f_H			
Distortion Yes/No?			

TABLE 23.14

Results for Experiment 23.9 for Different Values of Input Coupling Capacitors C_{in}

C_{in}	22 μF	1 μF	100 mF
Measured mid-band voltage gain			
Measured f_L			
Measured f_H			

TABLE 23.15

Results for Experiment 23.9 for Different Values of Output Coupling Capacitors C_O

C_O	22 μF	1 μF	100 mF
Measured mid-band voltage gain			
Measured f_L			
Measured f_H			

TABLE 23.16

Results for Experiment 23.9 for Different Values of Emitter Capacitors C_e

C_e	2100 μF	1 μF	1000 μF
Measured mid-band voltage gain			
Measured f_L			
Measured f_H			

Procedure: Restore the original values as shown in Figure 23.17. Modify the circuit shown in Figure 23.17 to add a capacitance, C_P in parallel to R_L (this can represent the load capacitance and/or the wiring capacitance). Select a value C_{P1} and perform an *ac analysis* to obtain the new cutoff frequencies to be included in Table 23.17, and repeat for another value of C_{P2}.

Examine the effect of the transistor capacitance between base and collector, C_{BC}: Change the transistor model and accommodate new values for capacitance between base and collector. In the transistor model the parameter for capacitance between base and collector is the parameter *CJC*.

Modify the model for the transistor Q2N2222 and run an ac analysis to obtain new values for cutoff frequencies and complete Table 23.18.

Provide an analysis and comments on all your results:

TABLE 23.17

Table of Values Results for Experiment 23.9 for Different Values of Output Parallel Capacitors C_P

Output Parallel Capacitance C_P	Original Values for Capacitor C_P, that is, $C_P = 0$ F	C_{P1}	C_{P2}
Measured f_L			
Measured f_H			

TABLE 23.18

Results for Experiment 23.9 for Different Values of Capacitors between Base and Collector C_{BC}

Capacitance between Base and Collector C_{BC} (CJC)	Nominal Value as in Part Previous Results	0.1 pF	100 pF
Measured f_L			
Measured f_H			

23.3.5   ASSIGNMENT 5. FET AMPLIFIER ANALYSIS AND DIFFERENTIAL AMPLIFIER

Aim: The aim of this session is to analyze the behavior of an amplifier using FET and differential amplifiers to determine the effects of the different components on the response of the amplifier; in particular, the CMRR response.

23.3.5.1   Experiment 23.10

23.3.5.1.1   FET Common Source Amplifier

For the amplifier shown in Figure 23.19, with the appropriate ac input voltage, perform a Pspice *ac analysis*.

Draw schematic for the circuit shown in Figure 23.19 and run an *ac analysis* with an ac source as input (magnitude less than 50 mV, phase 0°, no dc). Use probe to plot the voltage gain against frequency and complete Table 23.19. Use probe to plot the voltage gain (in decibels) against frequency (decades) and obtain the cutoff frequencies. (Remember definition of cutoff frequency: 3 db down from mid-band frequency.)

Sketch voltage gain (dB) versus frequency (log scale, decade) in Figure 23.20 including the cutoff frequencies.

Provide comments on the results:

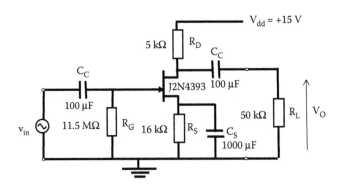

FIGURE 23.19   Circuit Diagram For The Fet Amplifier Used In Experiment 23.10.

TABLE 23.19

Table of Values Results for FET Amplifier

Used in Experiment 23.10

FET Amplifier	Value
Peak-to-peak input voltage	
Peak-to-peak output voltage	
Measured mid-band voltage gain	
Measured low-frequency cutoff, f_L	
Measured high-frequency cutoff, f_H	
Bandwidth (BW)	

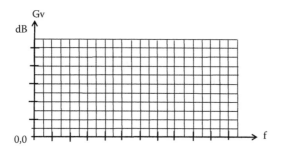

FIGURE 23.20   Sketch voltage gain G_v (in dB) versus frequency for the FET amplifier shown in Figure 23.19 for Experiment 23.10.

23.3.5.2   Experiment 23.11

23.3.5.2.1   Differential Amplifier

For the amplifier shown in Figure 23.21, with an input voltage V_{in} of 1 mV at 1 kHz, perform a Pspice *transient analysis*.

Draw a schematic circuit for this differential amplifier and run a *transient analysis* with a *VSIN* source as input with 1 mV of magnitude at 1 kHz with *VOFF* = 0. Use print step of 10 μs and final time 2 ms. Use probe to plot the output voltage v_{O1} against time.

Sketch in Figure 23.22 the single output voltage of the differential amplifier v_{O1} versus time.

Notes: When using transient analysis, trace variables must be plotted separately in order to determine the peak values. When using transient analysis, use peak-to-peak/2 to factor out any dc value and average the negative and positive peaks.

Calculate the single-ended voltage gain = v_{O1}/v_{in} = _____

Sketch in Figure 23.23 the single output voltage of the differential amplifier v_{O2} versus time and calculate the voltage gain.

Calculate the single-ended voltage gain = v_{O2}/v_{in} = _____

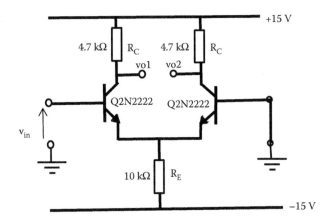

FIGURE 23.21 Circuit diagram for the differential amplifier used in Experiment 23.11.

Sketch in Figure 23.24 the differential output voltage between v_{o1} and v_{o2} versus time and calculate the differential-mode voltage gain

Differential-mode voltage gain = v_{O1-O2}/v_{in} = _____(Double-ended gain)

Provide an analysis and comments on your results:

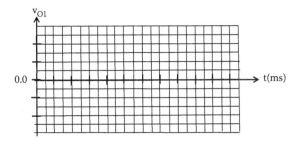

FIGURE 23.22 Sketch of output voltage v_{O1} against time for the differential amplifier used in Experiment 23.11.

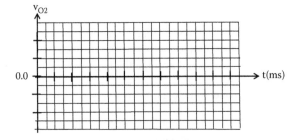

FIGURE 23.23 Sketch of single output voltage of the differential amplifier v_{O2} against time for Experiment 23.11.

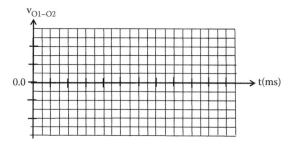

FIGURE 23.24 Sketch of the output voltage between v_{o1} and v_{o2} against time for Experiment 23.11.

Common mode rejection ratio: Modify the circuit in Figure 23.21 as shown in Figure 23.25.

Differential-mode voltage gain: Apply the following voltages: $v_{in1} = 1$ mV as magnitude at 1 kHz, VOFF = 0, and $v_{in2} = 0$ V (the difference of the inputs is v_{in1}, since v_{in2} is zero).

Draw the schematic circuit for this differential amplifier and run a *transient analysis* with a VSIN source as input, use print step of 10 µs and final time 2 ms. Use probe to display the output voltage v_O against time.

   Calculate the differential-mode voltage gain = v_O/v_{in1} = _____
   (v_O/v_{in1} is automatically differential-mode gain because v_{in2} amplitude is zero.)

Common-mode voltage gain: Apply the following voltages: $v_{in1} = v_{in2} = 1$ mV as magnitude at a frequency of 1 kHz (now the same voltage input is in both inputs) to measure the common-mode voltage gain.

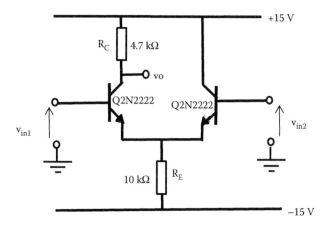

FIGURE 23.25 Circuit diagram for the modified differential amplifier with only one resistor R_c.

Run a *transient analysis* with a VSIN source as input, use print step of 10 μs and final time 2 ms. Use probe to display the output voltages and obtain the common-mode gain.

$$\text{Common-mode gain} = v_O/v_{in} = \text{_____}$$

Calculate the CMRR

$$\text{CMRR} = \frac{\text{Differential-mode voltage gain}}{\text{Common-mode voltage gain}} = \text{_____}$$

Current mirror: The CMRR depends on the common-mode gain. To provide a large CMRR the common-mode gain should be as small as possible. For a large CMRR, the resistor R_E should be large. One way of increasing the value of R_E without affecting the biasing of the amplifier is to replace R_E by a circuit known as *current mirror* shown in Figure 23.26.

Figure 23.27 shows the differential amplifier circuit of Figure 23.25 with R_E replaced by the current mirror shown in Figure 23.26.

Draw the schematic diagram for this amplifier and run a transient analysis for circuit in Figure 23.27. Repeat the measurement of the common-mode voltage gain and differential-mode voltage gain for this new amplifier and calculate the common-mode gain and the CMRR.

Use probe to display the output voltages and obtain the differential-mode voltage gain and common-mode gain.

Differential-mode gain = _____

Common-mode gain = _____

Calculate the CMRR

$$\text{CMRR} = \frac{\text{Differential-mode voltage gain}}{\text{Common-mode voltage gain}} = \text{_____}$$

Provide an analysis and comments on your results:

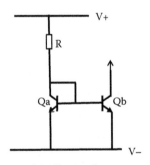

FIGURE 23.26 Circuit diagram for a current mirror to replace R_E in a differential amplifier.

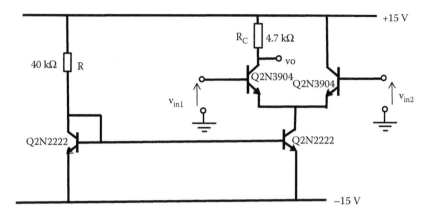

FIGURE 23.27   Circuit diagram of an amplifier with a current mirror as emitter resistance.

23.3.6   ASSIGNMENT 6. ACTIVE FILTER–POWER AMPLIFIER

23.3.6.1   Experiment 23.12

Aim: To analyze different responses of active filters and explore crossover distortion in a power amplifier.

23.3.6.1.1   Low-Pass Filter Single Pole

Note: Use appropriate input signals and choose appropriate Op-Amp with the required biasing.

Figure 23.28 shows an active filter with a single low-pass RC network that provides a roll-off of −20 dB/decade above the cutoff frequency (f_c):

This low-pass filter single-pole design has a Butterworth response with voltage gain $G_v = 1.586$ and with a cutoff frequency f_c equal to approximately 8 kHz.

Draw a schematic diagram for circuit 23.28 and run Pspice—ac analysis—with appropriate input voltage to show that the design meets the desired output response (i.e., voltage gain V_o/V_i versus frequency response, f_c, and appropriated roll-off characteristics).

Draw a graph of voltage gain V_o/V_i (dB) versus frequency response in Figure 23.29 for this low-pass filter and complete Table 23.20.

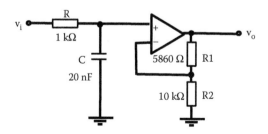

FIGURE 23.28   Circuit diagram for an active filter with a single low-pass RC network.

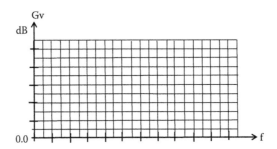

FIGURE 23.29 Graph of voltage gain V_o/V_i (dB) versus frequency response for the low-pass filter used in Experiment 23.12.

TABLE 23.20
Low-Pass Filter Single-Pole Response for Experiment 23.12

	Values
Voltage gain at mid-frequency	
Cutoff frequency	
Roll-off	

Cascade this filter to an identical stage and analyze the response in the same way as the low-pass filter single pole.

Provide an analysis on your results. Compare and comment on the two responses:

23.3.6.2 Experiment 23.13

23.3.6.2.1   The Sallen and Key High-Pass Filter

The Sallen and Key is one of the most common configurations for a *second-order* (two-pole) filter. A high-pass version of the Sallen and Key filter is shown in Figure 23.30.

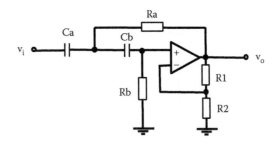

FIGURE 23.30   Circuit diagram for a Sallen and Key high-pass filter used in Experiment 23.13.

TABLE 23.21

Components Value for Design of Sallen and Key Filter of Experiment 23.13

Component	Value	Component	Value
Ra		Cb	
Rb		R1	
Ca		R2	

The value of cutoff frequency f_c and voltage gain for this filter are given by

$$f_c = \frac{1}{2\pi\sqrt{RaRbCaCb}} \tag{23.1}$$

$$G_v = \frac{R1}{R2} + 1 \tag{23.2}$$

Design a Sallen and Key high-pass two-pole filter with a cutoff frequency $f_c = 10$ kHz. You may assume Ra = Rb and Ca = Cb. You may also use the same values for R1 and R2 as single-pole design.

Calculate the values of components for your design and complete Table 23.21.

Use Pspice ac analysis with appropriate input to show that the design meets the desired output response (voltage gain V_o/V_i versus frequency, f_c, roll-off, etc.).

Draw a graph, Figure 23.31, of the voltage gain V_o/V_i (dB) versus frequency response for this Sallen and Key high-pass filter and complete Table 23.22.

Provide an analysis on your results:

23.3.6.3   Experiment 23.14

Filter response characteristics: Butterworth, Chebyshev, and Bessel filter have the same basic format, but the component values and amplifier gains differ for each type. When cascaded to form a higher order filter the gain values may differ for each section.

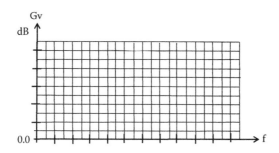

FIGURE 23.31   Graph of voltage gain V_o/V_i (dB) versus frequency response for the Sallen and Key high-pass filter used in Experiment 23.13.

TABLE 23.22
Sallen and Key High-Pass Filter
Response for Experiment 23.13

	Values
Voltage gain at mid-frequency	
Cutoff frequency	
Roll-off	

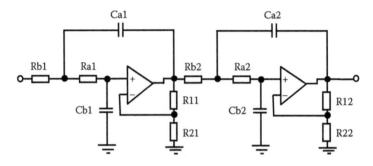

FIGURE 23.32 Circuit diagram for two Sallen and Key low-pass filter cascaded to create a fourth-order low-pass filter used in Experiment 23.14.

TABLE 23.23
Component Values for a Sallen and Key Low-Pass Filter
for Butterworth, Bessel, and Chebyshev Response

	Butterworth	Bessel	Chebyshev
Ca1	265 nF	187 nF	563 nF
Cb1	265 nF	187 nF	563 nF
Ra1	1 kΩ	1 kΩ	1 kΩ
Rb1	1 kΩ	1 kΩ	1 kΩ
R11	1.678 kΩ	1.678 kΩ	18.48 kΩ
R21	20 kΩ	20 kΩ	20 kΩ
Ca2	265 nF	167 nF	275 nF
Cb2	265 nF	167 nF	275 nF
Ra2	1 kΩ	1 kΩ	1 kΩ
Rb2	1 kΩ	1 kΩ	1 kΩ
R12	24.692 kΩ	15.17 kΩ	35.64 kΩ
R22	20 kΩ	20 kΩ	20 kΩ

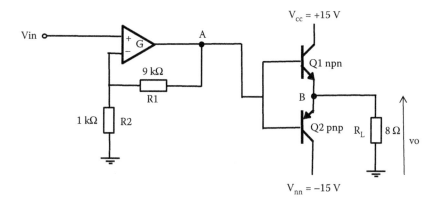

FIGURE 23.33 Circuit diagram for power amplifier class B used in Experiment 23.15.

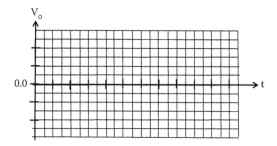

FIGURE 23.34 Output voltage V_o for the power amplifier class B used in Experiment 23.15.

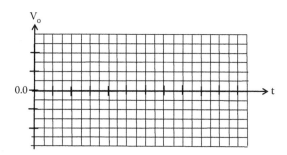

FIGURE 23.35 Output voltage V_o for the power amplifier class B for compensation of cross over distortion in Experiment 23.15.

This exercise follows from the design implemented in Chapter 20 (Section 20.8) of three fourth-order low-pass filters with a cutoff frequency equal to 600 Hz to provide Butterworth, Chebyshev, and Bessel response. Figure 23.32 shows two Sallen and Key low-pass filters cascaded to form a fourth-order filter.

The values of the components calculated in Chapter 20 are provided in Table 23.23.

Draw a schematic diagram for circuit shown in Figure 23.32 using the values of components indicated in Table 23.23 for a Butterworth response and run Pspice ac analysis, with the appropriate input, to show that this design meets the desired output low-pass filter response (voltage gain V_o/V_i versus frequency, f_c and roll-off response). Repeat this experiment for a low-pass filter with Bessel and Chebyshev filter response using values from Table 23.23.

Draw a graph of voltage gain V_o/V_i (dB) versus frequency response for the Butterworth, Bessel, and Chebyshev filters.

Provide an analysis on your results. Compare and comment on the three responses:

23.3.6.4   Experiment 23.15

23.3.6.4.1   Power Amplifier Class B

The circuit shown in Figure 23.33 is a class B amplifier, in which each BJT conducts for approximately half of the signal cycle. The class B output stage is driven by a differential amplifier. The feedback network consisting of R1 and R2 returns part of the output voltage to the inverting input of the differential amplifier. Examine the crossover distortion that happens when conduction is changing from one transistor to the other. Use Pspice with $V_{in} = 0.2$ V at 100 Hz with the appropriated biasing of the Op-Amp. Select a transient analysis to observe the voltage signal output of the differential amplifier at point A and at the output voltage V_o, as a function of time.

In Figure 23.34, draw the output voltage of the amplifier V_o as function of time for two cycles.

Now change the connection of R1 from point A in circuit shown in Figure 23.33 to point B and run a transient analysis to observe the output voltage signal V_o, as a function of time. In Figure 23.35, draw the new compensate output voltage V_o as a function of time.

Provide an analysis on your results. Compare and comment on the two responses:

23.4   KEY POINTS

- Pspice is powerful tool to perform quick simulation of circuits.
- Using a computer-aided package reduces the time dedicated to design.
- A computer-aided package such as Pspice provides the possibility to assess the behavior of a circuit when components change their values according to their tolerances. This makes it possible to check response for a "worst case" analysis.

REFERENCES

Hermiter, M. E. 1999. *Schematic Capture Microsim Pspice.* Pearson Education, Upper Saddle River, NJ, Prentice Hall. ISBN-13: 978013814043.

Keown, J. 1998. *Microsim Pspice and Circuit Analysis.* Prentice-Hall College Div. ISBN-13: 978-0132354585.

Index

A

AC, *see* Alternating current (AC)
Accumulation, 117–118
AC current, *see* Alternating current (AC)—voltage
Active components of circuit, 29; *see also* Passive components
   ideal voltage source, 30
   practical voltage source, 30–31
   voltage sources in parallel, 31–32
   voltage sources in series, 31
Active filter–power amplifier, 391
   filter response characteristics, 393–396
   low-pass filter single pole, 391–392
   power amplifier class B, 396
   Sallen and Key high-pass filter, 392–393
AC voltage gain at high frequencies, 227, 232
   FET common drain amplifier, 232–233
   FET common source amplifier, 227–228
AC voltage gain at medium frequencies, 223, 229, 231–232
   circuit diagram for MOSFET, 224–226
   FET common drain amplifier, 230
   FET common source amplifier, 224
   MOSFET common drain amplifier circuit, 230
Add Trace command, 370
Admittance y parameters, 148–149
Alternating current (AC), 60; *see also* Direct current (DC)
   application of Kirchhoff's Law to, 64–75
   circuits, 60
   networks, 374–377
   origin of phasor domain, 62–64
   signal as sinusoidal waveform, 61
   sweep, 374–377
   voltage, 61
   waveforms, 61
Alternating voltages, 61
Ampere (A), 12
Amplification device, BJT as, 106
   base transport factor, 109–111
   emitter injection efficiency, 108–109
   punch-through, 111
   schematic of current in transistor, 107
Amplifiers frequency response
   frequency response of amplifier, 180
   half-power gain, 179–181
   high frequency, CE amplifier at, 185–187
   low frequency, CE amplifier at, 182–185
   total frequency response, 187–188

Amplifiers overall gain in cascade, 201
   AC equivalent circuit, 202
   circuit diagram of n stages, 202
   voltage gain for $(n-1)$ stages, 202–203
   voltage gain of last stage n, 202
   voltage gain of source, 203–205
Amplitude and phase, 35, 64, 374
Analog computing, 360
Analog devices at high power, 358–359
Analog electronic applications, 323
   audio amplifier, 334–340
   automatic switch on of lamp in dark, 342–344
   automatic switch on of lamp in presence of light, 344–346
   components selection and circuit elements, 324–327
   DC power supply, 330–334
   delay switch, 347–348
   555 Timer, 349–351
   humidity detector, 346, 351–352
   multivibrators, 348–349
   Sallen and Key second-order Butterworth low-pass filter, 341–342
   simulation in circuit applications, 323–324
   smoke detector, 348
Analog electronics, 355
   analog computing, 360
   analog devices at high power, 358–359
   analog signals, 1–2
   biotechnology, 360
   nanotechnology, 359
   new analog building block working voltage under 1 V, 359
   physical stimulus, 1
   reconfigurable analog circuitry, 356–358
   revival in music industry, 359–360
Analog signals, 1–2
Analog systems, 2–3
   application and design, 3
   customer requirements, 3–4
   distortion, 6–8
   electronic design aids, 8–9
   noise, 6–8
   system design approach, 4–5
   technology choice, 5–6
   top-level specifications, 4
Analysis/Examine Output, 369
Analysis/Simulate Pspice, 369
Anode (A), 83
Anti-phase FB, *see* Negative feedback (NFB)

397

Astable multivibrator, 348–349
Asymptotic curves Bode plot, *see* Idealized Bode
    plots
Attenuation voltage, 203
Audio amplifier, 334
    class A amplifier design, 334–340
Auto bias, 137; *see also* Fixed bias
    equivalent circuit, 138
    examples, 139–140
Auto bias amplifier, 159
    AC equivalent circuit, 160
    current gain, 161–162
    power gain, 162–165
    voltage gain, 161
Automatic switch on of lamp
    in dark using BJT, 342–343
    in dark using Op-Amp, 343–344
    in presence of light using 555 Timer,
        352–353
    in presence of light using BJT, 344–345
    in presence of light using Op-Amp, 345–346
Avalanche
    breakdown, 84
    diodes, 84
    region, 100

B

Band-elimination filter, *see* Band-stop filter
Band-pass filters, 305
Band-reject filter, *see* Band-stop filter
Band-stop filter, 314–315
Bandwidth (BW), 179, 256
Base (B), 95
    transport factor, 107, 109–111
BC546A transistor, 337–338
Bessel filter, 318–319
Bias equalization, 290–292
Biasing, 213–220
    JFET, 217–220
    MOSFET, 213–217
    transistor, 129–130
Biasing, BJT
    auto bias, 137–140
    collector-feedback bias, 141–143
    fixed bias, 133–137
    load line, 129–133
    sources, 143–144
Bias point analysis, 366, 369, 371–373
Biotechnology, 360
Bipolar junction transistor (BJT), 95; *see also*
    Field effect transistors (FET)
    active-mode biased NPN transistor structure,
        97
    as amplification device, 106–111
    automatic switch on of lamp in dark using,
        342–343

    automatic switch on of lamp in presence of
        light using, 344–345
    C–B junction, 97
    common base connection, 102–103
    common emitter configuration, 103–106
    data for NPN transistor, 100
    efficient BJT as amplification device,
        106–111
    gain parameters, 102
    input characteristics, 98
    NPN and PNP BJT transistors configurations,
        96
    operating point stability, 377–380
    output characteristics, 98–100
    rating and selection of operating point,
        101–102
    testing transistors, 106
    transistor configurations, 98
BJT, *see* Bipolar junction transistor (BJT)
BJT amplifier analysis, 380
    AC analysis, 382–383
    coupling capacitors effects, 384
    DC analysis, 381
    feedback resistance effect, 384
    Miller effect, 384–386
Bode diagrams, TF and, 237, 240–243
    construction of Bode plots, 247
    examples of passive components, 237–238
    first-order high-pass, 239–240
    first-order low-pass, 238–239
    idealized Bode plots, 245–246
Bode plots, 237, 239, 243, 247–250
    magnitude, 243
    single low-pass frequency response, 244
    TF, 244–245
Breadboards, 327
Bridge rectifier, 87
Built-in voltage of diode, *see* Potential barrier
Butterworth filter, 319–320
BW, *see* Bandwidth (BW)

C

CABs, *see* Configurable analog blocks (CABs)
CAD, *see* Computer-aided design (CAD)
Capacitance, 21, 25–26
    effect on low-frequency response, 184–185
    electric flux, 23
    parallel plate capacitor, 22
Capacitors, 21, 153, 179, 326
    fabrication, 21
    in parallel, 22, 24
    in series, 24–25
Cartesian to polar form, 69
Cathode (K), 83
C–B junction, *see* Collector–base junction
    (C–B junction)

CBSC, *see* Comparator-based switched capacitor
        (CBSC)
CE amplifier, *see* Common emitter amplifier
        (CE amplifier)
Channel, 114
Charged atom, 16
Chebyshev filter, 320–321
Circuit analysis; *see also* Alternating current
        (AC); Direct current (DC); Electric
        circuits
    steady state solutions, 35
    transient solutions, 35
Circuit building and realization, 327–328
Class A amplifier design, 334; *see also* DC power
        supply
    activities/comments, 340
    auto-bias circuit, 336, 338
    C_{c1} and C_{c2} calculation, 339–340
    circuit diagram for, 340
    circuit selection, 336–337
    decide on transistor, 337
    design, 335
    I_B calculation, 338–339
    operation point selection, 335–336
    R_1 and R_2 calculation, 338
    specification, 335
Clipping, 8
CMRR, *see* Common mode rejection ratio
        (CMRR)
Cognitive radio (CR), 357
Collector (C), 95
Collector–base junction (C–B junction), 97
Collector current, 138, 151
Collector-feedback bias, 141–143
Common base amplifier, 196, 198, 199
    current gain, 199
    under h parameters, 197–198, 200
    medium frequency AC equivalent circuit,
        198, 200
    power gain, 199–200
    voltage gain, 198
Common base connection, 102–103
Common base current gain, 103
Common collector amplifier; *see also* Common
        emitter amplifier (CE amplifier)
    AC medium frequencies equivalent circuit,
        192
    circuit diagram for, 192
    current gain, 194
    input impedance, 194–195
    potential divider, 195
    power gain, 195
    voltage gain, 191–194
Common drain amplifier, 229
    AC voltage gain at high frequency, 232–233
    AC voltage gain at medium frequency,
        229–232

    input impedance at high frequency, 233–234
    output impedance at high frequency, 234–235
Common emitter amplifier (CE amplifier), 154,
        182, 197; *see also* Common collector
        amplifier
    frequency response, 205–211
    at high frequency, 185–187
    at low frequency, 182–185
    overall gain, 201–205
Common emitter configuration, 98, 103–106, 146
    base characteristic for BJT, 147
    collector characteristic for BJT, 148
    equivalent circuit using h parameters, 146
    experimental setup, 147
Common mode rejection ratio (CMRR), 276,
        277–278, 389, 390
    differential amplifier with emitter resistance,
        277
    Op-Amp, 276
Common-mode voltage gain, 389
Common source amplifier, 223
    AC voltage gain at high frequencies, 227–228
    AC voltage gain at medium frequencies,
        223–226
    input impedance at high frequency, 228–229
Comparator-based switched capacitor (CBSC),
        355
Comparator, Op-Amp as, 299
Compensation circuit, 285
Components selection and circuit elements,
        324–325
    building and realization of circuit, 327–328
    capacitors, 326
    inductors, 326
    npv, 326–327
    resistors, 325–326
    testing, 329
    transistors, 325
    troubleshooting, 329–330
Computer-aided design (CAD), 323–324
    software, 8
Computer-aided simulation
    practical assignments using Pspice, 370–396
    Pspice, 365–370
Conductance (G), 14–15; *see also* Mutual
        conductance model
Conduction band, 78
Conductivity in semiconductor materials, 77–78
Configurable analog blocks (CABs), 357
Construction of Bode plots, 247
Corner frequency, 246
Coulomb (C), 12
Coupling capacitors, 191, 205, 208
    effects, 384
CR, *see* Cognitive radio (CR)
Current divider, 59–60
    rule, 161, 182, 194

Current gain, 161–162, 166, 169–170, 194, 199, 200; *see also* Power gain; Voltage gain
 h parameter model, 154–155
 parallel current NFB, 263–264
 parallel voltage NFB, 261–262
 series current NFB, 260
 series voltage NFB, 257–258
 three-parameter model, 169–170
 T-model two-parameters, 173
 Y parameters, 166
Current mirror, 143, 281–282, 390–391
Currents signals, 61
Current–voltage characteristics (I–V characteristics), 123
 diodes current–voltage characteristics, 81
 forward bias, 81
 reverse bias, 82–83
 saturation, 124–126
Cut off, 1, 132, 349; *see also* Frequency cutoff
 region, 132

D

DC, *see* Direct current (DC)
DC power supply, 330; *see also* Class A amplifier design
 filter smoothing stage, 331–332
 rectifier stage, 331
 regulator stage, 332–334
 transformer stage, 330–331
Decibels (dB), 179
Degenerative FB, *see* Negative feedback (NFB)
Delay switch, 347–348
Depletion, 118–119
 layer, 81
 MOSFET, 213–214
 n-channel MOSFET biasing circuit, 214
Dielectric layer, 21
Differential amplifiers, 271, 296, 298–299, 387–391; *see also* Operational amplifier (Op-Amp)
 CMRR, 276–278
 differential-mode voltage gain, 274
 effect on noise, 274–276
 Op-Amp as comparator, 299
 single input voltage, 271–273
 summing amplifier, 300–301
 voltage gain, 297
Differential-mode voltage gain, 274, 389
Diffusion current, 81
Digital electronics, 1, 95, 323, 355
DIL, *see* Dual in line (DIL)
Diodes, 77
 applications, 85
 current–voltage characteristics I–V, 81–83
 full-wave rectifiers, 87
 half-wave rectifiers, 86–87

p–n junction, 80–81
 rectification, 85–86
 semiconductor material, 77–80
 single-phase bridge rectifier circuit, 87–92
 symbol and package, 83
 testing, 94
 types, 83–85
 voltage doubler, 93–94
 as voltage limiter, 92–93
Direct current (DC), 14, 36; *see also* Alternating current (AC)
 analysis, 370, 381
 circuits, 36, 42
 current divider, 59–60
 KCL, 36–41
 KVL, 36, 41–52
 mesh current method, 53
 networks, 371–373
 voltage divider, 58–59
Distortion, 6, 7, 257
 clipping, 8
 harmonic, 257
 pure sinusoidal original signal, 7
 sinusoidal waveform, 8
Documentation, 6
Doping, 79–80, 84, 108
Drain current, 121–126, 215, 219
Drain-feedback bias, 216–217
Drain saturation current, 213
Drift current, 81, 82
Dual in line (DIL), 283

E

Electric charge (Q), 12, 15–16
Electric circuit, 16, 32, 33–34
 active components of circuit, 29–32
 capacitance, 21–26
 circuit analysis, 11
 electric charge and current, 15–16
 electric circuits/networks, 32–34
 electric current in circuit, 16–17
 electrons movement, 16–17
 elements electric circuit, 32
 inductors, 26–29
 passive components, 17
 resistance, 17–21
 units, 11–15
Electric current, 11, 15–16
 in circuit, 16–17
 flow, 16
Electric flux, 23
Electromotive forces (emfs), 16, 17, 31, 32
Electron, 15, 32, 78
 contribution to current, 108
 current, 96, 97, 107

diffusion current density, 110
  flow, 16
Electronic systems, 2–3, 237
  approaches to design, 5
emfs, *see* Electromotive forces (emfs)
Emitter (E), 95
Emitter–base junction, 150
Emitter bias, *see* Auto bias
Emitter current, 138
Emitter follower, *see* Common collector amplifier
Emitter injection
  efficiency, 108–109
  ratio, 109
Energy (E), 1, 78–79
  bands in semiconductor materials, 77–78, 80
  storage, 29
  unit, 12–13
Enhancement MOSFET, 214
  channel voltage divider bias circuit, 215
  drain-feedback bias, 216–217
  example, 215–216
  FET equation, 215
Equivalent impedance, 186, 227
Esaki diodes, *see* Tunnel diodes
External loop, 52

F

Feedback (FB), 251
  in amplifiers, 251, 252
  NFB, 251–268
Feedback resistance effect, 384
Fermi level, 79–80
FET, *see* Field effect transistors (FET)
FET amplifier analysis, 386
  differential amplifier, 387–391
  FET common source amplifier, 386–387
Field effect transistors (FET), 5, 113, 223, 295;
    see also Bipolar junction transistor
    (BJT)
  as amplifiers, 223
  biasing, 213–220
  BJTs, 113
  circuit symbols, 115
  common drain, 229–235
  common source amplifier, 223–229
  current–voltage characteristics, 123–126
  fabrication, 113–114
  IGFETs, 114–115
  input characteristics, 123
  JFET, 115, 120–123
  MOS, 115
  output characteristics, 123
  static characteristics, 123
  structure, 114
  transfer characteristics, 123
Field programmable analog array (FPAA), 356–358

Field programmable gate arrays (FPGA), 356, 357
Filter(s), 305
  band-pass filter response, 314
  band-stop response, 314–315
  Bessel filter, 318–319
  Butterworth filter, 319–320
  Chebyshev, Bessel, and Butterworth filter
    response, 316
  Chebyshev filter, 320–321
  circuit diagram of inverting amplifier, 306
  design using standard tables, 316
  fourth-order response, 315
  high-pass filter response, 311–313
  low-pass filter responses, 308–311
  negative feedback voltage gain, 307
  response characteristics, 316, 393–396
  Sallen and Key fourth-order low-pass filter,
    317
  voltage gain, 306
First-order high-pass filter, 311–312
555 Timer, 349–351
  automatic switch on of lamp using 555 Timer,
    352–353
  humidity detector using, 351–352
  internal block diagram, 350
  package and pin out, 350
  two-tone musical instrument using, 352
Fixed bias, 133, 134–137; *see also* Auto bias
  amplifier AC analysis, 159
  biasing, 133
  circuit, 134
Force (F), 12
Forward bias, 81, 82
  base–emitter junction, 151
  current, 81
Four-terminal
  devices, 113
  network, 145
FPAA, *see* Field programmable analog array
    (FPAA)
FPGA, *see* Field programmable gate arrays
    (FPGA)
Frequency cutoff, 206–208, 211
  equivalent circuit at high frequency for
    source stage, 210
  high-frequency cutoffs for stages in cascade,
    210
  high frequency for stages in cascade, 210
  input circuit first stage, 206, 209
  of last stage, 205–206, 208–209
  low-frequency cutoffs for n stages in cascade,
    207
  low frequency of n stages, 207
Frequency response, 205
  high-frequency cutoff, 208–211
  low-frequency cutoff, 205–208
  series voltage NFB effects, 255–257

Full-wave bridge rectifier, 331
Full-wave rectifiers, 87, 88
    with smoothing capacitor, 89

G

Gain; *see also* Current gain; Power gain; Voltage
        gain
    parameters, 102
    second second-order filter, 320
Gallium arsenide (GaAs), 84
Gate (G), 113, 116, 120
    current, 216
    depletion of holes, 117
    insulated-gate FETs, 114–115, 116
    junction gate FETs, 115
    voltage, 219
Gunn diodes, 84
Gyrator, 326

H

Half-power gain, 179
    decibel notation, 179
    frequency cutoff, 180
    voltage gain of amplifier, 181
Half-wave rectifiers, 86–87
Henry (H), 26
High-frequency cutoff, 187, 208, 384–386;
        see also Low-frequency cutoff
    frequency cutoff input circuit first stage,
        209–211
    frequency cutoff of last stage, 208–209
    high-frequency cutoff for 1 to (n – 1) stages,
        209
    for 1 to (n – 1) stages, 209
High-pass filters, 305, 311
    first-order high-pass filter, 311–312
    Sallen and Key high-pass filter, 313
    second-order filters, 312–313
High-power transistors, 325
Hole, 78
    accumulation, 117
    contribution, 108
h parameter model, *see* Hybrid parameter model
        (h parameter model)
Humidity detector, 346
    using 555 Timer, 351–352
    two-tone musical instrument using 555
        Timers, 352
Hybrid parameter model (h parameter model),
        145, 153, 158; *see also* Three-
        parameter model; T-equivalent two-
        parameter model
    CE amplifier equivalent circuit, 154
    common emitter configuration, 146–148
    current gain, 154–155

equivalent circuit auto bias amplifier, 160
    input impedance, 155–156
    output impedance, 156–157
    power gain, 157–158
    small-signal practical CE amplifier, 158–165
    transistor amplifier analysis using, 153
    for transistor in CE configuration, 158
    voltage gain, 156

I

ICs, *see* Integrated circuits (ICs)
Ideal current source, 31
Idealized Bode plots, 245, 247
    Bode plot asymptotic phase, 246
    magnitude asymptotic, 245–246
Ideal voltage source, 30, 31
IGFETs, *see* Insulated-gate field effect transistors
        (IGFETs)
Impedance, 64–65, 374
    capacitance, 66–67
    inductor, 65–66
    resistance, 67–68
Indium phosphide (InP), 84
Inductance, *see* Self-inductance
Inductive proximity sensors, 29
Inductors, 26, 179, 326
    with air-filled coil, 26
    energy storage, 29
    inductive proximity sensors, 29, 30
    in parallel, 28–29
    permeability, 27
    in series, 27–28
Initial condition of component, 35
In-phase FB, *see* Positive FB
Input characteristics, 98, 123
Input current, 203
Input impedance, 194
    at high frequency, 228–229, 233–234
    h parameter model, 155–156
    inverting amplifier, 289
    noninverting amplifier, 295
    parallel current NFB effects, 264
    parallel voltage NFB effects, 262
    series current NFB effects, 260
    series voltage NFB effects, 254
    T-model two-parameters, 173–174
Input offset voltage, 284
Input voltage, 203, 227
Insulated-gate field effect transistors (IGFETs),
        114, 115
Insulators, 78
Integrated circuits (ICs), 85, 93
Internal distortion, series voltage NFB effects
        on, 257
Internal resistance, 30
Internal voltage of diode, *see* Potential barrier

International System of Units (SI system), 11
Inverse FB, *see* Negative feedback (NFB)
Inverting amplifier, 285; *see also* Noninverting
        amplifier
    bias equalization, 290–292
    input impedance, 289
    output impedance, 290
    virtual Earth, 287–289
    voltage gain, 285–287
I–V characteristics, *see* Current–voltage
        characteristics (I–V characteristics)

J

JFET, *see* Junction gate field effect transistors
        (JFET)
Joule (J), 12–13
JUGFET, *see* Junction gate field effect transistors
        (JFET)
Junction gate field effect transistors (JFET), 115;
        see also Metal–oxide–semiconductor
        field effect transistor (MOSFET)
    biasing, 217–220
    operation, 120–123

K

Kilowatt hour (kW h), 13
Kirchhoff's current law (KCL), 36, 36–41, 54–58
    circuit diagram, 37, 38
Kirchhoff's laws, 35, 141, 161, 199, 295, 323–324
    application to AC circuits, 64–75
Kirchhoff's voltage law (KVL), 36, 41–52
    circuit diagram, 42
    current divider, 59–60
    mesh current method, 53
    voltage divider, 58–59

L

Light-emitting diode (LED), 83, 84
Linear Bode plot, *see* Idealized Bode plots
Load, 17
Load line, 129, 382
    biasing transistor, 130
    cut off, 132
    example, 131–132
    saturation, 132–133
Long-pair amplifier, *see* Differential amplifiers
Long-tailed pair amplifier, *see* Differential
        amplifiers
Low-frequency cutoff, 183–184, 205; *see also*
        High-frequency cutoff
    frequency cutoff for 1 to (n – 1) stages, 206
    frequency cutoff input circuit first stage,
        206–208
    frequency cutoff of last stage, 205–206

Low-pass filters, 305, 308
    Sallen and Key low-pass filter, 310–311
    second-order low-pass filters, 310
    single first-order low-pass filter, 308–310
    single pole, 391–392

M

Magnitude of transfer voltage gain, 243
Majority charge carriers, 79
Mesh current method, 53
Metal (M), 116
Metal, oxide insulator, and semiconductor
        (MOS), 116
    operation, 116–117
    technology, 114
    transistor aspect ratio, 124
Metal–oxide–semiconductor field effect
        transistor (MOSFET), 113, 115;
        see also Junction gate field effect
        transistors (JFET)
    accumulation, 117–118
    biasing, 213–217
    depletion, 118–119
    inversion, 119–120
    MOS IGFET, 115–116
    threshold voltage, 120
Mid-frequency range, 179
Miller effect, 185, 229, 384–386
Miniaturization, 356
Minority charge carriers, 79
MOS, *see* Metal, oxide insulator, and
        semiconductor (MOS)
MOSFET, *see* Metal–oxide–semiconductor
        field effect transistor (MOSFET)
Multivibrators, 348
    astable multivibrator, 348–349
    555 Timer, 349–351
Mutual conductance
    model, 151–152
    parameter, 151

N

Nanotechnology, 359
Negative feedback (NFB), 251
    block diagram of system with, 252
    parallel current NFB, 263–268
    parallel voltage NFB, 260–262
    series current NFB, 258–260
    series voltage NFB, 253–254
Negative temperature coefficient (NTC), 346
Neural processing accelerators (NPUs), 360
Neutron, 15
Newton (N), 12
NFB, *see* Negative feedback (NFB)

Noise, 6, 7
  clipping, 8
  sinusoidal signal, 7
  sinusoidal waveform, 8
  voltage gain effect on, 274–276
Nominal preferred values (npv), 6, 326
Noninverting amplifier, 292; *see also* Inverting
      amplifier
  input impedance, 295
  output impedance, 295–296
  voltage follower, 296
  voltage gain, 292–295
Notch filter, *see* Band-stop filter
NPUs, *see* Neural processing accelerators (NPUs)
npv, *see* Nominal preferred values (npv)
NTC, *see* Negative temperature coefficient
      (NTC)
n-type material, 114
  semiconductor material, 77

O

Offset null circuit, 284
"Off the shelf" factor, 133
Ohm (Ω), 13
Ohmic region, 98, 122
Ohm's law, 11, 17–18, 36, 59, 68
Op-Amp, *see* Operational amplifier (Op-Amp)
Open-loop gain, 285
Open circuit voltage gain, 149
Operational amplifier (Op-Amp), 276, 281;
          see also Differential amplifiers
  automatic switch on of lamp, 343–346
  basic Op-Amp component, 282
  characteristics, 283
  circuit symbol, 283
  compensation circuit, 285
  frequency response, 301–303
  gain, 285
  ideal operational amplifier characteristics, 284
  inverting amplifier, 285–292
  modern Op-Amp system, 281–282
  noninverting amplifier, 292–296
  offset null circuit, 284
  slew rate, 285
Operation point (Q), 101, 129, 335
  rating and selection, 101–102
Opto-isolator, 93
Output characteristics, 98, 123
  avalanche region, 100
  BJT output characteristic, 100
  at one base current, 99
Output current, 194
Output impedance, 174
  at high frequency, 234–235
  h parameter model, 156–157
  inverting amplifier, 290

load resistance effect, 175
  noninverting amplifier, 295–296
  parallel current NFB effects, 264–265
  parallel voltage NFB effects, 262
  series current NFB effects, 260
  series voltage NFB effects, 255
  source resistance effect, 175–176
  T-model two-parameters, 174–176
Output voltage, 156, 191
Oxide insulator (O), 116

P

Parallel capacitor, 209
Parallel current NFB, 263; *see also* Series current
      NFB
  current gain, 263–264
  effects on input impedance, 264
  effects on output impedance, 264–265
  voltage gain, 265–268
Parallel voltage NFB, 260; *see also* Series
      voltage NFB
  block diagram of a system with, 261
  current gain, 261–262
  effects on input impedance, 262
  effects on output impedance, 262
  voltage gain, 261
Partitioning, 4
Passive components, 17; *see also* Active
          components of circuit
  capacitance, 21–26
  inductors, 26–29
  resistance, 17–21
PC, *see* Personal computer (PC)
PCB, *see* Printed circuit board (PCB)
pd, *see* Potential difference (pd)
Peak inverse voltage (PIV), 82
Peltier diodes, 84
Period, 61
Permeability, 27
Permittivity, 22
Personal computer (PC), 324, 365
Phase change
  at high frequency, 187
  at low frequency, 184–185
Phase–lead angle, 243
Phase response, 245
Phase–shift angle, 243
Phasor, 62, 63, 64
  diagrams, 70–75
  domain, 62–64
Photodiodes, 85
Phototransistor, 93
Pinch-off
  point, 121
  voltage, 124
PIV, *see* Peak inverse voltage (PIV)

p–n junction, 77, 80–81
Polar–Cartesian forms, 69
Positive FB, 251
Potential barrier, 81
Potential difference (pd), 17, 32, 35
Potential divider, *see* Voltage divider
Potential divider bias, *see* Auto bias
Power (P), 13
    amplifier class B, 396
Power gain, 162, 179, 195; *see also* Current gain;
        Voltage gain
    common base amplifier, 199–200
    T-model two-parameters, 177–178
    three-parameter model, 170–171
    transistor amplifiers analysis, 157–158
    Y parameters, 166–168
Practical assignments using Pspice, 370–371
    AC networks, 374–377
    active filter–power amplifier, 391–396
    BJT amplifier analysis, 380–386
    BJT operating point stability, 377–380
    DC networks, 371–373
    FET amplifier analysis and differential
        amplifier, 386–391
Practical current source, 31–32
Practical voltage source, 30–31
Printed circuit board (PCB), 324, 327, 328
Probe, 370, 381
Proton, 15
Pspice, 324, 365, 366
    AC networks, 374–377
    active filter–power amplifier, 391–396
    BJT amplifier analysis, 380–386
    BJT operating point stability, 377–380
    circuit creation/schematics, 367
    circuit diagram, 366
    connecting components, 367–368
    DC analysis, 370
    DC networks, 371–373
    FET amplifier analysis and differential
        amplifier, 386–391
    names to elements, 368–369
    practical assignments using Pspice, 370–371
    running simulation, 369–370
    subprograms, 365–366
p-type material, 113
p-type semiconductor material, 77
Punch-through, 111

Q

Quiescent point, *see* Operating point

R

Reactance, 68–69
Reconfigurable analog circuitry, 356–358

Rectification, 85–86
Regenerative FB, *see* Positive FB
Relative permeability, 27
Resistance (R), 13–15, 17, 18, 21
    Ohm's law, 17–18
    resistor in parallel, 19–20
    resistors in series, 19
Resistive region, *see* Ohmic region
Resistor(s), 11, 17, 180, 325–326, 367
    in parallel, 19–20
    in series, 19
Reverse bias, 82–83
    current, 81
Reverse breakdown process, 82
Root mean square (RMS), 86

S

Sallen and Key filter, 310
    high-pass filter, 313, 392–393
    low-pass filter, 310–311
    second-order Butterworth low-pass filter,
        341–342
Saturation, 124–126
    current, 82, 83
    FET current–voltage characteristics, 125
    region, 132–133
    region of I–V curve, 122
Schematics tool, 367
Schottky diodes, 85
SDR, *see* Software defined radio (SDR)
Second-order filter(s), 312–313, 318–321
    low-pass filters, 310
Self-bias, *see* Auto bias
Self-inductance, 26
Semiconductor (S), 116
    conductivity in, 77–78
    diodes, 83
    doping, 79–80
    energy bands in, 77–79
    materials, 77
    technology, 355
Series current NFB, 258, 259; *see also* Parallel
        current NFB
    current gain, 260
    input impedance, effects on, 260
    output impedance, effects on, 260
    voltage gain, 259–260
Series voltage NFB, 253; *see also* Parallel voltage
        NFB
    current gain, 257–258
    frequency response, effects on, 255–257
    input impedance, effects on, 254
    internal distortion, effects on, 257
    output impedance, effects on, 255
    voltage gain, 254
Short circuit current gain, 170

SiC devices, *see* Silicon carbide devices
(SiC devices)
Siemens (S), 148
Silicon (Si), 77
    diodes, 82
Silicon carbide devices (SiC devices), 358
Silicon dioxide (SiO_2), 113, 115, 116
Silicon-on-insulator devices (SOI devices), 359
Simulation in circuit applications, 323–324
Simulation Program with Integrated Circuit
    Emphasis (Spice), 324, 365
Simulator, 8–9
Single first-order low-pass filter, 308–310
Single input voltage, 271
    AC small-signal equivalent circuit, 273
    differential amplifier with, 272
    simplified T-model, 271
Single-phase bridge rectifier circuit, 87, 89–92
    conduction in bridge rectifier, 88
    example of rectifier diodes, 89
    positive half-cycle, 88
Single-stage amplifier, 201
Sinusoidal shape voltage, 61
SI system, *see* International System of Units
    (SI system)
Slew rate, 285
Small signal, 150–151
    analysis, 153
    auto bias amplifier, 159–165
    fixed bias amplifier, 158–159
    practical CE amplifier, 158
    small-signal practical CE amplifier,
        158–165
    transistor amplifier analysis, 153–158,
        165–168, 168–171, 172–178
Smoke detector, 348
SMT, *see* Surface mount (SMT)
Social and environmental implications, 6
Software defined radio (SDR), 357
SOI devices, *see* Silicon-on-insulator devices
    (SOI devices)
Solar cell, 85
Soldering, 328
Source, 17
Source follower amplifier, *see* Common drain
    amplifier
Spice, *see* Simulation Program with Integrated
    Circuit Emphasis (Spice)
Standard components, 325
Static switches, 356
Steady state solutions, 35
Strong inversion, 119–120
Summing amplifier, 300–301
Surface mount (SMT), 283
    components, 324–325
Switches (Sw), 357
System testing, 6

T

T-equivalent two-parameter model; *see also*
        Three-parameter model; Hybrid
        parameter model (h parameter
        model)
    AC-simple T-parameters transistor model,
        149–150
    emitter resistance, 150–151
Testing, 329
    transistors, 106
T-model two-parameters, 172; *see also*
        Y parameters
    current gain, 173
    input impedance, 173–174
    output impedance, 174–176
    power gain, 177–178
    transistor amplifier analysis, 172
    voltage gain, 172–173
TF, *see* Transfer function (TF)
Thermal voltage, 149
Three-parameter model, 149, 168; *see also*
        Hybrid parameter model (h parameter
        model); T-equivalent two-parameter
        model
    current gain, 169–170
    equivalent circuit of amplifier using, 168
    power gain, 170–171
    transistor amplifier analysis using, 168
    voltage gain, 168–169
3 dB frequency, *see* Corner frequency
Threshold voltage, 120
Top-level
    design, 5
    specifications, 4
Total frequency response, 187–188
T parameters, 145
    AC-simple T-parameters transistor model,
        149–150
    CE amplifier using, 172–173
Transfer function (TF), 237, 240–243
    Bode plots, 243–245, 247–250
    construction of Bode plots, 247
    examples of passive components, 237–238
    first-order high-pass, 239–240
    first-order low-pass, 238–239
    idealized Bode plots, 245–247
Transient analysis, 387, 390
Transient solutions, 35
Transistor(s), 1, 95, 129, 153, 281, 325
    admittance y parameters, 148–149
    amplifier analysis, 153
    configurations, 98
    h parameters, 145–148
    modeling, 146
    mutual conductance model, 151–152
    T-equivalent two-parameter model, 149–151

three-parameter model, 149
two-port network, 146
Troubleshooting, 329–330
Tunnel diodes, 84
Two-port network, 145, 146
Two sources biasing, 143–144

U

Unidirectional signals, 60
Units, 11
   charge, 12
   conductance, 13–15
   electric voltage, 13
   energy, 12–13
   force, 12
   physical quantities in SI system of units, 12
   power, 13
   resistance, 13–15

V

Valence band, 78
Veroboards, 327, 328
Very-large-scale integration (VLSI), 113
VIEWPOINT symbol, 369
Virtual Earth, 287–289
VLSI, *see* Very-large-scale integration (VLSI)
Volt (V), 13
Voltage divider, 58–59
   bias, 219–220
Voltage doubler, diode as, 93–94
Voltage follower, noninverting amplifier, 296
Voltage gain, 156, 161, 166, 168–169, 191,
      193–194, 198, 267–268; *see also*
      Current gain; Power gain
   CE amplifier using T parameters, 172–173
   coupling capacitors, 191
   equation, 201

equivalent circuit, 192
function, 167
at high frequency, 185–186
h parameter model, 156
last stage n, 202
at low frequency, 182–183
NFB on amplifiers, 266
for (n – 1) stages, 202–203
output voltage, 191
parallel current NFB, 265
parallel voltage NFB, 261
series current NFB, 258
series voltage NFB, 253, 254
source, 203–205
T-model two-parameters, 172–173
three-parameter model, 168–169
Y parameters, 166
Voltage limiter, diode as, 92–93

W

Watt (W), 13
Working point, *see* Operating point

Y

Y parameters, 148, 165; *see also* T-model
      two-parameters
   current gain, 166
   equivalent circuit auto bias amplifier using,
      165
   power gain, 166–168
   transistor amplifier analysis using, 165
   voltage gain, 166

Z

Zener breakdown, 84
Zener diodes, 84

Printed and bound by CPI Group (UK) Ltd, Croydon, CR0 4YY

01/11/2024

01782617-0012